A Review of Science and Technology During the 1972 School Year

Science Year

The World Book Science Annual

1973

Field Enterprises Educational Corporation
Chicago London Paris Rome Stuttgart Sydney Tokyo Toronto

Copyright © 1972
By Field Enterprises Educational Corporation
Merchandise Mart Plaza, Chicago, Illinois 60654
All rights reserved.
ISBN 0-7166-0573-2
Library of Congress Catalog
Card Number: 65:21776
Printed in the United States of America

The publishers of *Science Year* gratefully
acknowledge the following for permission
to use copyrighted illustrations. A full
listing of illustration acknowledgments
appears on pages 442 and 443

137   © 1969 by R. E. Dickerson and
      I. Geis. Reproduced by permission from
      The Structure and Action of Proteins
      by Richard E. Dickerson and Irving Geis,
      Harper & Row
257   Drawing by W. Miller © 1972
      The New Yorker Magazine, Inc.
283   A. I. McMullen, Microbiological Research
      Establishment, Porton Down, England.
      Crown copyright reserved.
286   Reprinted by permission from *The
      Christian Science Monitor* © The
      Christian Science Publishing Society.
      All rights reserved.
309   Cartoon by Al Kaufman © 1972
      by The New York Times Company.
      Reprinted by permission.
327   Sydney J. Harris © 1972
      Saturday Review, Inc.
332   Drawing by Chon Day © 1972
      The New Yorker Magazine, Inc.
359   Drawing by Weber © 1971
      The New Yorker Magazine, Inc.
376   Drawing by Richter © 1972
      The New Yorker Magazine, Inc.

Preface

You may not be able to detect it, but *Science Year* has a new look. With this edition, The Science Annual joins the growing numbers of newspapers and magazines that are typeset through the use of a computer.

The new typesetting system takes advantage of the speed and accuracy of modern electronics. It is contrasted to the age-old method of sending a typed manuscript to the typesetter where an operator, seated at a keyboard, retyped the manuscript. This enabled a machine to cast hot lead into type, one character at a time, which produced printed galleys or pages. With each retyping, errors were introduced which had to be found and corrected.

In the new system, the edited article is typed on paper and a magnetic tape is created in our editorial office. The impulses recorded on this tape represent each of the characters in the article. After the editor corrects the paper proof, the typist retypes only the corrections and "merges" them with the original magnetic tape. Coded instructions for typesetting are added to the tape and it is sent to the typesetter.

The typesetter converts the tape to another tape that can be mounted on the computer tape drives. Using information such as the size of the type or the length of the lines, the computer breaks the manuscript into lines, splitting words and adding hyphens where necessary. The computer then generates a printout—pages of text with symbols that represent instructions such as "capital letter," or "italic." The editor checks this printout for errors, adjusts the text length to fit the allotted space, and returns the printout to the typesetter.

The typesetter adds the corrections to his tape, and then uses it to generate a paper tape punched with rows of carefully aligned holes. This tape is fed into a photocomposition machine where five glass disks carrying the photographic images of 1,200 different characters whirl at 2,400 revolutions per minute in front of a focused light. The paper tape commands the light to flash whenever the proper character is before it, exposing characters at a rate of 40 per second on photographic film in lines and columns of text. When this film is developed, it is the source of the words which, along with the photographs and other illustrations, become the familiar pages of *Science Year.*

This sounds as if it were a very complicated and time-consuming process. There is, of course, much additional preparatory work required in the editorial office. But the total operation takes less time, and saves money as well. In addition, there is greater accuracy.

The most important result of the new system is an improved look to the pages of *Science Year.* The type is neater, cleaner, and more uniform than was possible with "hot type." To maintain the appearance of the book as originally designed, we chose a photographic type face that would match that of the old type as closely as possible. You might be interested in comparing this edition with last year's to see if you can detect the difference. If you can, we believe you will note an improvement in the appearance of *Science Year.* [Arthur G. Tressler]

Contents

10 **Assault on the Green Kingdom** by Darlene R. Stille
The inexorable growth of industry, agriculture, and recreation
facilities is spelling disaster for many species of plants.

24 **Escalating the War on Cancer** by James D. Watson
A Nobel prize-winning biologist warns that the new federal funds will not be
well spent unless the focus is changed from clinical to fundamental research.

33 **Special Reports**
Fifteen articles, plus a *Science Year* Trans-Vision®, give in-depth
treatment to significant and timely subjects in science and technology.

34 **Tuning in on the Grizzly** by Frank C. Craighead, Jr.
and John J. Craighead
Two brothers, both wildlife biologists, have completed a long-range
study of grizzly bears to determine if they have a future in man's domain.

50 **Reprieve for the Dwindling Ospreys** by David R. Zimmerman
Replacing pesticide burdened eggs with clean ones from another
area may bring the Long Island birds back from the brink of extinction.

60 **Mars Is an Active Planet** by William J. Cromie
Evidence from Mariner 9 suggests that geological activity
on Mars is fairly recent and, indeed, may be still going on.

76 **Beyond the Black Hole** by John A. Wheeler
A renowned Princeton physicist spins a futuristic yarn in which two space
travelers encounter a black hole, then reflect on its meaning to physical theory.

90 **The Search for a Human Cancer Virus** by Lee Edson
Scientists are closing in on evidence that there
is a link between viruses and human cancer.

104 **The Curious Ancestry of Our Cells** by Jerald C. Ensign
Bacteria and blue-green algae may be the still-living ancestors of
two components of modern cells—mitochondria and the chloroplasts.

116 **A Matter of Fat** by Jean Mayer
A noted nutritionist recounts what we have learned from studies
of obesity, and stresses the vital role of exercise in weight control.

132 **The Nutritious Trace** by Walter Mertz
Scientists are using new detection methods to discover very small amounts
of chemical elements that the human body requires for a balanced diet.

146 **The Making of a Champion** by Daniel F. Hanley
The chief physician for the U.S. Olympic committee relates how
new ideas about training and diet help athletes break records.

162 **Earth's Heat Engines** by Kenneth S. Deffeyes
Using data from satellite orbits, geologists at Princeton have plotted a
number of upwelling plumes around the earth that may cause continent drift.

177 **The Force That Moves Continents**
A *Science Year* Trans-Vision®
Transparent overlays unveil layers of the earth to show
tectonic plates and how a plume causes them to move.

188 **Science for the People of China** by John Barbour
American scientists visiting China learned how science and
technology is used there to enhance the welfare of the Chinese people.

204 **New Peril to Our Oceans** by John A. McGowan
The fallout of volatile material from the skies may prove to be a
greater hazard to the oceans than pollutants that reach them directly.

216 **Supertanker** by Wade Tilleux
The enormous ships being built to haul oil across the seas have
revolutionized engineering design and ship-handling techniques.

224 **Our Earliest Ancestors** by F. Clark Howell
A decade of anthropological work in East Africa has given us a
startling new insight into the origins of man and how he has developed.

238 **CERN: Experiments in Physics and People** by Robert H. March
The success of the high energy physics laboratory in Switzerland has
been achieved by the unique cooperation of physicists from many nations.

253 Science File
A total of 44 articles, alphabetically arranged by subject matter, plus 7
Science Year Close-Ups, report on the year's work in science and technology.

381 Men of Science
This section takes the reader behind the headlines of science and into the
research laboratories to focus on the men and women who make the news.

382 **Humberto Fernandez-Moran** by Richard A. Lewis
Born in Venezuela and educated in Europe, this University of Chicago
scientist is continuing his search for the ultimate structure of life.

398 **Christopher Kraft** by William J. Cromie
With cool, logical thinking, this aeronautical engineer guided the many
manned space flights that led to the incredibly successful moon landings.

414 **Awards and Prizes**
Science Year lists the winners of major scientific awards and describes
the men and women, their work, and the awards they have won.

423 **Deaths**
A list of noted scientists and engineers who died during the past year.

425 Index

Editorial Advisory Board

Harrison Brown is chairman of the *Science Year* Advisory Board and a member of *The World Book Year Book* Board of Editors. He is professor of geochemistry and professor of science and government at the California Institute of Technology, and also a member of the governing board and foreign secretary of the National Academy of Sciences. Dr. Brown is a graduate of the University of California and received his doctorate in nuclear chemistry from Johns Hopkins University in 1941. He is a noted lecturer and writer.

Adriano Buzzati-Traverso is Assistant Director-General for Science of UNESCO in Paris. He specializes in population genetics and radiation biology. Dr. Buzzati was born in Milan, where he received his D.Sc. degree in 1936. From 1948 to 1962, he was professor of genetics and head of the Institute of Genetics at the University of Pavia. Until 1969, he was director of the International Laboratory of Genetics and Biophysics at Naples, Italy.

Barry Commoner is professor of plant physiology at Washington University, St. Louis. He is also director of Washington University's Center for the Biology of Natural Systems, which studies the effects of contamination on the environment. Dr. Commoner received an A.B. degree at Columbia University in 1937, and his advanced degrees in cellular physiology from Harvard University. He is a member of *The World Book Encyclopedia* Advisory Board.

Gabriel W. Lasker is professor of anatomy at the School of Medicine, Wayne State University. He received his A.B. degree from the University of Michigan in 1934, an M.A. degree in 1940 and a Ph.D. degree in 1945 in physical anthropology from Harvard University. A past president of the American Association of Physical Anthropologists and former chairman of the Council of Biology Editors, Dr. Lasker is on the Editorial Council of the American Anthropological Association. He is editor of *Human Biology* and the author of *The Evolution of Man.*

Dr. Walsh McDermott is Livingston Farrand Professor of Public Health and Chairman of the Department of Public Health at the Cornell University Medical College. He received his B.A. degree from Princeton University in 1930 and his M.D. degree from Columbia University's College of Physicians and Surgeons in 1934. In 1967, Dr. McDermott was named chairman of the Board of Medicine of the National Academy of Sciences. He is editor of the *American Review of Respiratory Diseases.*

Roger Revelle is Richard Saltonstall Professor of Population Policy and director of the Center for Population Studies at Harvard University. He received a B.A. degree from Pomona College in 1929 and a Ph.D. degree from the University of California in 1936. He is former chairman of the U.S. National Committee for the International Biological Program. Much of his current work involves studies on the crisis in world hunger and malnutrition.

Allan R. Sandage is astronomer and a member of the observatory committee of the Mount Wilson and Palomar Observatories. He received a B.A. degree at the University of Illinois in 1948 and a Ph.D. degree in astronomy at the California Institute of Technology in 1953. Dr. Sandage is a member of the Royal Astronomical Society and the National Academy of Sciences. He is perhaps most noted for his discovery of quasi-stellar objects in the far reaches of space. He found the first "quasar" in 1960, and the first X-ray star in 1966.

Alvin M. Weinberg is director of the AEC's Oak Ridge National Laboratory in Tennessee. He is a graduate of the University of Chicago, 1935, and earned his doctorate in physics there in 1939. After three years of researching and teaching mathematical biophysics at Chicago, he joined the Metallurgical Laboratory (part of the Manhattan Project) in 1941. He moved to Oak Ridge in 1945, becoming director of the Physics Division in 1947, research director in 1948, and director of the laboratory in 1955.

Contributors

Auerbach, Stanley I., Ph.D.
Director, Ecological Sciences Division
Oak Ridge National Laboratory
Ecology

Barbour, John, B.A.
Science Writer
Associated Press
Science for the People of China

Belton, Michael J. S., Ph.D.
Associate Astronomer
Kitt Peak National Observatory
Astronomy, Planetary

Bromley, D. Allan, Ph.D.
Professor and Chairman
Department of Physics
Yale University
Physics, Nuclear

Budnick, Joseph I., Ph.D.
Professor of Physics
Fordham University
Physics, Solid State

Craighead, Frank C., Jr., Ph.D.
Adjunct Professor of Biology
State University of New York
President of the Environmental
Research Institute, Moose, Wyo.
Tuning in on the Grizzly

Craighead, John J., Ph.D.
Leader, Montana Cooperative
Wildlife Research Unit
Professor of Zoology and Forestry
University of Montana
Tuning in on the Grizzly

Cromie, William J., B.S.
President and Editor
Universal Science News, Inc.
Mars Is an Active Planet;
Christopher Kraft

Davies, Richard P., B.S.
Free-Lance Medical Writer
Close-Up, Medicine

Deason, Hilary J., Ph.D.
Former Director of Libraries
American Association for the
Advancement of Sciences
Books of Science

Deffeyes, Kenneth S., Ph.D.
Associate Professor of Geology
Princeton University
Earth's Heat Engines

Donohue, Jerry, Ph.D.
Professor of Chemistry
University of Pennsylvania
Chemistry, Structural

Drake, Charles L., Ph.D.
Professor of Earth Sciences
Dartmouth College
Geophysics

Edson, Lee, B.S.
Free-Lance Writer
The Search for a Human Cancer Virus

Ensign, Jerald C., Ph.D.
Associate Professor of Bacteriology
University of Wisconsin
The Curious Ancestry of Our Cells;
Microbiology

Evans, Earl A., Ph.D.
Professor and Chairman
Department of Biochemistry
University of Chicago
Biochemistry

Gingerich, Owen, Ph.D.
Astrophysicist, Smithsonian
Astrophysical Observatory
Close-Up, Astronomy, Planetary

Goss, Richard J., Ph.D.
Professor of Biological
and Medical Sciences, Brown University
Zoology

Gray, Ernest P., Ph.D.
Chief, Theoretical Plasma Research
Applied Physics Laboratory
Johns Hopkins University
Physics, Plasma

Griffin, James B., Ph.D.
Director, Museum of Anthropology
University of Michigan
Archaeology, New World

Hanley, Daniel F., M.D.
Resident Physician
Bowdoin College
The Making of a Champion

Hawthorne, M. Frederick, Ph.D.
Professor of Chemistry, University
of California, Los Angeles
Chemistry, Synthesis

Hayes, Arthur Hull, Jr., M.D.
Associate Professor of
Pharmacology, Cornell
University Medical College
Drugs

Herschbach, Dudley, Ph.D.
Professor of Chemistry
Harvard University
Chemical Dynamics

Hilton, Robert J., D.D.S.
Associate Professor
Operative Dentistry
Northwestern University
Medicine, Dentistry

Hines, William
Science Correspondent
Chicago Sun-Times
Space Exploration

Howell, F. Clark, Ph.D.
Professor of Anthropology
University of California, Berkeley
Our Earliest Ancestors

Irvin, Robert W.
Automotive Writer
The Detroit News
Close-Up, Transportation

Isaacson, Robert L., Ph.D.
Professor of Psychology
University of Florida
Psychology

Jahns, Richard H., Ph.D.
Professor of Geology and Dean
School of Earth Sciences
Stanford University
Geology

Kessler, Karl G., Ph.D.
Chief, Optical Physics Division
National Bureau of Standards
Physics, Atomic and Molecular

Kristian, Jerome, Ph.D.
Astronomer, Hale Observatories
Astronomy, Cosmology

Lee, Tsung-Dao, Ph.D.
Professor of Physics
Columbia University
Close-Up, Physics, Elementary Particles

Lewin, Arie Y., Ph.D.
Associate Professor, Coordinator
Social Policy Program
Graduate School of Business
New York University
Close-Up, Education

Lewis, Richard S., B.A.
Managing Editor
Bulletin of the Atomic Scientists
Humberto Fernandez-Moran

Lo, Arthur W., Ph.D.
Professor of Electrical Engineering
Princeton University
Computers

Lockwood, Linda Gail, Ph.D.
Assistant Professor of Natural Sciences
Teachers College, Columbia University
Education

Maran, Stephen P., Ph.D.
Head, Advanced Systems and
Ground Observations Branch
Goddard Space Flight Center
Astronomy, Stellar

March, Robert H., Ph.D.
Professor of Physics
University of Wisconsin
CERN: Experiments in Physics and People;
Physics, Elementary Particles

Mayer, Jean., Ph.D.
Professor of Nutrition, Harvard
University School of Public Health
A Matter of Fat

McGowan, John A., Ph.D.
Associate Professor of Oceanography
Scripps Institution of Oceanography
New Peril to Our Oceans

Meikle, Thomas H., Jr., M.D.
Dean, Cornell University Graduate
School of Medical Sciences
Neurology

Merbs, Charles F., Ph.D.
Associate Professor of Anthropology
University of Chicago
Anthropology

Mertz, Walter, M.D.
Chief, Vitamin and Mineral
Nutrition Laboratory, U.S.
Department of Agriculture
The Nutritious Trace

Novick, Sheldon
Editor, *Environment Magazine*
Environment

O'Neill, Eugene F., M.S.E.E.
Executive Director
Toll Transmission Division
Bell Telephone Laboratories
Communications

Price, Frederick C., B.S.
Managing Editor
Chemical Engineering
Chemical Technology

Price, John Bryan, Jr., M.D.
Associate Professor of Surgery
College of Physicians and
Surgeons, Columbia University
Medicine, Surgery

Rainey, Froelich, Ph.D.
Professor of Anthropology
Director, University Museum
University of Pennsylvania
Close-Up, Archaeology

Rodden, Judith, M.Litt.
Research Archaeologist
Archaeology, Old World

Romualdi, James P., Ph.D.
Professor of Civil Engineering
Director of Transportation
Research Institute
Carnegie-Mellon University
Transportation

Sagan, Carl, Ph.D.
Professor of Astronomy
Director, Laboratory for Planetary
Studies, Cornell University
Close-Up, Space Exploration

Shapley, Deborah, B.A.
Staff Writer,
Science
Science Support

Spar, Jerome, Ph.D.
Professor of Meteorology
New York University
Meteorology

Steere, William Campbell, Ph.D.
President, New York
Botanical Garden
Botany

Sutton, H. Eldon, Ph.D.
Professor of Zoology
University of Texas, Austin
Genetics

Tilton, George R., Ph.D.
Professor of Geochemistry
University of California, Santa Barbara
Geochemistry

Treuting, Theodore F., M.D.
Professor of Medicine
Tulane University School of Medicine
Medicine, Internal

Watson, James D., Ph.D.
Director, Cold Spring Harbor
Laboratory
Professor of Molecular Biology
Harvard University
Escalating the War on Cancer

Weber, Samuel, B.S.E.E.
Executive Editor
Electronics Magazine
Electronics

Wheeler, John A., Ph.D.
Joseph Henry Professor of Physics
Princeton University
Beyond the Black Hole

Wittwer, Sylvan H., Ph.D.
Director, Michigan Agricultural
Experiment Station
Michigan State University
Agriculture

Woltjer, L., Ph.D.
Rutherfurd Professor of Astronomy
Chairman, Astronomy Department
Columbia University
Astronomy, High Energy

Zimmerman, David R.
Free-Lance Science
and Medical Writer
Reprieve for the Dwindling Ospreys

Contributors not listed on
these pages are members of
the *Science Year* editorial staff.

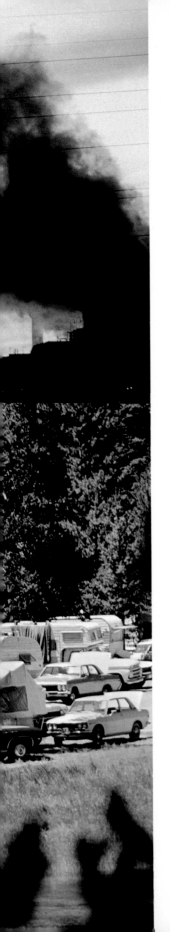

Assault on The Green Kingdom

By Darlene R. Stille

**In city, countryside, and wilderness alike,
man-made dangers lay siege to our plant life**

The threat to wildlife in the United States has become a widely recognized problem, and many people are concerned about the fate of endangered species of animals. But, few realize that a threat to our plant life also exists. Through man's initial invasion and development of the continent, many of our native plant species have become extinct. Others now face new, insidious forms of danger.

When the first Europeans landed on the shores of America, they found a land of seemingly endless beauty and resources. But, they were followed by waves of settlers, who swept across the land, cutting down trees in the eastern forests, plowing under the prairie soil in the Midwest, and raising sheep and cattle in the Far West, where overgrazing virtually destroyed the native grasses. Meanwhile, cities began to grow. The land they stand on has now been buried under concrete, and the suburban ring continues to expand, further destroying habitats for local plant life. What remains of our swamps and coastal marshlands is being "reclaimed" for agricultural or industrial purposes; water is being diverted and drained away for commercial development.

In urban and industrial areas, the plants that have survived the bulldozer now face a far more sinister threat—pollution. Automobile exhausts and toxic wastes from industry have badly injured and even killed certain species of plants. Pollution damage to farm crops causes huge financial losses each year.

The threat to our plant life has not been confined solely to developed areas, however. As the population grew it also became more mobile; today, people invade the wilderness by the millions. Unfortunately,

11

Lichens, partnerships of algae and fungi, have almost completely disappeared from cities, possibly because of sulfur dioxide and the low humidity and soil moisture in these areas of concrete and cement.

Parmelia sp.

some careless wilderness visitors trample plants, injure trees, and tear up the soil with minibikes, snowmobiles, and dune buggies. Even well-intentioned tourists can harm plant life by trampling vegetation in delicate ecosystems or by picking or digging up rare species.

The problem of endangered plants is complex and difficult for scientists to assess. They know little about the extent of the damage and danger. Some rare species that are threatened in the few locations in which they grow are clearly in danger of extinction. They are regarded as universally endangered species. But, for the most part, the problem lies in locally endangered species. A plant that may be on the verge of vanishing from one area may be fairly abundant in another. As a result, the peril to local plant life tends to be overlooked.

Aside from the irreplaceable aesthetic loss when a plant species disappears from an area, there is the larger ecological concern. Since all living things are dependent upon one another, no species can be removed from an ecosystem without creating a chain reaction. Also, a disappearing plant is a signal that something is wrong—a polluted environment, a dangerously low water table, not enough open space, or too many people concentrated in one area.

Botanists have long been concerned with the various dangers to plants. For example, a group of U.S. Department of Agriculture scientists have been studying the types of pollution that affect certain species of plants and the susceptibility of individual plants within a species. They are trying to develop hardier plants that can survive in a polluted atmosphere until the sources of pollution are eliminated.

In the case of plants that face extinction, some conservation groups advocate the establishment of an international seed bank to preserve plant specimens for future generations. The United Nations Food and Agriculture Organization regards seed banks as vital to the prevention of worldwide famine. Because of the increasing use of high-yield hybrid grains, primitive food plants are rapidly disappearing. With them goes the gene pool that was thousands of years in the making—starting when man began to cultivate the soil. The numerous primitive strains developed special characteristics enabling them to fight off diseases or survive changes in local environments. Left with relatively few kinds of high-yield plants that are vulnerable to pests and disease, future generations face widespread famine unless the resistant primitive varieties are preserved for crossbreeding purposes.

Protecting plant life in wilderness areas or preserving plant habitats can be complicated by legal or political problems. Business and conservation interests often clash. However, legislative action, encouraged by conservation groups, has prevented the destruction of some plant habitats and given legal protection to some endangered species. In our overcrowded national parks, the establishment of well-marked nature trails can discourage people from wandering over large areas, trampling the vegetation. To deal with the annually increasing number of visitors, some ecologists have recommended eliminating campgrounds from the national parks and also limiting the number of people allowed in the parks at any given time. A program of rationing the wilderness is already in limited use in Yosemite National Park in California.

Horned bladderwort (*Utricularia cornuta*)

On the valuable land near Lake Michigan, the specter of development threatens to wipe out bladderworts, *above and below,* that grow in ponds between dunes.

Water lotus (*Nelumbo lutea*)

Floating bladderwort (*Utricularia inflata*)

Because of pollution and other changes in Lake Erie, half of the 40 aquatic flowering species that grew about 80 years ago in Put-In-Bay, Ohio, on the western lakeshore have vanished. The water lotus, *left,* is now rare.

13

Douglas fir (*Pseudotsuga menziesii*)

Ponderosa pine (*Pinus ponderosa*)

Air pollution affects plant life in and near urban or industrial areas. Ozone in photochemical smog, made by sunlight acting on hydrocarbons and nitrogen oxides, has killed or injured more than a million ponderosa pines in the San Bernardino National Forest, about 70 miles from Los Angeles. Ecologists blame fluoride emissions from an aluminum reduction plant in Montana for damage to Douglas fir and lodgepole pine in Glacier National Park and Flathead National Forest.

Lodgepole pine (*Pinus contorta* var. *latifolia*)

White lady's-slipper (*Cypripedium candidum*)

Oval-leaved milkweed (*Asclepias ovalifolia*)

Fringed gentian (*Gentiana procera*)

Farming, industrial developments, and the growth of cities destroyed the Midwestern prairies. This loss of habitat, coupled with a drop in the water table, threatens the existence of many prairie plants, such as the yellow fringed orchid.

Yellow fringed orchid (*Habenaria ciliaris*)

Red mangrove and flower (*Rhizophora mangle*)

The drainage and diversion of water through canals has dangerously lowered the water table in southern Florida. Now, fires rage out of control in the Everglades area because the water that normally covers the rich organic muck dries up. The soil itself burns; fire destroys bald cypress trees and the roots of smaller plants. Also, the Florida coast is under constant development, and mangroves that once held the ocean at bay are being replaced by retaining walls.

Bald cypress (*Taxodium distichum*)

Pink lady's-slipper (*Cypripedium acaule*)

The uniquely beautiful Red River Gorge in Kentucky began attracting increasingly large numbers of visitors in the 1960s. At that time, the U.S. Army Corps of Engineers planned to build a dam that would have flooded the entire area. Protests by conservationists caused the corps to build the dam elsewhere, but also created great public interest in the gorge. Ironically, the area is now threatened by a torrent of tourists. Under this unusual ecological strain, a number of plants native to the Red River Gorge area, particularly lady's-slippers, have begun to decline.

Yellow lady's-slipper (*Cypripedium calceolus* var. *pubescens*)

Moss campion (*Silene acaulis*)

Rydbergia (*Hymenoxys grandiflora*)

In Rocky Mountain National Park, visitors damaged alpine tundra by littering, gathering rocks, and trampling moss campion and other cushion plants. This exposed the soil to wind erosion. Establishing well-marked nature trails eased the problem, but some damaged areas that contain much moisture may require up to 1,000 years to recover.

Roseroot (*Sedum rosea*)

The story of endangered plants has yet to be fully told. One of the main obstacles to evaluating the situation is the lack of information on what species are endangered, where, and why. Local botanical societies, conservation groups, or concerned individuals can help in assessing the scope of the damage by reporting any threatened plant life in their vicinity to the National Parks and Conservation Association in Washington, D.C. The association is a member of the International Union for Conservation of Nature and Natural Resources, publishers of the *Red Data Book* listing endangered species of animals and plants throughout the world.

However, saving locally endangered plants generally calls for local action. And, the problem brings one fundamental question of priorities down to the community level: What is more valued, man's notions of progress, or the preservation of the natural environment?

Escalating
The War
On Cancer

By James D. Watson

The new cancer money should be invested in laboratories directed by young scientists and dedicated to fundamental research in animal cells

A desire to banish human cancer has been a prime goal of human consciences for a very long time. The collection of diseases that the term cancer encompasses have been all too often unsparingly fatal, as well as being marked by long periods of pain and anguish. But now, as a result of developments in modern medicine, the picture is radically changing. A number of cancers, particularly those that can be diagnosed early, can be cured. And there is the hope that future scientific advances may soon bring us to a situation where the majority of cancers, if properly treated, can be eradicated.

Behind this new state of affairs is the realization that the science of biology is now advancing very fast, and soon we may be able to pinpoint the essen-

tial differences between normal cells and their cancerous counterparts. For most of its history, biology had been essentially a descriptive science and biologists could not ask incisive questions about the molecular nature of the parts of the cell. But beginning some 50 years ago, the chemical nature and functioning of various key biological molecules began to yield to systematic analysis. Greatly promoting the emergence of biochemistry as a major scientific discipline was the direct application of many new ideas and methods from the fields of chemistry and physics. The availability of key radioactive isotopes, and the use of X-ray diffraction techniques for the structural analysis of macromolecules, for example, were signal events in the study of the chemical nature of living cells.

Correspondingly, a number of scientists trained in physics or chemistry set out to give biology a quantitative vision that before had only been a fringe component. Key progress came quickly in understanding the nature of viruses, particularly those that multiply in bacteria. Especially important was the realization that the genetic components of all viruses were nucleic acid molecules. And when in 1953 the structure of DNA was found to be a double helix, composed of two polynucleotide chains with complementary sequences of bases, it became possible to phrase key genetic questions in precise chemical terms. By 1961, the main pathways of the flow of genetic information (DNA to RNA to protein) were worked out. Moreover, only five years later, in 1966, all the key features of the genetic code became clear.

With these discoveries, life, as such, lost much of its previous mystical quality. Today, virtually all biologists believe that the means either exist or soon will be developed to understand, at the molecular level, all the key features that give a cell the capacity to grow and divide.

We likewise think that this is now the time to seriously study cancer cells and to find out what features give them the capacity for uncontrolled growth, and why they do not stay in their normal cellular environment, but rather spread throughout the body. In this work we wish to find out why the cancer cells seem to simultaneously acquire two abnormal properties: (1) dividing when they should not; and (2) losing their ability to recognize their normal cellular neighbors. Clearly, a key feature in most, if not all, cancer cells are specific changes at the cell surface. So we must ask whether a change at the cell surface leads to loss of control at DNA replication, or whether the cancerous process starts by some molecular disturbance of the DNA replication process which sets into motion a chain of events that culminates in the production of abnormal cell surfaces.

We must also ask what triggers the changes that mark the transformation of a cell from normal to cancerous. Is it a spontaneous event whose frequency we cannot control, or are there external agents that cause it? Now we suspect that many, if not most, cancers arise because of the presence of a variety of outside agents which collectively we call carcinogens. Many act in very specific ways inducing one unique type of cancer. Others increase the frequency of many cancer types. For example, lung cancer is usually caused by cigarette smoking.

Infection by many viruses also can lead to the development of cancer cells (see THE SEARCH FOR A HUMAN CANCER VIRUS). Work with animals has revealed that many common viruses have the potential to cause cancer when infecting a suitable host animal. In establishing viruses as carcinogens the importance of the immunological response as a means for fighting cancers first became clear. Viral infections usually lead to the development of tumors only when their host's immunological response cannot function normally.

This means that when a cancer cell arises due to viral action, its specific cell surface usually becomes recognized as "foreign." Thus it generates an immunological response that eventually will kill it and all its descendants. And we guess that the same picture holds when cancer cells arise due to other causes—that is, a cell would not be cancerous unless it possesses an abnormal cell surface. The effective antibodies are a specific class which are bound to small lymphocytes. They are usually called "cell-bound antibodies" and their respective immunological response, a "cell-mediated immunological response."

The question then must be posed: How do tumors arise in animals which possess functional immunological systems? Is it because the affected host cannot make the specific antibodies needed to recognize the surface component of the respective cancer? Or does some blocking factor, or factors, prevent the cell-bound antibodies from killing the cancer cells? There are now hints that this latter alternative is correct and, once we understand this blocking phenomenon, it may be possible to somehow increase the effectiveness of the cell-bound antibodies and so destroy most growing cancers.

At the same time, we must continue to work on how infection of cells by certain viruses leads to their transformation into cancer cells. In the past, we have largely tended to think of viruses as objects which multiply within cells to form many thousands of identical progeny. In so multiplying, most viruses so greatly disturb the normal functioning of their host cells that the cells inevitably die. In contrast, when an infecting tumor virus transforms a normal cell into a cancerous one, its first step is not to start multiplying, but rather to insert its chromosome into one of its host's chromosomes. The functioning of one or more of these viral genes while inserted into a host chromosome leads to viral-induced cancer. Now we would like to know what these viral genes do. Do they code for specific enzymes which affects the cell surface, or is their action on some aspect of DNA metabolism?

We do know, however, that the genes of many cancer viruses are usually not in an active form after they become integrated into a host chromosome. In fact, many "healthy" cells contain inactive cancer virus chromosomes. Only when some specific triggering event occurs, such as the presence of a carcinogen, do these "cancer genes" start working and change a normal cell into a cancerous cell.

The author:
James D. Watson is Director, Cold Spring Harbor Laboratory, Long Island, N.Y., and professor of molecular biology at Harvard University. He shared the 1962 Nobel prize in physiology and medicine for discovering the structure of DNA.

The fact that a cancer virus chromosome can be present without causing cancer opens up the possibility that cancer viruses can be transmitted through healthy eggs and sperm. Many people now suspect that most cancer viruses are not infectious in the ordinary sense of passing cancer through free virus particles. For example, breast cancer may be caused by a cancer virus that is largely, if not entirely, passed through the chromosomes of germ cells, and not through contact of a healthy person with a diseased person.

It is thus clear that there now exists a variety of experimental approaches, each of which might suddenly reveal one or more profound insights into the nature of the cancer cells and the ways we may prevent them from multiplying. We must not, however, have any illusion that the task will be easy or that any guarantee can be made for real success even on the fundamental scientific level over the next decade. Congress and the President have decreed that the Conquest of Cancer is now a national goal. In December, 1971, the National Cancer Act was signed into law. It greatly increased the amounts of money available for cancer research—perhaps $1 billion a year soon will be available. This does not insure, however, that we will act sensibly. There is no a priori reason why the now boldly proclaimed "War on Cancer" will be any more successful than the other highly touted "wars" of the past decade—the "War on Poverty," the "War Against Crime," the "War Against Drugs," or the "War in Southeast Asia."

D oing "relevant" research is not necessarily doing good research, and it will be a tragic mistake if we do not face up to how hard our objective may be to achieve. Therefore, I think we must look very carefully at the current way we spend our money for cancer research, see what we are getting, and ask whether just more of the same is needed or some really radically new approaches should be adopted. I think we should focus on a number of facts in particular.

■ Historically, most cancer research has been done in clinical environments, either directly in hospitals or in research institutes operated adjacent to and in conjunction with major clinical facilities. While such institutes realize the necessity for long-term research aimed at a fundamental understanding of the cancer cell, their association, per se, with the agonies of the cancer patient mean that they will constantly be looking for research which might yield immediate short-term benefits. Most of them underwent their major periods of expansion when chemotherapy was first being tried out during the 1950s. Optimism then existed that anticancer drugs would soon be discovered that would quickly rid the world of all the major cancers. All such institutes thus bear the organizational stamp of an era which hoped, if not believed, that cancer would not be a major problem a decade later.

Unfortunately, in the meanwhile, no broadly effective anti-cancer drugs have been found though some real cures are being obtained through injection of specific anti-cancer drugs into patients suffering from several rare forms of cancer. One example is Burkitts lymphoma

"... there now exists a variety of experimental approaches, each of which might suddenly reveal one or more profound insights into the nature of the cancer cells ..."

—a cancer of children that largely occurs in Africa. Moreover, there are recent indications that some forms of childhood leukemia also may be cured—this time by the combined use of several anti-cancer drugs. But many other major cancers have proved almost completely refractory to all known drugs. Where chemicals have any effect at all, the remissions that they cause last, at best, only several months. So, the growing feeling exists that the perfect anti-cancer drug may never be found by the current random-search approach in which each year we continue to screen thousands of different chemical compounds for some sort of anti-cancer activity.

So today, much of the effort of major centers for cancer research goes to attempts to heighten the immunological response against tumors or toward marginal improvements in radiation and surgical therapy. Because of these very immediate humanitarian objectives, the major research thrust of almost all major "cancer institutes" does not go toward developing the fundamental knowledge that eventually might be needed for rational attempts at chemotherapy. Although excellent molecular biology, cell biology, and immunology have been carried out within them, it would be very deceptive to believe that these institutions are now, or could become, a significant force in the development of modern biology.

"... laboratories must be created which will allow more people to work together on common objectives and which will provide a degree of instrumentation ..."

■ There is a mistaken feeling that the rash of recent spectacular successes, such as the working out of the genetic code, means that all the key science has already been done, and that applied research is what is needed to solve the cancer problem. In fact, almost all of our "hard" facts still concern bacterial cells and work at the molecular level with the much more complex higher animal cells is essentially in its infancy. I would guess that from 10 to 15 years will be needed to put fundamental animal-cell biology on a firm molecular basis.

■ The very complexity of the cell in higher animals in comparison to the bacteria cell means that much more effort is likely to be needed to solve a given problem than was necessary to work out a key feature of the bacteria cell. This means that laboratories must be created which will allow more people to work together on common objectives and which will provide a degree of instrumentation not yet found in most major American laboratories.

■ No one now wants to face up to the dilemma that virtually all the major universities are financially unable to come up with plans for coordinated research which reflects real expansion. Now most new efforts are papier mâché organizations designed to let the National Cancer Institute (NCI) dole out even more money for pre-existing research programs. I see no sign that any of our major universities, except the Massachusetts Institute of Technology, are in a rush to join the cancer crusade at the fundamental research level.

Yet, traditionally, almost all the cleverest scientists in the United States have been located in its universities, where they can influence the education of the next generation of scientists. The best of our younger biologists will thus home to university environments, even if it means they will either not be involved at all in the cancer program or, if they do work with cancer cells, they will be doing so with, in effect, one hand tied behind their back.

■ We are happily passing out increasingly large amounts of cancer money to profit-making industrial organizations set up specifically to receive governmental cancer largesse. In the past, such companies have never been able to attract the best minds, and I see no indication that the picture is changing. So it seems very likely that, as a result of the exceedingly bountiful amounts of money now pouring out through the various windows of the NCI, the average quality of the scientists engaged in cancer research will go down, not up.

■ Increasingly immense sums of cancer money are being distributed as "contracts" to carry out research which the NCI decrees important to its various programs on the origin, nature, and cure of cancer. Scientists who receive such money are limited to projects which the NCI thinks will pay off and cannot use such money to try other ideas. For legal reasons, contracts must be renewed every year and so the NCI can watch closely what is happening with its money and quickly terminate work not in accord with official policies. This approach might be the best one if we were full of real leads for solving any major cancer. But to my knowledge, there does not exist even one good idea, either inside or outside the NCI, for curing any of the major cancers that have proven refractory to known therapy.

"... our nation may very likely squander a great opportunity to transform cancer research ... into a serious discipline."

■ I am struck by the reticence of most leaders in the cancer field to judge most of our large-scale clinical research programs in terms of whether they are likely to have any significant effect on mortality from cancer. Perhaps this reflects their embarrassment over the wildly optimistic claims of some 15 years ago when chemotherapy was our great hope. But the failure to make hard-boiled assessments makes sense only if either we think the "cure" will come independent of what we do or, deep down in our hearts, we suspect that within our lifetimes the major unconquerable cancers will remain with us, no matter how many bright people we bring into cancer research. Given this latter assessment, it makes sense to refrain from criticizing others while you yourself have nothing to offer, especially when you know that the public wants to hear that their money is buying a series of innovative research programs that will empty the cancer wards.

■ The National Cancer Act by itself is unlikely to significantly affect the way the government cancer bureaucracy hands out its money. Although the legislation created two outside advisory bodies—a three-man National Cancer Panel, with direct access to the President, and an 18-man National Cancer Board—neither will have any operational responsibilities. The increase in major advisers, in the absence of any increase in their real power, may only make matters more diffuse and make everyone even more dependent upon the director and his staff.

Thus, our nation may very likely squander a great opportunity to transform cancer research from a collection of disorganized approaches into a serious discipline. Before the money was available, there was no opportunity to upgrade the existing patterns of cancer research. But with money flowing at a flood pace, we can create a brand of cancer re-

"... create ... 15 to 20 new laboratories whose principal responsibility will be the working out of the fundamental principles of animal cell biology ... located so that they can interact with the best of existing biological and chemical research, and can attract many of the best students ..."

search where success depends upon intellectual innovation and the production of meaningful ideas, especially since many of our best younger biologists now see that the challenge of understanding the cancer cell may be more exciting than any other field.

These young biologists are a very different breed from those who came into cancer research in the 1940s and 1950s. Then, the most creative people thought that working on animal cells was beyond their capabilities and they chose to work on the much simpler bacteria. But now there are innumerable younger scientists in their late 20s or early 30s who wish to master the biochemistry and genetics of the cancer cell. Even more important, they think they have a chance to succeed, provided that they are given free rein. But seeing the chaotic way cancer money is doled out, they wonder whether they can do anything to set things straight, especially since by tradition all real policy decisions about how research money gets spent have been made by men in their 40s and 50s, if not 60s.

They must realize, however, that at this critical moment there is no group of wise decision makers who will run the ball for them. A vast planning effort by the NCI in the winter of 1971-1972 brought together hundreds of specialists in the cancer field. The result was a National Cancer Plan, which will soon be forgotten. It runs some 1,000 pages in length and is so gentlemanly in calling for all ideas worth exploring that it will never serve the purpose of focusing research on problems that have a good chance of being solved.

As a result, much, if not most, of the new cancer research money will fall out to good intentions, rather than to clearly established priorities established by tough-minded assessments of the current prospects for modern biology. And we will further delude the public, if not ourselves, into believing that our current research programs involve the best scientific talent our country possesses. Instead, we will be witnessing a massive expansion of well-intentioned mediocrity.

We could avoid this, however, if the NCI would quickly make four important decisions:

(1) To create, over the next 6 to 8 years, 15 to 20 new laboratories whose principal responsibility will be the working out of the fundamental principles of animal cell biology. They should be located so that they can interact with the best of existing biological and chemical research, and can attract many of the best students who want to choose science for a career. Most would thus be found in academic environments, often as integral parts of departments of major universities. The creation of such laboratories in most cases would have to involve the construction of new buildings, specifically designed for work with animal cells and their viruses. I would guess the cost of each such building to be some $8 to $10 million, and that some $3 to $4 million per year would be needed to cover salaries and research expenses.

These new laboratories will be needed in addition to the new Cancer Centers specified by the 1971 cancer law. Such centers are to be set up with the requirement that, besides providing clinical care, they also do research. But, for reasons discussed earlier, clinical research centers will be largely engaged in immediate clinical objectives, and so it would be a tragic mistake to believe that they will carry out the necessary fundamental research. Hence the need to support centers having no clinical responsibilities. Approximately $200 million will be needed to build and equip the 15 to 20 centers and some $100 million a year must be allocated to keep them operating at full efficiency.

(2) To establish 150 to 200 lifetime positions for scientists who will work in the new laboratories of fundamental research. Long-term salary stability will be necessary for the universities to create the new academic positions—professorships—needed for true expansion. Unless the younger scientists now starting work with cancer cells know that success with their research will be rewarded with permanent academic jobs, they are unlikely to take the high risks associated with very difficult problems. Science students are all too aware of the massive unemployment of scientists who formerly were connected with the space program. So increasingly, they are choosing medical careers, where money will never be lacking, over the uncertainties of a research career where salaries are only guaranteed for a few years at most.

"… establish 150 to 200 lifetime positions for scientists who will work in the new laboratories of fundamental research."

The cost of maintaining these new academic positions would at most be some $10 million a year, or only about one per cent of the total money which soon will go to cancer research. All these jobs should not be funded simultaneously but gradually within the next decade, as the new research laboratories are being formed. Knowledge that such jobs will soon exist would instantly transform our financially stagnant academic world into one which can realistically join the cancer program.

(3) To thoroughly overhaul the current contract programs starting with creation of some high-powered review committees composed of some of the best of the younger scientists connected with cancer research. Optimally, none should now be supported by contract money, because their advice might substantially alter the way future contracts are awarded. Now almost all contract money goes toward: (a) buying research aimed at the origin of cancer—in particular, the possibility that viruses are involved, or (b) developing new anticancer drugs.

No general overviews of either program have ever been done by outsiders, despite a massive shift toward contract research and the anticipation that soon some $200 million per year will be spent this way. With sums this large, the outside scientific community is bound to ask questions about their effectiveness. Even before these programs became so widespread, feelings existed that much of such money was being badly misspent. Until real appraisals are done by scientists known both for their candor and for their ability to get science done, there will be a strong tendency in many scientific circles to associate all cancer research with the weakest of the contract programs, and to dismiss almost the whole NCI effort as a giant pork barrel. This would be a tragic mistake for, as I have tried to show above, the term "cancer research" can

be synonymous with much of the better biology, if not most science, done in the world today.

(4) To initiate a series of imaginative, advanced training courses to promote quickly the development of specialized technical skills that seem essential for a successful attack on the cancer cell. Now we are for the most part supporting cancer research by individuals specifically trained to work in a field of biology unrelated to cancer. Unfortunately, they are seldom given the chance to master the fundamentals of their new research areas. Among experimental disciplines desperately in need of more thoroughly trained talent are animal virology; protein chemistry, in particular the investigation of the glyco-protein molecules found in cell membranes; glycolipid structure and biosynthesis; the culture of animal cells; higher cell genetics; and chromosome structure. Now we essentially let these disciplines develop by chance, hoping that if someone is clever he will move toward an exciting vacuum. But given the decision to almost instantly increase the money spent by some three times, we must take positive training steps, or be prepared to accept undistinguished second-rate, if not third-rate, research.

There will, of course, be many people who will object to these suggestions. Among their reasons will be the accusation that they will cost far too much money and will deprive many cancer patients of the possibility of curing their disease. But all the new research efforts I have outlined above, together with pre-existing commitments toward fundamental research, will not cost more than $200 million per year— only about 20 per cent of the NCI budget projected for the next five years. Money spent for long-term goals thus only marginally can affect either current or future clinical programs.

The second objection may be that not enough good younger scientists are in the pipeline to staff the new buildings that I want constructed. This presupposes that most young people can only follow orders and that what is lacking in cancer research is trained lackeys, rather than senior scientists whose capacity for innovation is limited only by the number of servants they can get hold of. If that were true, their ideas could be farmed out to large industrial-type operations and the wanted results would roll in. But in my experience, the superstars of a given era acquired their reputations when they were young and in their later years seldom come up with anything truly inventive. When something really novel emerges from one of their labs, credit almost always should go to the younger scientist who had the clever idea.

Moreover, I strongly feel that most people seldom live up to their true potential, and that if they are given the chance to think big they often rise to the occasion. This is what happens in real wars where one's life, and sometimes one's country's future, lies in doubt. Then we quickly give real power to the young and promote them to high ranks. Generals sitting in big chairs in Washington have never won battles and never will. Only the young can do it. And with this war, unlike with our squalid mess in Southeast Asia, they have the will to win.

Special Reports

The special reports and the exclusive *Science Year* Trans-Vision®
give in-depth treatment to the major advances in science. The
subjects were chosen for their current importance and lasting interest.

34 **Tuning in on the Grizzly** by Frank C. Craighead, Jr. and
John J. Craighead

50 **Reprieve for the Dwindling Osprey** by David R. Zimmerman

60 **Mars Is an Active Planet** by William J. Cromie

76 **Beyond the Black Hole** by John A. Wheeler

90 **The Search for a Human Cancer Virus** by Lee Edson

104 **The Curious Ancestry of Our Cells** by Jerald C. Ensign

116 **A Matter of Fat** by Jean Mayer

132 **The Nutritious Trace** by Walter Mertz

146 **The Making of a Champion** by Daniel F. Hanley

162 **Earth's Heat Engines** by Kenneth S. Deffeyes

177 **The Force That Moves Continents**
A *Science Year* Trans-Vision®

188 **Science for the People of China** by John Barbour

204 **New Peril to Our Oceans** by John A. McGowan

216 **Supertanker** by Wade Tilleux

224 **Our Earliest Ancestors** by F. Clark Howell

238 **CERN: Experiments in Physics and People** by Robert H. March

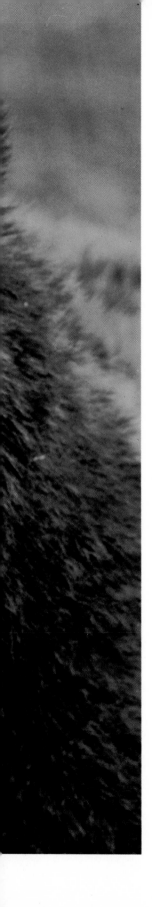

Tuning in on The Grizzly

By Frank C. Craighead, Jr. and John J. Craighead

Sporting collars with radio transmitters attached, grizzly bears have broadcast information that may help them to survive in Yellowstone National Park

Ⓦe approached the female grizzly on foot, at the end of a long day's hike through a snowstorm. Cautiously we followed the radio signal to within 20 feet of her den. Apprehensive that she might suddenly emerge but elated over our success, we stood savoring the moment, engulfed by swirling snow.

That day, bear and man were brought together in Yellowstone National Park's wilderness for a project breakthrough. We knew we had a tool and a technique that could enable us to locate grizzly bear dens and, for the first time, gather data on the denning behavior of the grizzly. The testing phase for our equipment was over, and we were ready to begin our ecological studies in earnest.

Our story began in 1959, when we, along with some of our colleagues and students, undertook a long-range study of the grizzly bear (*Ursus arctos horribilis*) in Yellowstone National Park. The grizzly was considered an endangered species, but scientists knew so little about the biology of these bears that they could not suggest how to halt the decline of the species. They were not even sure how many grizzly bears lived in Yellowstone National Park.

Our goal was to accumulate the scientific information needed to learn how this bear interacts with man. We hoped this information might help ensure the survival of the Yellowstone grizzly population in the face of man's accelerated invasion of its habitat.

To observe the grizzly in its natural state, we employed some of the most ancient and some of the most modern tools of science in man's oldest and perhaps most complex laboratory—the natural environment. Primarily, we combined the age-old method of observing free-roaming animals with the inventions of the electronic age.

Today, between 200 and 250 grizzly bears live in Yellowstone National Park and the adjoining national forests (the Yellowstone ecosystem). In 1971, more than 2 million people visited Yellowstone National Park, and each year the number will surely mount. Grizzly bears do not mix readily with humans, and when two aggressive, unpredictable animals—grizzly bear and man—attempt to share the same area, there is bound to be conflict. Even though the grizzly possesses enormous physical strength, he is no match for the ways of modern man.

Less than 200 years ago, this animal competed with the North American Indian for a wide variety of wild foods in a spacious, uncrowded environment. Wherever the grizzly existed, it was either the dominant species or a coequal with aboriginal man. To many Indian tribes, the grizzly was a sacred animal, hunted as a test of strength. The Indians prized the grizzly's claws as a symbol of great skill in hunting. They also believed that eating raw bear heart could give them the courage and physical strength of a grizzly bear. However, with the settlement of the West, grizzlies began to disappear. Grizzly bears were indiscriminately shot, trapped, and poisoned by white explorers, homesteaders, hunters, and ranchers.

The number of grizzly bears in the continental United States has rapidly declined since the early 1800s. Before the coming of the white man, grizzly bears ranged from the Pacific Ocean to the Mississippi River and from Mexico to the Arctic Circle. In California alone, there were once an estimated 10,000 grizzlies, but by 1924, all had disappeared. Probably less than 800 or 900 grizzlies now exist in the entire continental United States. They are generally found only in the high mountain country and the wilderness areas of large national parks and forests. Only in Alaska and the western part of Canada are grizzly bears still relatively abundant.

The authors:
Frank C. Craighead, Jr., is senior research associate at the Atmospheric Sciences Research Center, State University of New York, Albany, and president of the Environmental Research Institute, Moose, Wyo. John J. Craighead is leader of the Montana Cooperative Wildlife Research Unit, U.S. Bureau of Sport Fisheries and Wildlife, and professor of zoology and forestry at the University of Montana, Missoula.

To understand and thus intelligently contribute to the management of grizzly bears within the 5-million-acre Yellowstone ecosystem, we had to determine how many grizzlies lived there and find answers to questions about their birth rate, death rate, age structure, and behavior. We also had to learn how they relate to man and his use of the Yellowstone environment.

Envision how you would learn the vital statistics of life and death as well as the daily and seasonal activities of 200 human beings if you could not communicate with them or even identify individuals. Also, consider how you would study an animal whose habitat is 85 per cent densely timbered terrain. Within this habitat, the grizzly is active mostly at night. In addition, it spends half of the year sleeping in a cave hidden by a deep blanket of snow.

We solved the identification problem by developing techniques for capturing and color marking bears so that we could observe them as individuals in the population. We captured them in culvert traps or shot

The authors prepare to pull a grizzly, which has been immobilized by a muscle relaxant, out of a culvert trap to measure it and color mark it with ear tags.

them in the open with syringe darts containing immobilizing drugs fired from a rifle. Our trap consisted of a piece of metal culvert, 3 feet in diameter and 8 feet long, mounted on a trailer frame. It had a grate at one end and a door at the other that slammed shut when a grizzly entered to take our bait of raw meat or bacon. Through holes in the sides of the culvert trap, we injected the bear with a muscle relaxant. After the grizzly was safely immobilized, we attached numbered metal ear tags and color-coded plastic ear markers. We also tattooed the bear's number under one of its forearms.

Finding and following the elusive grizzlies posed greater difficulties. We had to develop a research technique that would permit us to locate and observe these animals both day and night as well as follow them over extensive areas of their terrain. To accomplish this, we collaborated with engineers to design a radio-tracking system. Our field requirements for studying the ecology of the grizzlies were translated into tracking-system specifications on signal range, weight, power, battery life, and frequency (32.02 megahertz).

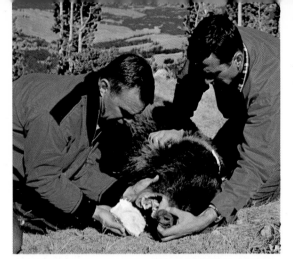

Dental impressions are used, *top*, to determine the ages of the bears. An electrocardiogram, *above*, shows the bear's heart action. A spring scale, *right*, is used to weigh young grizzlies.

In the fall of 1960, we tested a prototype transmitter with a power output of 10 milliwatts, and by the summer of 1961, using a 100-milliwatt transmitter, we were ready to put a radio on our first grizzly. We captured and immobilized a female, bear No. 40, and outfitted her with a special collar containing a radio transmitter and weighing about 2 pounds. Then we released her and began testing our system. We tracked this female bear intermittently for a month, picking up her pulsed signal, or "beep," on a portable directional receiver at distances of up to three-quarters of a mile. Later modifications greatly improved this wildlife telemetry system, but nothing could detract from that first thrill when we followed a signal to the crest of a hill and saw the radio-collared grizzly plodding along, completely unaware of our presence. We then realized that we could arrange a meeting between bear and man at times and places of our own choosing.

With our receivers coupled to stationary and mobile directional antennas we could simultaneously monitor four grizzlies and discriminate between them. In order to distinguish between bears, each transmitter was pulsed at a different rate. We used fixes, the intersections of two or more radio bearings, to plot the locations of the animal on a map. The system also had the potential of transmitting physiological data, such as body temperature, at the same time that it was sending information on the location of the bear.

We first concentrated on obtaining data on sex and age, growth and development, and reproductive condition. Not only did we mark the captured grizzlies, but we also measured and weighed them. We found that a large adult female weighs from 350 to 450 pounds; a large male, from 800 to 1,100 pounds. We even took an electrocardiogram of a grizzly and discovered that its heart action was similar to that of a dog. But, some of the most important data we collected had to do with age.

One of our most valuable "diagnostic" instruments for determining the health of the grizzly bear population in Yellowstone is the age structure. We knew the ages of some of the bears we had marked, such as cubs and yearlings. But, the ages of older grizzlies we captured and marked had to be determined by precise, scientific means.

To do this, we used dental stone to make foot and tooth impressions. We measured the impressions to determine the ages of bears up to 3 and 4 years old. Older bears' ages were found by extracting their small fourth premolar. When the roots of these teeth were sectioned and

A groggy grizzly bear, wearing plastic ear tags and a radio collar, is released back into his environment. His new adornments will enable researchers to monitor his future activities.

stained, we could count the cementum layers, just as one counts the growth rings of a tree. By tabulating all of the annuli, or "rings," we could assign an accurate age to each adult we captured. The bear was then released back into the population, minus his fourth premolar.

Information from thousands of observations and from systematic censuses of marked bears whose ages were definitely known, taken over a 12-year period, plus the data we obtained from determining the ages of adult bears, told us the size and the age structure of the population. On the average, 19 per cent of the population are cubs; 13 per cent, yearlings; 11 per cent, 2-year olds; 13 per cent, 3- and 4-year olds; and 44 per cent, adults. Adults ranged from 5 to 25 years old. These data told us that the Yellowstone grizzly bear population was stabilized or perhaps increasing at a very slow rate. In other words, the population was in delicate equilibrium.

While we were accumulating population statistics, we were also studying other important aspects of grizzly bear life, such as denning behavior. The den of bear No. 164 was the first of 12 that we located by radio tracking. We found that all bears excavate their dens; they do not use natural shelters. All dens are dug in isolated sites with the entrance facing north. Most are at the base of a large tree and lined with boughs.

The grizzlies use the claws on their forefeet to dig their dens. They sever roots that get in their way with their teeth. The claws, which may be 3 to 4 inches long, or longer, are one of the characteristics that distinguish the grizzly from the black bear.

A directional antenna to receive signals from radio-collared grizzlies helps researchers track bears over rough terrain up to 20 miles away.

Grizzlies enter their dens during snowstorms. This effectively, if not intentionally, erases their tracks and all other signs of their presence. This behavior trait could help protect the bear from his major enemy, man, when the grizzly is most vulnerable.

Bears are not true hibernators. During the winter sleep, a bear's body temperature drops only by 5° to 11° F. Because the temperature is still relatively high (90° to 96° F.) and the metabolic rate is correspondingly high, bears can be easily roused. Nevertheless, the grizzly apparently neither eats nor eliminates while in its winter den.

While in their winter dens, pregnant females give birth to from one to four cubs. The average litter is two cubs. Birth generally occurs early in February. The newborn cubs weigh from 12 to 18 ounces and are blind and hairless. Around mid-May or early June, the mother and cubs leave the den and move to spring feeding areas. By this time the cubs have fur and weigh from 10 to 12 pounds.

When they are about 6 months old, the cubs start feeding on vegetation and meat. In September or October, when the cubs weigh from 100 to 125 pounds apiece, they accompany their mother to an area where she digs a new winter den. Seldom is a den reused. About the middle of November, during a heavy snowfall, the family enters the roughly 5 by 4 by 8-foot excavation.

The following spring, the female may wean her now year-old cubs, or she may suckle them through another season, den with them, and wean them as 2-year-olds. She weans them just prior to the breeding season in June and July. Until she has weaned her offspring, the female will not accept a mate.

One season, while tracking No. 40, we came upon female No. 101 and her cubs. Bear No. 40 led us to where No. 101 was digging a den. We learned that two females and their cubs will sometimes travel together for months, and that females will adopt orphaned cubs.

The male, however, has no role in the family life. Adult males are avoided or even attacked by females with cubs because it is not uncommon for male bears to kill the helpless young. On many occasions, we watched female grizzlies with cubs attack large males that approached too closely or exhibited aggressive behavior. A female with cubs reacts suddenly when confronted with an aggressive male. She alerts her cubs, but they often flee in different directions, making it impossible for her to retreat with them in an orderly fashion. She appears confused. With head and neck extended and ears laid back, she rotates, stepping first in one direction, then in another. Suddenly she wheels and attacks the male. Her attack may last only seconds, and then she quickly retreats to search for the cubs.

During the breeding season, the female will mate with more than one male. The large males compete with each other for the females. Fur flies and deep wounds may be inflicted. It is not uncommon for a male's ear to be ripped off, a lip torn, or a jaw fractured.

The grizzlies' complex social structure is most evident during the breeding season. Dominant males win their positions in the social hierarchy by aggressively competing for the females. Aggressiveness is also a factor in determining a female's rung on the dominance ladder.

This socially useful arrangement of dominant and subdominant animals tends to reduce conflict among the grizzlies. Subdominant animals display certain submissive signs and postures to avoid fighting

At his headquarters in Yellowstone National Park, John Craighead gets a fix on a grizzly from the intersection of two radio bearings and plots the location of the bear on a map.

Their densely timbered, rugged habitat makes observing grizzlies a challenging task.

with dominant animals. Submissive behavior can also end a fight before lethal wounds are inflicted. The most common submissive posture is lowering and turning the head to the side while slowly retreating.

At no time did we see grizzlies staking out definite territories. We never observed them defending an area, whether a range or a den site, against other grizzlies. A show of dominance, when and where needed, seems to substitute for any clear-cut defense of a definite territory.

However, grizzlies do have home ranges and smaller seasonal ranges within the home range. We plotted these ranges to determine how much space a bear needs. During an 8-year period of research, we studied 24 radio-collared grizzlies. For example, we watched bear No. 40 for portions of eight consecutive years and tracked her to four different dens. Her plotted movements showed her lifetime home range to be about 30 square miles.

We plotted the positions of all 24 grizzlies on maps, and by drawing connecting lines through peripheral fixes, we were able to outline sea-

A radio-collared bear signals the location of its den, *left,* even though snow erased its tracks. A grizzly family heads for dinner, *below,* along one of the bear trails found near sources of food.

After feeding on elk carcass, *below,* a grizzly covers it with soil and debris. The grizzly will later return to its cache, which can provide food for up to a week.

The fiercely protective mother grizzly is never far away from her playful cubs. Females may have from one to four cubs, but the average litter is two.

sonal, annual, and home ranges of the bears. These varied from 12 square miles for the home range of a young female to 168 square miles for the summer range of a young mature male. The ranges of some adult males were more than 1,000 square miles. Factors such as age, sex, food supply, and den sites influence the size of a grizzly's range. Most ranges overlap and are shared by more than one bear. For example, weaned yearling grizzly No. 202 established his range within the boundaries of his mother's and also dug his first and second winter dens within his mother's home range.

Foraging treks appear to be part of the grizzlies' normal pattern of movements. Grizzly No. 76 made a 50-mile excursion in seven days, circling Mount Washburn and crossing the Yellowstone River. Female No. 150 and her three cubs roamed a range of 27 square miles. But, for a full month before hibernating, they foraged in a restricted area of about 1 square mile. Such information gave us clues to how small an area could support a grizzly family. We found that, on the average, 29 square miles supports one grizzly in the Yellowstone ecosystem.

Because we could close in with our tracking system and observe grizzlies undetected, we were able to learn more about their diets. In early spring, when they emerge from their winter dens, grizzlies feed on carcasses of elk and other large herbivores that have died during the winter. After feeding on this carrion, the grizzly generally covers the remaining portion of the carcass with soil and debris and returns to feed on it for several days or a week. It does not vigorously defend its food against other grizzlies but normally shares it with them. Grizzlies, however, will defend food against black bears and coyotes, and also against man. Grizzly No. 158 led us to a carcass where 24 other grizzlies had assembled and were feeding at one time. Their behavior was governed by social status, with subdominant bears deferring to the dominant ones.

Grizzly bears are not efficient predators on large game. Their role as a carnivore is first as a scavenger, second a predator on small prey, and last a killer of large prey animals.

During the summer, the grizzlies congregate in meadows and river bottoms to dig roots and tubers and graze on sedges, clover, and grasses. They also eat berries, feeding almost exclusively on them when they are abundant. We monitored grizzlies in open areas feeding on field mice and pocket gophers when the populations of these small rodents were high. During seasons when pine nuts were plentiful, we moved with the bears to the timbered ridges where the white bark and limber pines grew. In spring, we watched them strip the bark from lodgepole pines to consume the edible cambium, the layer of soft tissue between the bark and wood of trees.

Although a relatively inefficient carnivore, the grizzly bear is an extremely efficient omnivore. In the process of evolving omnivorous feeding habits, the grizzly also evolved some useful social habits. He can function as a "loner" or within a complex social structure. This dual behavior has enabled this great bear to exploit a wide range of food sources, but in modern times it has also brought the grizzly trouble by bringing it into conflict with man. The Yellowstone grizzly has become a threat to park visitors, and this, in turn, is a threat to the grizzly.

Since it is natural for grizzlies to congregate where food is abundant, the garbage dumps in Yellowstone Park seasonally attracted large numbers of grizzly bears. Some of these mobile, naturally inquisitive animals wandered into campgrounds by accident and fed on food scraps they found lying around.

A cub rummaging in campground trash cans could grow up to be a problem bear. Adult bears that learn to forage in campgrounds lose their fear of man.

A portable culvert trap used to remove problem bears awaits this adult grizzly feeding at a park garbage dump.

Grizzlies that learn to forage in campgrounds become more aggressive than the bears that feed only at dumps, and thus are more dangerous to man. We have found that such animals learn to associate food getting with humans and soon lose their natural distrust and wariness of man and human scent. Such animals may coexist with people for extended periods of time. However, sooner or later, these man-conditioned animals are startled by humans at close range or, in attempts to get food, completely disregard humans. This may occur in a campground, on a trail, or in the back country, and may result in human injury or death. But, for a grizzly bear to lose its natural shyness or fear of man requires cooperation and encouragement, and the initiative is usually with man.

Bear No. 202, for example, expanded his home range to include a campground. Though he was not an aggressive bear, he learned in the campground to associate food with man and soon lost his wariness of park visitors. His campground forays become bolder and bolder until he had to be killed for safety's sake.

One way of dealing with these man-conditioned bears is to transport them by truck, boat, or helicopter to remote regions of the park. We attached radio collars to some of these transported bears and followed their movements. The grizzlies demonstrated time after time that they have strong homing instincts. Almost invariably, those transported and released within the park returned quickly to the original point of capture. We radio tracked bear No. 170 for 34 miles through timbered, mountainous terrain. In 62 hours, she returned to her home range. When these troublesome bears keep returning to areas heavily populated with park visitors, they must be either killed or shipped to zoos.

To keep grizzlies from becoming conditioned to man, the bears and man must be kept apart in Yellowstone. This means that the bears

Troublesome bears are often trucked to remote park areas. But, most of these grizzlies soon find their way back to campgrounds.

must be slowly re-educated to use entirely natural foods in wilderness areas of the park. Not only must garbage dumps be eliminated, but the campgrounds must be "sanitized." For example, bear-proof garbage cans will help discourage grizzlies from foraging in camping areas. But campers must also cooperate by not leaving food out in the open.

Within the park, the ranger staff controls troublesome grizzlies. From 1931 to 1970, 140 grizzlies were killed. This is an annual average of 3.5 bears over a 40-year period. However, 22 grizzlies were eliminated from the Yellowstone population in control measures in 1970. Twenty-one met death from other causes. In 1971, the total known mortality inside and outside Yellowstone Park was 45 grizzlies, making the known mortality for the two-year period 88 grizzlies. While some new cubs may have gone unrecorded, approximately 42 were born during the same period. This indicates a mortality rate twice as great as the birth rate, and this in itself would appear to be serious. But, the recorded death rate does not tell the entire story. Statistics compiled

Grizzly Bear Behavior— And Yours

If while vacationing in Yellowstone National Park you come upon a large bear with dark-brown to light-blond hair, long claws, and a distinctive hump above its shoulders, you will have met a grizzly. But, your chances of meeting one are small; your chances of being killed or injured by this huge, but usually man-shy bear are even smaller.

Grizzly bear attacks on man are rare, but in recent years, fatal attacks in Glacier National Park have generated public apprehension. However, even in Yellowstone, where the grizzly population is relatively high and the number of visitors increases each year, there is little chance of a bear-man confrontation.

Since 1900, only three persons have been killed by grizzlies in Yellowstone. Two deaths occurred in the early 1900s; one, in 1972. From 1931 to 1970, grizzlies injured 63 persons. During those 39 years, an average of 1 million tourists visited the park each year, so the injury rate from grizzlies was 1 person per 600,000 visitors. From 1959 to 1970, grizzly injuries averaged 2 a year, about 1 for every 900,000 visitors.

It is not always easy to determine why some of these attacks occur. The temperaments of individual grizzlies differ and seem to depend on their age, mood, or physical condition. A hungry or injured grizzly may attack more readily than a well-fed, healthy bear. Females weaning their offspring are very irritable. A female with cubs is nervous and defensive and quick to attack.

Grizzlies attack when they are startled, cornered, or injured, and at times without any apparent provocation. The victim is usually knocked down by blows from the grizzly's powerful forepaws or by the sheer weight and momentum of the charging bear. Grizzlies inflict injury primarily by biting the victim, usually on the head.

Our studies show that the grizzly has poor eyesight. Apparently it cannot identify an animal or man beyond 600 feet. However, it can detect motion beyond that distance and will circle downwind to investigate with its keen sense of smell.

If human scent is detected, a grizzly normally retreats. But, its behavior depends on the distance between bear and man at the moment the grizzly discovers him. There is a critical distance within which the bear will attack. Usually, it is less than 300 feet in open country.

You do not want to surprise a grizzly. So, if you are hiking in open alpine country, be alert and cautious when approaching the top of a hill or ridge. If you see a grizzly in the distance, circle upwind so that it can get your scent.

Grizzlies often sleep during the day in heavily timbered areas. A grizzly startled from sleep at close range (100 to 150 feet) may attack. To avoid surprising a grizzly, talk or whistle when walking in wooded areas.

If you should surprise a grizzly at close range, do not run or make noises that might sound threatening to the bear. You cannot outrun a grizzly; some can run as fast as a horse. Just stand still and talk softly until the bear retreats. If you can find a sturdy, low-limbed tree nearby, climb it. Grizzlies cannot climb trees, but they have been known to "tree" people for several hours. So, if time permits, take your pack up with you.

Should the bear attack and no tree is available, "playing dead" will probably minimize injuries. Resistance may prolong the attack, and movements or any attempt to flee is likely to bring about more attacks.

Do not feed a grizzly, intentionally or by accident. The smell of food attracts grizzlies, so be sure to carry dried or well-packaged foods when camping in the wilderness. And, keep your camp area clean. Before going to bed, cache your food and cooking pots at least 200 feet from where you sleep. Sleep under a tree that is easy to climb in case a bear should enter your camp.

Before leaving camp, clean it up and carry all trash out with you. Never bury trash. Burn your cans to eliminate odors, then flatten them, put them in a strong plastic bag, and carry them out.

Practicing caution and common sense in grizzly country can protect both you and the bears.

during our study indicate that additional grizzlies die from unknown and unrecorded causes. Consequently, the population loss for 1970 and 1971 is even more serious than the death statistics show.

Complex and sensitive biological factors maintain the balance in this small population of grizzly bears. The females first breed successfully at 4½ years of age, though many may not produce cubs until they are 8 or 9 years old. Only about 45 per cent of the adult females breed each year, and a relatively small number of productive females maintain the population level. Therefore, an increase in the death rate of adult females could jeopardize the population.

Furthermore, 27 per cent of the cubs born do not survive the first year and a half of life. So, not all of those estimated 42 grizzly cubs born in 1970 and 1971 may live to replace the 88 or more grizzly bears that died during the same period.

The effects of future environmental changes, hunting pressure, or management practices could be critical. Changes unfavorable to the population would be expected to show a decrease in the percentage of cubs in relation to other age groups. If this occurs, or if the ratio of adult females to adult males changes adversely or the population level drops rapidly, then we will be forewarned that the grizzly is indeed in danger of disappearing from the Yellowstone ecosystem.

To help us measure and assess future changes, we are building a computer model of the Yellowstone grizzly population. The model is based on the data collected from studying 264 color-marked and 24 radio-collared grizzlies over the past 13 years. It will enable us to project data and predict the future for the Yellowstone grizzlies.

There are some who ask: What good is a grizzly, an animal that is potentially dangerous to man? The national parks, they argue, are for people to enjoy in peace and without fear of attack from these animals. But, the grizzly is one of the many living things—plant and animal—that are intricately related and interdependent upon one another. It is this interdependence that makes continuing life on earth possible, and we must respect it and learn more about it to ensure our own survival. For man to eliminate every species that appears to be a potential danger is to betray a fundamental disrespect for life. And, if men have grown so intolerant and selfish in their dominance of the environment that they cannot coexist with some 200 grizzly bears, how can they hope to coexist with each other?

For further reading:

Craighead, F. C., Jr., and J. J. Craighead, "Knocking Out Grizzly Bears for Their Own Good," *National Geographic*, August, 1960.

Craighead, F. C., Jr., and J. J. Craighead, "Trailing Yellowstone's Grizzlies by Radio," *National Geographic*, August, 1966.

Olsen, J., *Night of the Grizzlies*, G. P. Putnam's Sons, 1969.

Leopold, A. Starker, "Weaning Grizzly Bears," *Natural History*, January, 1970.

Reprieve for the Dwindling Ospreys

By David R. Zimmerman

A young scientist's unique scheme is buying time for the birds in their struggle to survive chemical annihilation

A college student's anguish at watching the disappearance of wild birds he loves has given birth to a new conservation strategy—one that may be their salvation. The program may even be extended to save other bird populations now endangered by the chemical poisons that man has dumped into the environment.

In 1968, when he first realized that he might help, Paul Spitzer was 21 years old. The bird that stirred his compassion is the osprey, the diving, chocolate-brown and white, fish-catching bird of prey with a wingspan approaching 6 feet. It was once a familiar species that lived near bodies of water in most parts of the world. One colony bred abundantly around the marshy mouth of the Connecticut River, where it flows into Long Island Sound near Spitzer's hometown, Old Lyme, Conn.

Abandoned osprey nest symbolizes the tragedy of the vanishing Connecticut colony.

The author:
David R. Zimmerman is a free-lance science writer. An amateur bird watcher since he was a child, he has written several articles on endangered birds.

Today, Spitzer is a Cornell University graduate student working for his Ph.D. degree in ecology and a leader in the environmentalists' struggle to save U.S. wildlife from chemical death. His ideas and work and that of several young colleagues were the basis for convening the first North American Osprey Research Conference, at the College of William and Mary, Williamsburg, Va., in February, 1972.

Ever since Rachel Carson published *Silent Spring* in 1962, documenting that DDT was causing a decline in American birdlife, many deeply concerned young men and women have wanted to help stem the destruction. Spitzer may be the first scientist, of any age, to develop a way to do so. Ideas from young people can lead to important achievements, particularly when they are developed in collaboration with older, more experienced heads. For Spitzer, the older head was Roger Tory Peterson, the noted ornithologist and bird-guide artist. When Spitzer was only 10 years old, he and his family moved to Old Lyme, where the boy became Peterson's neighbor and admirer, and eventually his protégé and co-worker in the struggle to save the osprey.

Peterson, whose interest was soon picked up by the young Spitzer, is an osprey enthusiast. He had first come to Old Lyme to live near what was then one of the world's largest colonies of breeding ospreys. In about 1960, Peterson first noticed that the ospreys' young were becoming rare and suggested that pesticides were the culprit. Spitzer often accompanied Peterson on visits to the nests and was deeply moved by what he was seeing. His distress over the problem spoke eloquently when the younger man later observed, "The dying osprey is part of the agony of the dying river and marsh.... The salt marsh is beautiful and productive, part of the countryside, and part of New England's heritage, but [it is] the osprey [that] gives it animation...."

In the Connecticut River estuary, the birds declined from 71 active nests in 1960, to 13 in 1965, to fewer than 5—most of them barren—in 1972. What disturbed Spitzer most was the manner in which death came to the birds. A bird that is shot with a gun falls instantly dead. The pesticide-burdened flock of ospreys did not just die. The adults came back to their New England haunts from migration each spring, repaired their nests, mated, and laid eggs that failed to hatch. Each spring, they tried—and failed—again. Steadily, their numbers dwindled. The colony was threatened with extinction.

A process called biological magnification contributed to the ospreys' decline. Chemicals spilled into rivers, for example, are often concentrated in algae. In the operation of a food chain, each successive predator absorbs—and further concentrates—the poisons contained in its prey. After being passed from algae to worms or other small water animals to fish to ospreys, for instance, the chemicals are so concentrated that they may damage reproductive mechanisms.

Despite high concentrations of pesticide in their systems, the birds instinctively continued their habits. Only the delicate biochemical balances of their reproductive processes were affected, so that they laid

Osprey Colonies

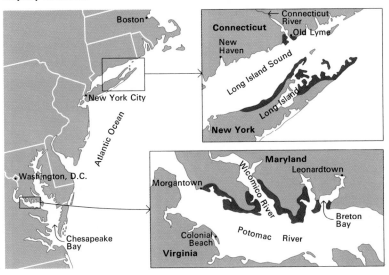

eggs with shells that were damaged or too thin to last through hatching, or with embryos that failed to develop. Scientists are concluding that many species of birds are similarly threatened. They believe that DDE, a breakdown product of the DDT that enters the ospreys' system through the fish that they eat, somehow interferes with the transport of calcium from the mother's long bones, where it is stored, to the shell. Thin, fragile shells with inadequate calcium are what Spitzer calls "the fingerprint of DDE." He now feels DDT may be twice guilty—as the chemical parent of DDE and as the source of the high toxicity that destroys embryos before hatching.

Polychlorinated biphenyls (PCBs), persistent and widely used industrial chemicals, also concentrate in fish-eating birds. Some authorities believe these also attack birds' reproductive ability. PCBs are chlorinated hydrocarbons closely related to DDT. In 1971, Spitzer found that Long Island Sound ospreys were more contaminated with PCBs than reports indicated for any other wildlife in North America.

The distinctiveness of this new, slow, chemically induced death inspired the germ of Spitzer's plan to save the birds: If the eggs alone were faulty, then replacing them with healthy eggs laid by "cleaner" birds in less-contaminated regions might keep the Connecticut River colony alive. Ospreys usually return to their nurseries when it is their time to become parents, and Spitzer hoped that fledglings hatched from foster eggs by Old Lyme ospreys would return to Old Lyme to breed. His scheme might preserve the colony, and its migratory traditions, until pesticides are brought under control and the birds are able to resume their normal reproductive pattern.

Peterson was impressed enough by Spitzer's proposed experiment to gain him an audience with academic and governmental conservation experts. The young college student was hired as a federal biological aide. In the spring of 1968, with the help of the U.S. Fish and Wildlife Service's Patuxent Wildlife Research Center in Maryland, he began transferring osprey eggs and chicks. He moved them from fertile nests near Chesapeake Bay to barren ones along the shores of Long Island Sound. Later, grants from the Audubon Society and the Northeast Utilities Company helped to keep the project going.

The eggs were carefully carried to Connecticut by boat, automobile, airplane, and then automobile and boat again. To protect their delicate contents, they were nestled into foam-lined picnic coolers or suitcases and warmed with hot-water bottles. At all times, the containers were held on the courier's lap so the shock absorbence of a human body further cushioned them from jolts, bumps, and engine vibrations along the way. The downy chicks were easier to transport, usually traveling in a cardboard box in a station wagon. The station wagon soon acquired a strong aroma from the fish-eating birds.

The osprey foster parents proved remarkably cooperative. When Spitzer approached their nests to replace the damaged eggs with viable transferred ones, they flew up to the sky, soaring, diving back down in anger, and uttering their high-pitched distress cry, *cree, cree, cree.* But after the exchange was made and Spitzer had withdrawn, they invariably returned. As if nothing had happened, they began brooding their foster eggs and, later, feeding their foster chicks.

Work in the marsh is not easy. To reach a nesting tree standing in shallow water, Spitzer dons canvas shoes and wades to its base. Climbing to the nest—as high as 80 feet up a dead tree—can require special equipment. Spitzer occasionally wears a pair of lumberman's climbing irons strapped to his legs. Sometimes he can use a ladder.

The nest itself, a large structure of twigs and small branches, may be hard to get at, even though the outraged ospreys leave promptly. Spitzer usually finds it flaring wide above him, almost too wide and shaky to reach around and into.

Sometimes the original eggs had crumpled before replacements were available. But Spitzer and his colleagues—a group of Peterson's friends and neighbors—found a way to hold the ospreys to their nest and maintain their parental impulses and behavior for days or even weeks: They loaded the nest with a hard-boiled hen's egg from someone's lunch.

Accepted, hatched, and nurtured, the transferred young grew up as Long Island Sound ospreys. Before their first flight, Spitzer attached brightly colored bands and streamers to their legs so he could identify them later. More than 50 of them thus far have fledged (matured enough to fly) in five seasons. Presumably, they headed south on their maiden migrations. Thus, it seemed clear that the first part of Spitzer's experiment had succeeded.

The big question remained: Would they return to Long Island Sound to breed? Spitzer assumed that they would return to the place where they were raised, rather than to Chesapeake Bay, where they were conceived. Before the ospreys first return—as adults, usually in

Delicate osprey eggs are cradled in soft plastic and kept warm with hot-water bottles during the journey from Chesapeake Bay's shores to Long Island.

their fourth spring of life—the hazards of nature and man take a heavy toll: Normally, only about a third survive to breed. The egg transfers began in 1968, so 1972 and 1973 pose important tests of Spitzer's plan. Each year, he has examined every osprey nest, hoping the bright bands and streamers will still be attached to its occupants, identifying them as the birds whose adoption he brought about.

Spitzer's original transfer plan was based on one other critical assumption: That osprey colonies elsewhere could spare a supply of eggs or young each year. Initially, the Chesapeake Bay colony could. But pesticides and PCBs now threaten the future of the Chesapeake ospreys. Reproduction in many sections of the Chesapeake Bay area has fallen below the average of one fledgling per active nest each year that can sustain the birds there.

Fortunately, a solution may be at hand. Robert S. Kennedy, then a graduate student at William and Mary, conceived a way to manipulate the faltering Chesapeake Bay birds. His approach might help them to produce enough viable eggs to maintain their own population and provide a surplus for southern New England.

Kennedy, an ornithology student and coordinator of the osprey research conference, recalled a classic observation from the 1930s, when collecting wild birds' eggs was an accepted practice and collectors sought to increase their destructive harvest. Ospreys normally lay only one clutch of two or three eggs a year, but collectors reported they will produce a second, supplemental clutch if the entire first clutch is taken away as soon as the female lays it. The thievery could pay a dividend. Here, Kennedy realized, might be an untapped conservation re-

Paul Spitzer and a co-worker row to the more remote nests along the marshy shores of Long Island Sound.

source. His scheme—which he calls "double-clutching"—succeeded in its first trial in 1971. It calls for taking eggs from the nests of less-contaminated ospreys as soon as the birds lay their first clutches. Some birds in a region are less contaminated than others, at least partly because there are less pesticides in their immediate area. The "stolen" eggs are put in a poultryman's incubator. Artificial incubation of osprey eggs had only rarely succeeded. Kennedy's system quickly hatched several, primarily because he got the eggs back into an incubator within two or three hours. In addition, careful record keeping let him concentrate on nests that had produced healthy chicks a year earlier. As the first clutches matured in the incubator, the adults laid a second clutch, which they were allowed to keep.

When the first clutches hatch, the young birds are returned to nature. Some are given to local birds that, while too contaminated to lay viable eggs of their own, are still able to raise healthy young. Others are shipped elsewhere—to Spitzer in Connecticut, for example.

Double-clutching's potential was demonstrated when the first three pairs of clean, double-clutched ospreys produced an average of 1.67 young per pair in their incubated first clutches and 2 young per pair in their undisturbed second clutches. The total—3.67 young per productive pair—virtually doubled their natural production. Buoyed by success, Kennedy reported, "We are going to do this program in Virginia and, hopefully, are going to do it on a scale that is large enough so that we can provide as many eggs as are needed in New England and elsewhere." Birds of prey, particularly fish-eating birds, are endangered in many parts of the United States.

Even unhealthy birds may benefit. Kennedy feels that birds with pollution-damaged reproductive systems may rid themselves of part of the problem with early clutches of eggs. This would mean that subsequent clutches would have a better chance to survive. In one instance, his theory seems to have borne fruit. A pollution-tainted female that had produced only dead eggs finally laid one that hatched. It was from her second clutch, after Kennedy had removed her first of the season.

How successful Spitzer, Kennedy, and their young co-workers will be remains to be seen. Even with double-clutching, a supply of healthy birds seems essential to provide enough healthy eggs. It may already be too late to save the Long Island Sound ospreys. The breeding population of ospreys over that entire area has already fallen below 100 nests, less than one-tenth of the population before the use of pesticides and other chemicals became so widespread.

Nevertheless, Spitzer and Kennedy have introduced a whole new conservation strategy. When a bird population's future hinges on an annual production of dozens or even hundreds of eggs, it is obvious that such methods as double-clutching and egg transfer can be immensely important conservation techniques. True, their greatest promise is in assisting naturally small populations of large, easy-to-locate

A telescope tripod over his shoulder, Spitzer heads for the marshy area near the Connecticut River in search of osprey nests.

Spitzer carries foster eggs to a special platform, built to give ospreys a nest location free from marauding animals. In the nest, *center,* the transplanted eggs await adults' return.

birds, such as ospreys. But, because these large birds of prey are at the end of pesticide-concentrating food chains, they also are among those in the most imminent danger.

Other birds of prey are in even greater danger than the osprey. The population of the bald eagle is down to about 3,000 in North America, virtually none below the Canadian border. The peregrine falcon, one of the world's swiftest birds, is virtually extinct as a breeding bird in the contiguous 48 states, principally because of DDT and DDE attacks on its reproductive system. A scant handful of peregrines continue to nest in North America west of the Rocky Mountains. Ornithologist Tom Cade of Cornell University suggests that double-clutching be applied to increase the offspring of the survivors. Besides peregrines, other hawks and perhaps even eagles may be induced to lay two or more clutches each year. The method was even under consideration as a way to rebuild the population of rare monkey-eating eagles, which are found only in the Philippine Islands.

Late in the spring of 1972, Spitzer reported enthusiastically that he had found four cases of clear success in his fight for the ospreys. Mature birds from transplanted eggs had returned to Long Island Sound to begin raising their own families. Spitzer thinks that several more birds have returned, although their identification is less positive. If the pattern continues, the life and customs of the Long Island Sound ospreys may be sustained for a few years. If, during that extra time, the chemicals that attack the birds can be controlled, those years may mean the salvation of the dwindling colony.

Adult ospreys return promptly to the nest and begin brooding the new eggs as if they were their own.

Increasingly, the persistent pesticides that are the source of many of the damaging chemicals are being restricted. Their lingering effects, however, may be too persistent to be overcome so quickly.

How fruitful the techniques born of Paul Spitzer's concern will be—whether chemical threats can be eased within the time the new methods may buy—cannot now be predicted. But it is clear that Spitzer and others like him have gone far toward fulfilling a personal and scientific credo that he set for himself five years ago when he began his project.

"Perpetuation of ospreys in an area where they appear doomed has beauty, and an element of defiance, as well as scientific merit," he said. "If I, as a scientist, do not treat the osprey as something rare and precious, I cannot expect others to do so...."

For further reading:

Bent, Arthur Cleveland, *Life Histories of North American Birds of Prey*, Dover, 1961.

Carson, Rachel, *Silent Spring*, Fawcett, 1962.

Hess, John, editor, "Where Have All the Ospreys Gone?" *National Wildlife*, December, 1970.

Hickey, Joseph J., editor, *Peregrine Falcon Population: Their Biology and Decline*, University of Wisconsin Press, 1968.

"Osprey Refuge Planned," *National Parks*, September, 1969.

Peterson, Roger Tory, "Osprey: Endangered World Citizen," *National Geographic*, July, 1969.

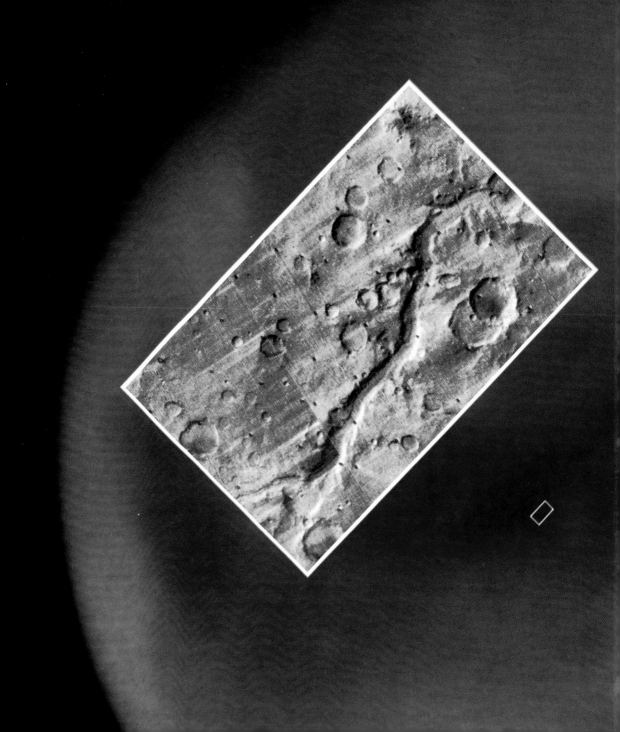

Mars Is an Active Planet

By William J. Cromie

Mariner 9 reports suggest the varied face of Mars was shaped by volcanic action, driving dust, and melting frost

The scientists had puzzled over the mysterious dark spots for days. Four huge, round blemishes on the face of Mars had been photographed by a camera aboard Mariner 9 just before the spacecraft went into orbit on the afternoon of Nov. 13, 1971. A violent dust storm cloaked the Martian surface, but earth-based radar measurements revealed that the blotches were in a high area, over 9 miles above the lowest points on Mars. The puzzling spots, the scientists reasoned, must be mountain peaks rising above thick, windblown clouds of dust.

On November 17, Mariner 9, the first vehicle to orbit another planet, was in a position to take a high-resolution photograph of one of the spots. Scientists gathered in the windowless mission-control

Televised Mariner 9 photos of Mars, set against an earth-based telescopic view, shows a valley whose meandering shape suggests water erosion.

61

center at the Jet Propulsion Laboratory (JPL) in Pasadena, Calif. JPL is a National Aeronautics and Space Administration (NASA) facility operated by the California Institute of Technology. The picture signals traveled as electronic pulses over the 50 million miles from Mars to a 210-foot-diameter antenna at Goldstone, Calif. There, they were amplified and sent along a chain of microwave relay stations over the San Gabriel Mountains to JPL.

The first photograph flashed on television screens in the control center in the late afternoon. "The picture was a dusty gray-white," Robert H. Steinbacher, a JPL planetologist, recalls. "We could see there was something there, but we could not see the details."

The photograph was immediately put through a computer process that enhances contrasts in shading. When the retouched picture appeared, "an audible 'Oooh!' went through the whole place," said Steinbacher. There before the scientists was one of the most important discoveries ever made about our solar system. The picture showed that Mars is a dynamic, geologically alive planet, not the cold, dead world most everyone had thought it to be.

In less than a minute, a still clearer, filtered version of the photograph came on the screens. It revealed a vast crater with scalloped edges, characteristic of a volcanic crater, or caldera, on Earth.

Some scientists doubted what they saw. For most, however, the photo was conclusive evidence of volcanic activity on Mars. Molten rock must have surged repeatedly out of, then subsided into, volcanic vents, collapsing the crater walls and creating the 40-mile circular pit.

"Seeing a caldera without large numbers of meteoroid impact craters in the hardened lava around it," Steinbacher pointed out, "meant Mars was active in recent geological times—and may still be active."

As the dust began to clear in December and January, the scientists were able to see lower and lower levels. "Then it was just one surprise after another," said Bradford Smith, a New Mexico State University astronomer and a member of the Mariner 9 scientific team. The big crater turned out to be the top of the largest volcano ever seen by man. Called *Nix Olympica* (Snows of Olympus), the volcano is probably 30,000 feet high with a base some 310 miles wide. "This is twice as big as the great volcanic pile that forms the Hawaiian Islands—the greatest one on Earth," says geologist Harold Masursky of the U.S. Geological Survey. Masursky led the Mariner 9 television experiment.

The other dark spots also turned out to be huge calderas. One of them, called South Spot or the Gordian Knot, is 75 miles in diameter. Unlike craters on Earth, it has a flat floor, terraced walls, and is surrounded by a series of arcing faults. "It is unique in our solar system," commented Steinbacher. By mid-1972, Mariner 9 had shown with thousands of pictures that the planet Mars is unique.

For centuries, men thought the red planet was like Earth. When Italian astronomer Giovanni V. Schiaparelli discovered *canali* (channels) on Mars in 1877, he thought they were natural waterways. But the

The author:
William J. Cromie is president and editor of Universal Science News, Incorporated, and a frequent contributor to *Science Year.*

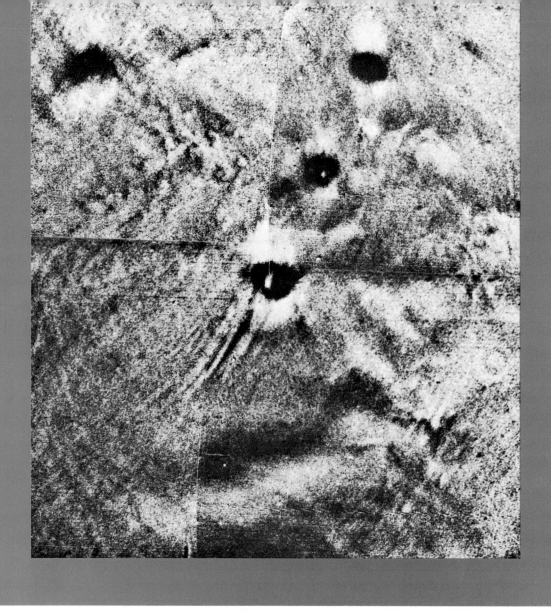

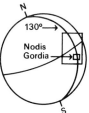

word was translated into English as "canals," and by the beginning of this century, Harvard University astronomer Percival Lowell found wide acceptance for his idea that the canals were constructed by intelligent beings. He believed the canals carried water from the polar caps to desert areas. Mars watchers had for centuries reported an annual wave of darkening that moved across the surface in spring and summer. Lowell attributed it to the growth of plant life.

In 1965, the world got its first close look at another planet when Mariner 4 flew within 6,100 miles of Mars and sent back 22 photographs. They showed a barren and primordial-looking surface, cratered by the impacts of billions of meteorites. The photographs so resembled those of the lunar surface that most scientists concluded that Mars was a cold, inactive body like the Moon. In 1969, Mariners 6 and 7 passed

Special processing of a photomosaic reveals some detail through the great dust storm. The lowest dark spot is a volcanic caldera, about 124 miles across, south of Nodis Gordia.

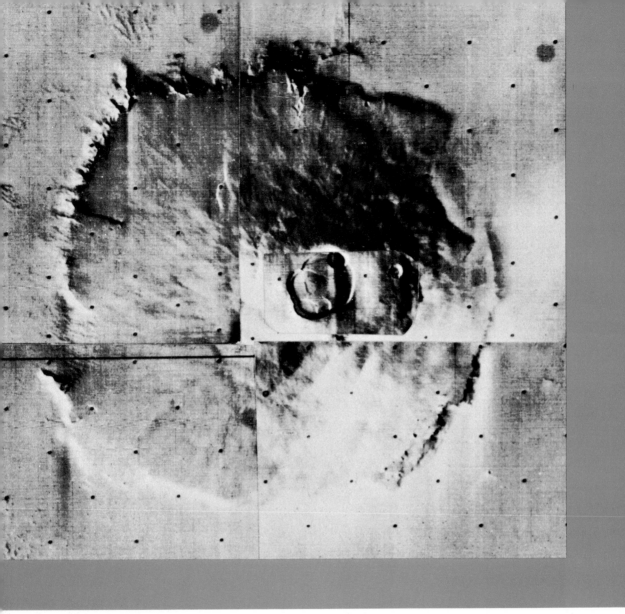

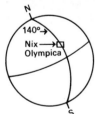

Gigantic Nix Olympica, a volcanic mountain, has a crater 40 miles in diameter and a base 300 miles across. A ring of steep cliffs drops off to the surrounding plain.

within about 2,000 miles of Mars, snapped about 200 pictures, and made complex scientific measurements. These fly-by missions uncovered some puzzling and unexpected features, but did little to change the impression that Mars was much like an oversized Moon.

At the western edge of the area covered by its cameras, Mariner 6 had photographed terrain so jumbled that geologists, despite all their special and precise technical vocabulary, could only call it "chaotic." A NASA publication described the region as "a mess of short ridges and depressions that could be called the 'badlands' of Mars."

"We just got a glimpse of the badlands in 1969," noted Smith. "It should have been evidence that Mars was more active than we thought, but we missed the clue. Then, in 1971, we spotted the high volcanic mountains poking up through the dust, crying to be discovered. When the dust cleared, we saw that this region of calderas and extensive faulting was just west of the badlands."

When Mariner 9 began its journey on May 30, 1971, earth-based telescopic views showed that the Martian surface was clear. But on September 23, a large cloud of yellow dust began rising. By November 13, little of the surface could be seen through the thick blanket of dust. Astronomers knew that dust storms covering millions of square miles of the surface can last for months. And this was the greatest dust storm in recorded history. Disappointment was widespread at JPL.

Two days after Mariner 9 began its first orbit, flight controllers at JPL fired small rockets to bring the spacecraft into an elliptical orbit around Mars with a low point of 862 miles and a high point between 10,000 and 11,000 miles. Mariner 9 settled silently into its new path to wait out the storm.

On its elliptical path, the spacecraft crossed the orbit of a Martian moon, Phobos, that circles the planet only 3,750 miles away. Outside Mariner's sweep, a second moon, Deimos, orbits at an average distance of 12,000 miles. In 1877, the tiny satellites' discoverer, American astronomer Asaph Hall, named them *Deimos* (Terror) and *Phobos* (Fear) after the attendants of Mars, the mythical god of war.

Mariner's close-up views, the first ever obtained of the satellites, showed dark, irregularly shaped bodies covered with meteorite craters. The tiny moons have gravities too weak to pull them into smoother, more spherical shapes.

Mariner 9 measurements found that Deimos, the outer moon, is about 7½ by 8½ miles in size. Phobos is larger as well as closer to Mars. It measures about 16 by 13 miles and survives on the edge of destruction. Phobos is nearly as close to Mars as a satellite can be without disintegrating under the latter's pull of gravity.

Craters are much more common on both moons than on Mars. This means their surfaces are older, and that some erosive process has worn away or covered up the oldest craters on Mars. The photographs also revealed that Phobos and Deimos are among the darkest objects in the solar system. This suggests they may have once been part of the same body. Perhaps the moons had once escaped from the asteroid belt between Mars and Jupiter, or perhaps they were remnants of the material from which Mars itself was formed.

Even while its target was hidden, Mariner 9 gave scientists unexpected new information about Mars. Tracking data showed that the spacecraft was not behaving as it should. Analyzing the data, scientists calculated that the planet had an equatorial bulge of more than a half mile at about 110° west longitude, and a similar, complementary bulge on the opposite side of the planet. The equator is slightly depressed at points midway between the bulges. The bulges provided an extra gravitational tug on Mariner.

No one knows why Mars is out of round this way, but Phillip A. Laing, a JPL physicist, offered one idea he called "pure speculation."

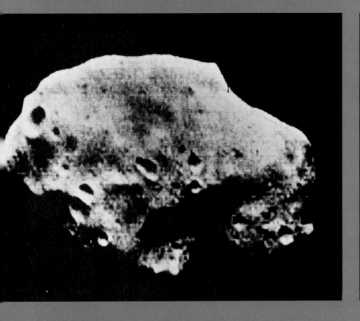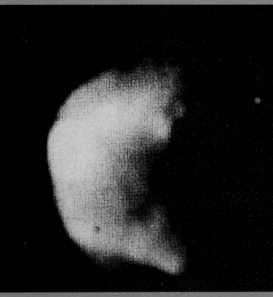

Mars's two small moons, lumpy, irregular Phobos, *above,* and Deimos, *above right,* are too small to be pulled into spherical shape. The profusion of craters on Phobos suggests that it may be very old.

"Mars may have originated as part of a double planet," he said. "If Mars rotated synchronously with its companion [the same side always facing the other planet], the gravitational pull would have elongated the Martian equator while the planet was still hot and its outer rocks soft." The surface of Mars nearest its companion would have been pulled toward the other planet, creating a bulge. The comparative weakness of that force on the far side would also create a bulge there. "After the crust hardened, the bulges would remain," Laing explained. "The companion then might have disintegrated or broken up into asteroids. This theoretical planet would have to have been quite close to Mars—somewhere between Phobos and Deimos. The two moons rotate synchronously with Mars, so they might be pieces of Mars's expired twin planet," he speculated.

Mariner scientists also found surprising local irregularities in Mars's gravity. "Observations of the motions of the moons caused us to anticipate a rough Mars, but it turned out to be more irregular than we expected," said Jack Lorell, an applied mathematician in charge of the mission's celestial mechanics experiment. These irregularities were mapped from the wobbles and dips they caused in the orbit of Mariner 9. Lorell said they were similar to Moon orbiters' fluctuations, attributed to the attractions of *mascons* (large concentrations of mass) beneath the surface. Denser than surrounding material, the mascons' gravity pulled on the lunar spacecraft and changed their orbits.

But, "Mars is much rougher gravitationally than either the Moon or the Earth," Lorell pointed out. "Also, high gravity areas on the Moon occur where topography is low, while on Mars they correlate with high features." Thus, the irregular gravity on Mars seems produced by masses of material piled on the surface rather than in or beneath it.

By the first week in January, much of the dust storm had subsided. Smith was able to report enthusiastically, "The clearing was astonishing." Throughout the modern JPL facility, scientists rejoiced over the clarity of the new pictures coming in to Pasadena from Mariner 9's television cameras. But the relative suddenness of the clearing itself presented them with a new puzzle.

"We thought," said Smith, "that quick, violent disturbances in September had stirred up large amounts of very, very fine material, and that this had been distributed around the planet by a global wind system. The rate of clearing from September to December indicated that the talclike particles would settle slowly and the surface would be obscured for a long time—perhaps months, even years. But the sudden and rapid clearing in late December and January showed us that the particles in the atmosphere settled fast and were therefore much larger than we had believed. They were probably on the order of 10 microns, instead of 1 micron or less as we first thought. (A micron is about .00004 inch.) This conclusion has a consequence that is a little difficult to accept—that the dust must have been continuously stirred up for three months over the entire planet. What kind of a wind system could keep blowing for so long on such a scale? We just don't have the answer."

There is one clue to the wind's origin. Global dust storms seem to occur when Mars is closest to the Sun. On its elliptical, 687-day orbit, the distance from Mars to the Sun varies from 128 million to 155 million miles. Solar energy is 44 per cent greater when Mars is closest to the Sun. "At the time of the closest approach, solar heating might cause atmospheric instabilities," Smith theorized. "Differences in temperature would build up too fast to be smoothed out by normal processes of heat transfer in the atmosphere. This could cause movements of the atmosphere—winds on the order of 225 miles an hour, strong enough to pick up the dust." But their duration was still puzzling.

Some Mariner scientists saw the dust storm as a boon. Masursky called the arrival during the storm's peak "a most fortunate happenstance. The first 30 or 40 days of the mission turned out to be a rare opportunity to study the dynamics of the Martian atmosphere."

Movement of dust creates constant changes on the face of Mars. Planetologists closely watched a region called Hesperia, which Earth-bound observers had seen change repeatedly from dark to light and back. Just before the 1971 dust storm, telescope observations showed part of this area to be dark. Then, Mariner 9 photographs showed a mixture of both light and dark areas there. Still later, Earth-based observations showed the region had turned light again. The scientists concluded that dust storms cause such changes.

Blowing dust also seems to be the best explanation for the seasonal wave of darkening that Lowell attributed to blooming vegetation.

The surface of Mars, as it might look from a Martian jet plane, includes landslide areas and valleys that could result from subsurface collapse. The area is about 200 miles wide.

None of the four Mariner spacecraft spotted any signs of plant life on the Martian surface or any trace of organic compounds.

The blowing dust also might explain a puzzling area in Mariner 6 photographs. Called Hellas, it is a vast, featureless, circular area a thousand miles across. "There should be craters there," said Smith. "They must either have been eroded away or covered up by dust."

Mariner 9 confirmed what previous spacecraft had found: Mars is a world void of oceans and rivers, dry as the Sahara, cold as Antarctica, and enveloped in a thin "air" of carbon dioxide. Since its axis is tilted about 25 degrees, Mars has seasons much like Earth. Earlier, Mariner 6 and 7 had registered a high temperature of about 60° F. in the equatorial regions and a low of –240° F. in the wintery "snowdrifts" near the

south pole. Mariner 9 recorded temperatures ranging from 65° F. at the equator to –200° F. at the north pole.

Mariner flights have shown that atmospheric pressure on Mars is only about one-half of 1 per cent of that on Earth. This pressure is not high enough to allow water to exist in a liquid state. Consequently, it neither rains nor snows on Mars. Water vapor and carbon dioxide condense out of the atmosphere as a light, solid frost. When the temperature rises, this frost sublimates directly into a vapor.

If there is no liquid water, what of the canals that, for nearly a century, were thought to crisscross Mars? Careful observers of earlier years had even mapped them. But today's scientists have decided that they just don't exist. Instead, they have decided the channels are really

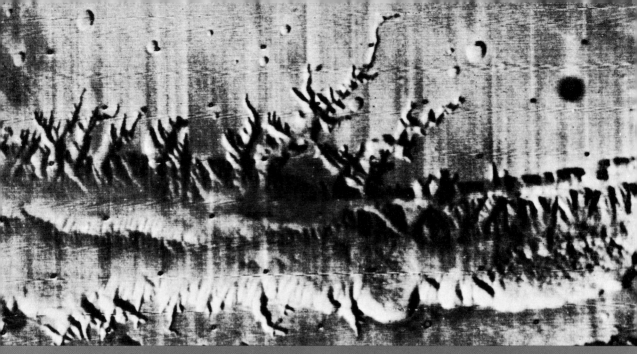

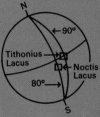

Branching canyons along a valley in the Tithonius Lacus area, *above,* and Noctis Lacus, *below,* resemble stream systems and hint at water erosion. But some scientists are convinced that the resemblance is only superficial.

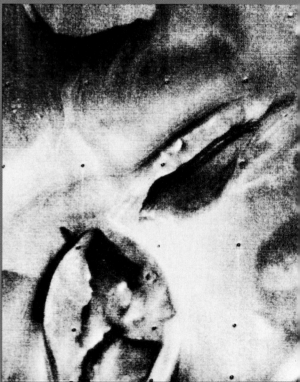

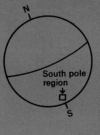

Mystifying geological processes have formed strange patterns in the south polar region of Mars. Dark patches at bottom are on the floors of surface depressions.

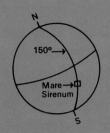

The first clear view of cracks in the Martian crust was found in Mare Sirenum. They apparently occurred as upper surface layers stretched. The widest is about a mile across.

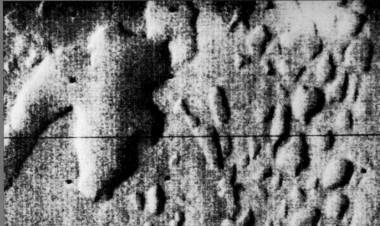

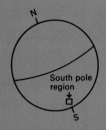

Strange hollows without terraces were unknown on Mars before Mariner 9. Thawing of ice under the surface may have caused it to collapse.

roughly aligned, closely spaced craters and dark spots that the eye tends to see as continuous lines.

Instruments detected less water vapor than scientists had expected. Most of it was found in the atmosphere around the south polar cap. Even there, it is "roughly one-thousandth of what we see in the Earth's atmosphere," according to Rudolph A. Hanel, a NASA scientist from the Goddard Space Flight Center's Laboratory for Planetary Atmospheres in Greenbelt, Md.

Apparently, when it is summer in one hemisphere, water vapor there moves toward the other. The warmer polar cap shrinks and the colder one grows. "Finding the highest amount here [near Mars's south pole] leads us to the conclusion that water vapor is being released from the south polar region as the cap retreats," Hanel said. "The absence of water vapor over the north polar cap indicated that the vapor there is condensed in the carbon dioxide. Therefore, water vapor must be

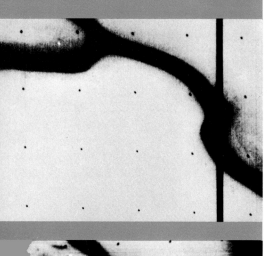

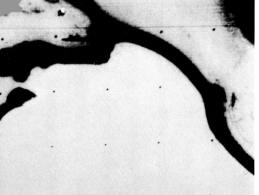

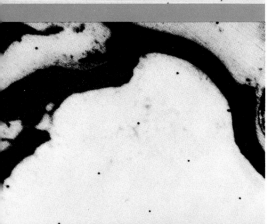

Pictures of Mars's south polar "dry ice" cap taken on November 19 and 28, and on December 1, *top to bottom,* indicate that its carbon dioxide frost cover was shrinking.

transported from the south polar cap onto the north polar cap either by dust storms or by the general circulation" of the atmosphere.

As it does on Earth and Venus, an extremely thin cloud of hydrogen surrounds the outer atmosphere of Mars. Mariner 9 found that the hydrogen cloud extends out more than 12,000 miles from the planet's surface. Scientists believe the hydrogen comes from water that is continuously being lost from Mars. Charles A. Barth, a University of Colorado physicist in charge of one Mariner 9 experiment, described the process: "Water comes up out of the interior of Mars, goes into the lower atmosphere, and is carried to the polar caps. When one polar cap thaws, the water is released into the atmosphere and much of it migrates to the other cap. Some of it, however, goes into the upper atmosphere—very little because the stratosphere of Mars is very cold, $-191°$ F. What little does get into the upper atmosphere is *photodissociated* (split into hydrogen and oxygen) by solar energy." Because of the weak gravity of Mars (about ⅜ that of the Earth), the hydrogen atoms escape into space easily," Barth said. "If you translate the escape rate of hydrogen atoms into other terms, it turns out that 100,000 gallons of water per day is lost from the atmosphere of Mars."

Lack of rain and the absence of oceans and rivers makes the surface of Mars drastically different from the Earth's. On man's home planet, water is the great sculptor. It carves mountains, splits plateaus with canyons, wears down volcanoes, shapes coastlines, fills low spots with sediment. But on Mars, wind rules. It piles dust into long, low ridges; it picks up the dust and uses it like sandpaper to wear down sharp and angular features in the rock. "Mars shows us what Earth would look like if it had no water," Smith explained.

Global dust storms and huge volcanoes make Mars unique from the Moon or the Earth. But they are not the only processes shaping Mars's face. On Jan. 6, 1972, high-resolution cameras recorded several large closed basins and small pits about 500 miles from the south pole. The small hollows are from 1 to 2 miles in diameter; two of the large basins measure about 10 miles across.

The pits and hollows may have been scooped out by wind, but many geologists lean toward the idea that they resulted from slumping of the surface crust after underground ice thawed. Many scientists think there are large accumulations of ice under the Martian surface. Hot gas and molten rock rising from the interior could produce enough heat to melt or evaporate it. "Most of the subsidence, or slumping, we see occurs in tectonically active areas where heat is available," Smith pointed out.

Smith, Steinbacher, and others believe this process could have produced a 46,000-square-mile area of deep canyons separated by flat-topped plateaus that was photographed in Mars's *Noctis Lacus* (Lake of Night) area. The canyon walls, which slope 10 to 15 degrees down to wide, smooth floors, have been intricately fluted by erosion, perhaps by wind. The canyons are from 6 to 12 miles wide and as much as 6,600 feet deep (comparable to the 5,500-foot-deep Grand Canyon in Arizona). "This type of slumping occurs in Alaska on a small scale when *permafrost* (frozen subsurface ground) thaws," Steinbacher noted. "The thing that is unique about it on Mars is the huge scale."

East of the badlands and north of the canyons, a vast chasm 75 miles wide and extending more than 3,100 miles along Mars's equator cuts through high plateaus. Geologists compare its size to the 2,000-mile-long Great Rift Valley that runs through eastern Africa. Atmospheric pressure changes across the chasm's width indicate to the scientists that the chasm is as much as 9,500 feet deep. Mars experts believe it was formed by cave-ins along faults, or lines of weakness, in the Martian crust. "Parallel to the chasm is a smaller fault with a number of little craters that we think are a line of volcanic vents," explained Masursky. Withdrawal of molten rock could cause slumping that would deepen and widen a fault.

Other Martian valleys are thought to be cut purely by lava flows. When lava cools, the top surface hardens first. If remaining molten material then drains away, a thin crust is suspended over a void. Meteorite impacts or other stresses can cause the roof to cave in, exposing the valley. Wind erosion would further modify its shape.

Such features demonstrate that Mars has, or once had, a hot, active interior. The absence of large numbers of meteoroid impact craters and the presence of sharp features in these areas indicate they formed in geologically recent times. "Their age," Smith said, "must be counted in millions, not billions, of years."

But some of the surface seems as ancient and dead as the Moon. "The most astonishing fact brought out by Mariner 9 is that Mars is so different from place to place," remarked geologist Bruce Murray of the Mariner scientific team. "Part of it looks like an old Moon and part of it looks like a new Earth. It's like two different planets."

On Earth, geological action is still taking place. From Moon rock samples, scientists conclude that such action there stopped over 3 billion years ago. Some geologists theorize that Martian evolution stopped some time later. Others, like Murray, believe that widespread geological activity on Mars is just beginning. All the planets are believed to have formed at the same time—about 4.5 billion years ago—

from the coming together of rocks, dust, and gas. Decaying radioactive minerals in this primordial debris caused heat to build up inside the planets. According to Murray's theory, the heat has built up more slowly within Mars than within the Earth because Mars is smaller and thus lost heat almost as fast as it accumulated. "Once heating reached a certain point, it triggered volcanism, faulting, and other processes that are changing Mars from a Moonlike to an Earthlike planet. The active areas—those that resemble Earth—are growing at the expense of the older, Moonlike areas," he said. "Mars is intermediate in evolution between Earth and the Moon because it is intermediate in size."

Controversy centers around a strange valley photographed from 1,033 miles away by Mariner 9 on Jan. 19, 1971. About 3 miles wide, it snakes along for some 250 miles, resembling a giant arroyo—a water-cut gully like those common in the dry, mountainous regions of the Southwestern United States. Feeding into it is a complex of tributaries much like those at the head of river systems on Earth. "There is a great temptation to explain them as due to liquid water," Smith noted. But water cannot exist as a liquid on Mars and there is very little water vapor present in the atmosphere. Only the polar caps might hold large amounts of water. "If there are rivers on Mars, that's where the water would have to come from," Masursky stated. "The presence of liquid water would improve tremendously the possibility that there has been, and is now, life on Mars," Smith added.

Smith, Masursky, and others theorize that every 25,000 years, because of a slow, cyclic change in direction of the planet's axis, temperatures may rise enough for rain to fall and rivers to flow on Mars. Higher temperatures would release the water frozen in a polar cap. The outpouring water vapor would increase the atmospheric pressure, perhaps enough for water to exist in a liquid state on the red planet. "Of course, this whole sequence is pure speculation," Smith emphasized. "The only reason for such a theory is as a possible explanation for signs of water erosion on Mars."

"You can take the most likely explanation to account for individual features and build a theory on it, but that does not mean it's right," countered Murray. "If there was once enough water to erode a canyon 250 miles long and a mile deep, then there should be evidence of running water on other parts of Mars. You should see signs of water courses on the slopes of Nix Olympica and other high volcanoes, for instance. We don't see this."

The new information from Mars does not answer the question of whether or not life exists there. Mariner 9 was not designed for that purpose. It was only to identify the best places to look for unearthly organisms, to ferret out warm, possibly wet locations, probably low spots where atmospheric pressure is highest. These would then be considered as landing sites for two Viking spacecraft that will examine the surface for signs of life. The Vikings are scheduled to be launched in 1975 and to land on Mars the following year. Soil samplers will scoop material from the surface and deliver it to an automated laboratory where it will be assayed for signs of life—such as bacteria or one-celled plants.

But the United States may not be the first country to search for Martian life. Russia has already made the first landing on Mars. In May, 1971, the month Mariner 9 left Earth, Russia also launched two unmanned probes, Mars 2 and 3. Mars 2 went into orbit on November 27, 14 days after Mariner 9. Mars 3 arrived on December 2 and sent an automated craft to make the first soft landing on Mars that same day. The lander began sending television signals via the mother ship 90 seconds after touchdown. Transmission stopped, however, 20 seconds later. Only part of one frame was sent, and no elements of contrast were received on Earth. The fragment was thus unreadable.

This much experience may help Russian scientists make a successful biological assay of Mars before Viking does, perhaps in late 1973. "It is quite possible they will be the ones to make the discovery of Martian life," Smith said simply. "It would depend on how good their instrumentation is."

Man has never yet set foot on another planet. Mars seems to be the logical place for him to take that giant step. If the American Viking or its Russian counterpart should find evidence of living things on Mars, man would undoubtedly mount a crash program to land on the planet that has most fascinated him for centuries. After the explorations of Mariners 4, 6, and 7, the probability of living things on a cold, dead world seemed very low. But now that Mariner 9 has shown that liquid water may exist on Mars, many scientists feel the probability of finding life has increased enormously. Others see no reason for such optimism, and feel that those who do are long on imagination but short on reason.

Murray pointed out that the achievement of all the Mariner 9 objectives would bring man's knowledge of Mars only about as far as his knowledge of the Moon was in 1959. And looking back, it is clear that his knowledge of the Moon in 1959 was far from definitive. Full understanding of Mars will require specimens of Martian material.

Thus, for many scientists on the Mariner 9 team, the search for life is not the only reason to go to Mars. "Living organisms or not," says Masursky, "finding that Mars is alive and active makes it a thousand times more interesting than before." Murray added: "The popular fantasies about canals and intelligent beings have now been replaced by knowledge that it is a unique planet with its own geological processes—one that can tell us much about the evolution of our solar system."

For further reading:

Leighton, Robert B., "The Surface of Mars," *Scientific American*, May, 1970. A report on Mariner 6 and 7 pictures.

Steinbacher, Robert H., *et al*, "Mariner 9 Science Experiments: Preliminary Results," *Science*, Jan 21, 1972. Eight early reports by members of the Mariner 9 team on different aspects of the program and its findings.

Wilson, James H., *Two Over Mars*, National Aeronautics and Space Administration, 1971.

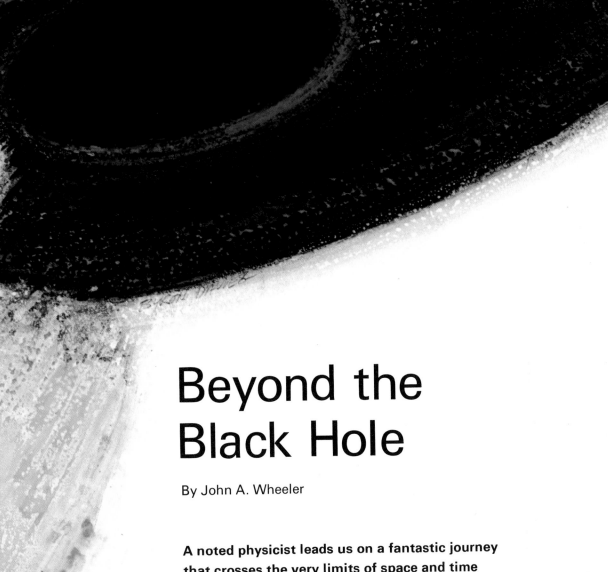

Beyond the Black Hole

By John A. Wheeler

**A noted physicist leads us on a fantastic journey
that crosses the very limits of space and time**

"Surely our rocket engine is off, Audrey."

"Of course it is, Fred. I checked the shut-off valve and fuel supply myself at 11 o'clock as I always do. What are you worrying about?"

"Audrey, we've lived with weightlessness all these 48 months, and now something is going wrong. I'm frightened. Remember Professor Frenchet's prediction? 'There's a stellar cluster you'll be passing through that contains a black hole,' he said. And we're passing through that cluster right now! Can't you feel something strange?"

"No. I'm still strapped in my chair. Otherwise, the least movement bounces me from one side of the cabin to the other."

"But look at what's happening to all the pencils and pliers and the other loose things. Some are collecting at one end of our cabin and some at the other. There's not one loose thing left in-between."

"Do you mean we're in a tidal force field that's pulling harder at the ends of the ship than at the middle?"

"Yes, Audrey, and that's bad enough. But it's only a symptom of something far worse, I'm afraid."

"Look here, Fred. Stop worrying about such strange things as black holes. Here we are in the most natural state of motion—falling through the emptiness of space. There's nothing to get in the way if we don't bump into some star. And I don't see a star nearby, so stop worrying."

"But that's just what I'm telling you. We *have* almost bumped into a star. Only it's a black hole. Its gravity has caught us without warning. A star that gave off light would have scared us off long before we got this close."

"This close? How close? Now you frighten me, Fred. Oh, something strange *is* happening. I'm being stretched!"

"We must be very close to the black hole. The force has pulled off my shoes! Quick, Audrey, open the fuel valve!"

"Oh, Fred, I can hardly ... I'm being pulled apart! There! Just did it."

"Audrey, we're in orbit around the black hole. Quick, before we both faint, what are we going to do? Back to Earth or on to Zeta Zeta?"

"Let's go home, Fred—right now. Push the E button—the Earth button."

"Done! Can you hear me above the main engine, Audrey?

"There, the engine has shut off.... Isn't it good to be back to normal weightlessness! That dreadful stretching has stopped. Are you all right, Audrey?"

"Yes, Fred, but look what I've made you do. All that time and money and effort are gone into nothingness. We've dumped the great dream overboard. We were to carry human life and human knowledge to the 'They' on Zeta Zeta. But not now. When will the chance ever come again? And Fred, what about us? When we were on course we had a future. It was an uncertain future, with the 'They' in it, but it was an exciting one. Now what future do we have? I'm terrified we'll never make it back to Earth."

"Don't talk that way, Audrey. You weren't pulled apart, and neither was I. What a narrow escape it was! Stop being terrified of the future. We've just had all the terror we can take in one dose.

"We would never have made it if Dave had not insisted on wiring in that automatic 'black hole circuit.' I always joked about it, but he put it in anyway. He told me never to turn off the fuel valve. If we did, and ran onto a black hole in our sleep, the rocket engine would not fire automatically to drive us from its pull, and we'd spiral into the hole.

"We overrode his instructions. We gave the rocket fuel only at the last moment. It was a miracle that its thrust, and the computer that programmed that thrust, could deposit us in a circular orbit, zooming around and around the black hole. A little mistake in thrust could easi-

The author:
John A. Wheeler, professor of physics at Princeton University, is a member of the National Academy of Sciences and former president of the American Physical Society. He was awarded the Albert Einstein Medal in 1965.

ly have plunged us down the hole, to destruction. Or it could have sent us hurtling out into space in some random direction, lost forever. How wonderful it was that Dave's circuit could do that steering and do it right! To do that job, it had to work out how massive the black hole was. Then, after piloting us into a circular orbit, it had to keep accurate track of our circling. Otherwise, it could never have decided when to blast us out of orbit when I pushed the E button.

"Audrey, let's find out more about what we've escaped. Let's give Dave's computer a closer look. Push the 'display' button."

"The panel shows 'm = 39,960 M.' "

"A black hole almost 40,000 times the mass of the Sun! No wonder it was able to reach out and pull us in!"

"It also says '$J/m^2 = 0.999$'."

"That means the black hole was spinning at almost its greatest possible speed. It must have been formed fairly recently. Listen, there's the buzzer. Do we want to let it go out or shall we push the 'delete' button? I had almost forgotten that all we say is recorded on tape and will soon go back to Earth by radio unless we erase it."

"We almost always do, Fred, but not today. It will be the most important message Earth has ever received from us!

"Fred, we haven't been listening to any newscasts from Earth in the last few months. They became too boring. But no matter how boring, it would be comforting to hear one now."

"But, Audrey, how can the news be anything but stale? It took 3,000 Earth years to get a radio message to us when the black hole crisis struck. It took us 3,000 Earth years plus only a day more to cover the same distance. Between the time we took off and the time of transmission of the latest radio message to reach us at the black hole, there's been only 24 hours of fresh news. I used to think we would have to live 3,000 years to travel 3,000 light-years. But at our ship's high speed it has taken us only 48 months on our clocks. We have had only one day's news bulletins spread over the trip. It's no wonder they are so repetitious! So why try to get another?"

"Because, Fred, now we're going back toward Earth. Instead of half an hour of Earth history being stretched into a month of our time, from now on it's just the other way around. A month of Earth history is compressed into a half hour. Each day we are going to be getting four years of news; each month, 125 years of history. In our entire 48-month trip back to Earth, we are going to be moving 6,000 years into the future. How fascinating! Quick, let's turn on the radio.... Fred, don't we owe them some news, too?"

"No, Audrey, forget it. From what we've said, it isn't even useful to send any more messages. We're traveling back toward Earth at very close to the speed of light. Our distress radio message from the black hole will reach Earth only 24 hours before we do. All the messages we could send during our return will reach the Earth bunched up into the last day before we arrive. Nobody there can possibly be prepared to record four years of messages in a day. Skip that futility! Let's put everything we can on tape. Let people study our record at leisure after we

"... now we're going back toward Earth. Instead of half an hour of Earth history being stretched into a month of our time, from now on it's just the other way around. A month of Earth history is compressed into a half hour."

arrive. Now we have a lot of work to do. We have to go over and over what has happened and what it means. Then we have to record it all. We have to do a lot of heavy reading about this in the microfiche library of all the world's science and history that we have aboard. What a job of analysis and reflection we have ahead of us to make that tape! Good thing we have four years to do it in."

■ ■ ■ ■ ■ ■ ■

"Fred, my part of the report will begin with history. In the reading I've been doing these past few months, I've discovered that the idea of a black hole is very old. In 1798, French mathematician Pierre Simon Laplace noted that a star as dense as the Sun, but extending out more than twice as far as the Earth, could not give off light. A year later, Laplace published the details of the reasoning that led him to this conclusion. To escape from the Moon, he explained, a projectile needs a speed of 1½ miles per second. To escape from the Earth, it must start at 7 miles per second; or from the Sun at 383 miles per second. Laplace asked how big an object would have to be so that a projectile starting off with the maximum known velocity, the speed of light itself, would fall back defeated. This was the first mention of what investigators later began to call a black hole.

"In 1939, U.S. physicists Robert Oppenheimer and Hartland Snyder analyzed the gravitational collapse of a spherical cloud of dust particles. They did not deal with matter of a fixed density and compose objects of different sizes, as Laplace had done. Instead, they considered what happens when a fixed amount of matter becomes denser and denser. In the process of compression, the pull of gravity at the surface of the cloud of dust becomes stronger and stronger.

"In the beginning, light emerging from an atom on the surface of the dust cloud successfully overcomes the pull of gravity and escapes into outer space. Light emitted later on cannot escape. Gravitation becomes too strong. The attracting mass has not grown any greater, but the distance of the light source from the center of the cloud has changed. According to Sir Isaac Newton's 'inverse square' law of gravitation, when this distance has fallen to, say, half its original value, the pull of gravity at the cloud's surface has increased fourfold. Thus, any cloud, however massive, will eventually become a black hole if it continues to shrink.

"Oppenheimer and Snyder found that Albert Einstein's 1916 general relativity theory predicts similar results. However, because that analysis deals with curved space, the distance used in this case is not that measured directly from the surface to the center of the collapsing cloud. Instead, one must calculate it by measuring all the way around the belly of the cloud, and then dividing that by 2π, or 6.28, to get the distance. If Earth's Sun, which has a radius of about 430,000 miles, were to collapse, light could escape until the distance, so defined, drops down to 1.83 miles. This critical distance is called the Schwarzschild radius. For a star a hundred or a million times more massive than the Sun, the Schwarzschild radius is exactly a hundred or million times

"... if we had fallen inside that black hole, forces would have stretched us and our spacecraft past our breaking points.... No one would have been any the wiser."

larger. The surface of the collapsing Sun falls from two Schwarzschild radii to one in 11.9 millionths of a second, and from one Schwarzschild radius to zero—complete collapse—in another 6.5 millionths of a second. The time required for an object of a million solar masses is a million times greater.

"According to Einstein's standard general relativity theory, the collapse forces all the matter within the Schwarzschild radius, and even the space immediately surrounding the matter, to infinite compaction. The prediction applies to the gravitational collapse of a star to a black hole. And it applies to any object falling into an already existing black hole."

"Audrey, let me interrupt. You haven't yet said anything about the two scales of time."

"Right, Fred. The rapid collapse to infinite compaction is measured on a clock that is falling with the cloud of dust. The fall looks completely different when it is followed by an observer viewing it from far away. He sees the cloud at its 'end' collapsing ever more slowly as it approaches the Schwarzschild radius. Light from the cloud's surface becomes redder and fainter. The cloud, and anything dropped in later, fades to invisibility at the Schwarzschild radius. The reason is simple—light from within the Schwarzschild radius cannot escape to a faraway observer.

"Fred, if we had fallen inside that black hole, no one on Earth would have heard of us again. None of our radio messages would have ever got outside the Schwarzschild radius. Tidal forces would have stretched us and our spacecraft past our breaking points. They would have squeezed us sideways to infinite density. No one would have been any the wiser."

"Let me read a little from my part of the report, Audrey. 'When we lifted off, no one had yet detected a black hole. Hopes to do so were rising higher and higher. Individual stars in our Galaxy, the Milky Way, were thought to collapse to form black holes during the normal course of their evolution. The dense, starry center, or nucleus, of the Galaxy also seemed a promising place to search for a black hole. There, a hole should be made by the falling together of many stars, one after another, over a long period of time. Such a black hole could well have a mass of a million or even a billion Suns. Now our little spaceship can report that there is at least one black hole, close to 40,000 times the mass of the Sun, at the center of a cluster of stars. This cluster has an estimated total mass of about 220,000 solar masses.' "

■　■　■　■　■　■　■

"I found something in our microfiche library that I want to read to you, Fred. The author reviews Einstein's reasoning that the universe must be 'closed'—no matter which direction light goes, it cannot escape. He cites the work of many investigators showing that such a universe will end in a fantastic collapse of all the stars and galaxies. He compares this to the collapse that goes on within a black hole. 'One can observe

the gravitational collapse of a star to a black hole from far away. One can safely observe an existing black hole from outside the Schwarzschild radius. But in the predicted gravitational collapse of the universe, there will be no outside space from which to observe. Everyone will be inside, caught up in the collapse—people and planets, rocks and radiation, stars and black holes will all fall together.'

"Isn't the author really saying that the gravitational collapse of a star—or the fall of a spaceship into an already collapsed star—is a kind of preview of the gravitational collapse of the universe itself? If that is so, Fred, you and I were closer to seeing how things will be at that climactic time than anyone else ever was.

"It seems so easy to say 'How can the end of the universe matter to me?' After all, cosmologist Allan Sandage of the Hale Observatories estimated 80 billion years as the time, based on his observations, from the birth of the universe, through its expansion and recontraction, to its final collapse. Only about 10 billion years have elapsed so far, and 70 billion years is certainly a long time to wait. But, in my mind's eye, I can see all the matter of the universe pouring together from all directions to that final catastrophe."

"Audrey, physicists describe the end of the universe, in fact its whole history, in a domain called superspace, not in the three-dimensional space we know. Superspace does not have four or five dimensions, it has an infinite number. In superspace, the entire universe is described by a single point. The infinite coordinates of this one point tell the size of the universe, the location of all the particles in the universe, and the curvatures of space at all locations near and far. Move this single point to a slightly different position, and all its coordinates will change a little. These changes reveal the change in the size of the universe, the change in the location of all its particles, and all the details of all alterations in the curvature of space. In this way, the history of the universe can be described by a succession of points in superspace, or, more accurately, when one considers that observers at different places each have their own time scales, by a leaf in superspace.

"The point in superspace at one end of the leaf of history represents the indefinitely small beginning of the universe—the 'Big Bang.' Each point beyond this tip of the leaf describes a momentary configuration of the universe during the course of its expansion and recontraction. The last point at the other end of the leaf represents the universe at its end, when it collapses to infinite compaction."

"But Fred, what happens *after* a star or the universe has collapsed? Is that point the end of everything? Or does something lie beyond the

"... the gravitational collapse of a star—or the fall of a spaceship into an already collapsed star—is a kind of a preview of the gravitational collapse of the universe itself ... in my mind's eye, I can see all the matter of the universe pouring together ... to that final catastrophe."

black hole? We're more than halfway back to Earth. We have explored, we have read, and we have written. But I am more on edge than ever to know what it all means."

"Let me tell the story as some physicists are beginning to view it, Audrey. As the point runs along the leaf in superspace from one end to the other, as the universe swells, then shrinks and collapses, one is reminded of a negatively charged electron moving in ordinary space toward an attracting positive charge. According to the prediction of classical mechanics, as the distance between the two charges shrinks, the attracting force between them becomes stronger. The electron's kinetic energy should rise without limit. If this prediction were correct, all matter, being made out of positive and negative charges, would undergo catastrophic 'electric collapse' in less than 10^{-15} second. Matter could not exist.

"Yet matter does exist, and is stable. During the early 1900s, this contradiction between prediction and observation marked a great crisis in the history of physics. The crisis forced physicists to recognize that classical mechanics is not a good picture of nature at extremely small distances. In reality, the electron is scattered by the positive charge and avoids the catastrophe of electric collapse. There is a certain chance, or probability, that it will be scattered in this direction, another probability it will go in that direction, yet another probability it will go in another direction.

"Experiments confirm this probabilistic picture of matter. Without this picture, we would not know how to predict the properties of atoms. Possessing it, we understand the reactions of chemistry. We tailor a transistor to its task. We master nuclear energy. We design new kinds of materials. We conceive a laser and make it lase. Whatever we call this picture—quantum principle or probabilistic character of nature—it is the overarching principle of physics in the 20th century.

"The impossibility of electric collapse must have deep consequences for gravitational collapse. For instance, general relativity theory cannot be correct in predicting that the matter in a star will collapse to a mathematical point of infinite density within a black hole. And it cannot be correct to think that the end of the universe is described in all accuracy by a single point at the end of the leaf in superspace.

"In both cases, as with the electron, the final stage of collapse has to be described by a kind of scattering. The inevitable catastrophe predicted by general relativity, in which a star or the universe shrinks to unimaginable dimensions, must be replaced by many possible outcomes, each having its own probability.

"Physicists know how to calculate the probability that an electron will be scattered at a certain angle by a positive charge. They do not know how to calculate the probability that a certain outcome will result from gravitational collapse. Gravitational collapse is the greatest crisis physicists face. Dynamics must go on. Yet, our knowledge of the dynamic object, the structure of space itself, is too primitive to allow us to forecast what happens at the very small dimensions found in the final stages of gravitational collapse."

"But why not experiment, Fred? If a black hole is an experimental model for the collapse of the universe itself, then why not simply drop things in it and find out how and where they scatter?"

"Audrey, if we stay outside the Schwarzschild radius, we see only the approach of the object to the black hole, never what happens to it inside the black hole. And if we are part of the collapse, we are destroyed before that stage in the compaction is reached. In neither case do we get the answer to our question."

"Fred, what could space possibly be like at these tiny dimensions?"

"Our conception of space—its geometry—is like elasticity, to paraphrase some of Russian physicist Andrei Sakharov's conclusions on this point. Under ordinary inspection, a bowlful of jelly seems to be elastic. With sufficient magnification, however, you could see that there is no such thing as elasticity inside the atoms in the jelly. Instead, there are only charged particles, emptiness, and motion.

"Similarly, Einstein's theory of general relativity gives a useful description of the geometry of space and time at any everyday scale of distances, and even at distances far smaller than an electron. However, the very concepts of space and time must lose their meaning at distances billions and billions of times smaller than an electron. Such ideas as 'distance,' 'direction,' 'before,' and 'after' lose all sense, just as 'elasticity' loses sense within the atom. But how do we describe the dynamics on this scale? What 'machinery' underlies space and time? To account for what goes on, we must have a concept as different from the general relativity theory as anything could be. Some physicists call it 'pregeometry.'

"Elasticity helps us understand how matter stretches, but not how it tears. Likewise, what happens in the final stages of gravitational collapse of a star or the universe cannot be understood in terms of how it smoothly collapses. It can only be understood in terms of pregeometry —and that tool is not available. Several investigators are studying the possibility that pregeometry may be nothing more than the accumulated result of a very large number of simple yes-or-no decisions."

"Fred, nobody has ever explored as vast a range of space and time as we. In imagination we can look up and down the course of history, from the beginning of the universe to the end of time. But Fred, what happens *beyond* the end of time? Can we say nothing about these new cycles of the universe, other than that each has its own probability?"

"If a throw of the dice, in effect, determines the scattering of an electron by a positive charge, then does it not also govern the properties of the universe in each new cycle of expansion and recontraction? Can the properties be anything but erased in the final stages of collapse and written in afresh, at new values, at the start of the new cycle? In the wake of a great storm, one is not surprised to see a new pattern of clouds left in the sky.

"The universe must be reprocessed through a knothole in the final crunch of gravitational collapse. A new cycle of expansion starts with every physical constant changed—new types of elementary particles, and a different number of these particles, interacting with different

forces. The new universe has a new radius at its maximum expansion. Its lifetime from start to end has a new value.

"Compare this reprocessing of the universe with what geologic time does to rock: The curvature of space near a black hole or any other center of gravitation is as far from revealing the actual structure of space as the sheerness of a cliff seen from a distance is from revealing the structure of the rock. A closer look reveals that the rock is stratified. Many a man passed by those layers without reading their hidden message. How preposterous it was to imagine that a snow-covered cliff a mile high was once buried two miles down, metamorphosized there by heat and enormous pressure into what we see today.

"Space, too, is stratified. Focus energy from opposite directions into a tiny region of a vacuum. Particles boil out of the emptiness. Positively and negatively charged electrons spring into existence, along with pairs of muons, kaons, protons, and a long list of other particles. Fire energy into another region of space and the identical particles will materialize. Radiation reaching us from distant galaxies reveals that this 'stratification of space' is exactly the same at points light-years away.

"All the experimental evidence has been marshaled most penetratingly by mathematician-physicist Freeman Dyson of the Institute for Advanced Study in Princeton, N.J. It shows that any relative change in the course of a year in the stratification of space—the properties of the particles—is far less than the relative change in the radius of the universe during a year. Thus, the stratification does not appear to change as the universe grows older. If we can say that the stratification of rock is frozen in by relief from heat and pressure, it also seems reasonable to say that the stratification of space is frozen in after the reprocessing of the universe."

"I can accept the idea of a frozen-in stratification, Fred. Indeed, I can't see any escape from it. But I have difficulty with the idea that the properties of the universe can be changed from one cycle to another at the whim of probability. Physics to me is the science of the definite. Here everything is jumping unpredictably to a new value in the final crisis of collapse. How can that be physics?"

"But Audrey, the transformation of the universe seems to many scientists an almost inescapable consequence of general relativity and the quantum principle, the two great concepts of 20th century physics. Yet, the transformation *is* fantastic to contemplate. How can the dynamics of a system so incredibly gigantic be switched at the whim of probability from one cycle that has lasted, say, 100 billion years to another that will last only 1 million years? At first, only the fact that the universe gets squeezed down to unbelievably small size allows one to accept such an incredible transformation. But then you look at the stratification of a towering cliff, or a bird never seen before, and you realize that the

"... the history of the universe can be described by a leaf in superspace ... almost all cycles of the universe will be barren of life ... Our cycle of the universe ... is built for man."

universe is full of incredible transformations—metamorphosis of rock; mutation of species; chemical transformation; nuclear disintegration; particle decay; and the reprocessing of the universe."

"Wait a minute, Fred. You speak of a cycle of the universe that lasts only a million years. But such a universe would have to be less than a million light-years in radius at the time of maximum expansion. There won't be room in it for anything larger than our Milky Way. And there won't be nearly time enough for stars to evolve. What sorry kind of a universe will that be?"

"You've put your finger on a really exciting point, Audrey. Most cycles of the universe won't permit life at all. Let me rephrase the reasoning of physicist Robert H. Dicke of Princeton University in these words, 'What good is a cycle of the universe with no one around to look at it? For observation to be possible there must be a mind, and for a mind to be possible there must be life. Moreover, to build life requires elements such as carbon, nitrogen, and oxygen. To produce these elements calls for thermonuclear combustion. But thermonuclear combustion requires several billion years of cooking time in a star. And a universe that is to last several billion years must expand to be several billion light-years in radius. So why is the universe as big as it is? Because we are here!' "

"How preposterous, Fred. I have always conceived of the universe as a gigantic machine. It was set up on some master plan that took no account of man. Then, life developed here and there as a kind of accidental afterthought, a cog too tiny to influence the course of colossal stars."

"Audrey, an astrophysicist at Cambridge University in England, Brandon Carter, calculated how stars would look in a universe whose physical constants are different from those in the present cycle of the universe. For instance, he considered an atomic constant, the reciprocal fine structure constant, equal to 137.0388. This constant measures how much faster light moves than the electron in the hydrogen atom. Carter found that a star's luminosity varies with the 20th power of this constant. A change of that constant by 1 per cent one way from 137 will turn all the stars into red stars, and no star like the Sun will be possible. A change of 1 per cent the other way will turn all the stars into blue stars, and again no star like the Sun will be possible. Carter gives reasons to suspect that a fine tuning of many other physical constants is also required before life can develop. According to this line of reasoning, almost all cycles of the universe will be barren of life, mind, and thought. If I try to express this idea in simple words, Audrey, I can only call it 'the biological selection of physical constants.' Our cycle of the universe is a very rare exception. It is built for man."

"You make me dizzy, Fred. When we took off from Earth, everyone wanted to find out where else there might be life. Radio transmitters and receivers were being used from time to time to send messages out into space and to try to pick up messages from space. Many astrophysicists said that with 100 billion stars in the Milky Way, the chances for our finding life were overwhelming. They kept telling us that there were millions of suns and planetary systems at least as favorable for life as our own. There was supposed evidence of signals coming from Zeta Zeta, and we volunteered to make contact with life there. Enormous effort went into building a launcher to get us up close to the speed of light quickly and painlessly. Special antiradiation medicine had to be developed for us. Now, what you tell me casts doubt on our voyage and on all my past ideas about the universe.

"Let's carry what you are saying to the logical extreme, Fred. It takes a very narrow squeak for a cycle of the universe to permit life at all, even at one place. If life had originated in more than one place, that would have meant that the universe was larger and longer-lived than necessary. The creation of life would be more 'expensive' than it needed to be. So the chances are overwhelming that Earth is the sole outpost of life in the universe, and we had no right to expect to find life on Zeta Zeta. Am I wrong in my reasoning?"

"What you say is all new and startling to me, Audrey. We could refuse to accept the idea that the universe is reprocessed, and that all the physical constants are refrozen at new values from time to time. Then we would be left with all the mystery of what happens beyond the end of time. And we would have no explanation of what gave the physical constants the values they have. I always found it thrilling to contemplate how physicists found their way through the mystery of the atom— not total collapse, but chance scattering. That surely is the right line here, too. But then I see the force of your reasoning: Mankind—the mankind of our Earth—may well be alone in the universe—our universe, a universe made for man."

■　■　■　■　■　■　■

"Audrey, it's only four more days now until we land! The 'They' we are going to meet on Earth will have made many discoveries quite new to us. We may at first be overwhelmed, but we must never forget that we have a new perspective on the universe. Our present tentative answers may be wrong. But in science it is much more difficult to ask the right questions than to look for the right answers."

"Fred, how fortunate we have been. Our encounter with the black hole brought us to the brink of extinction, closer to gravitational collapse than anyone has ever been before. Our accident revealed to us the fate of stars and of the universe itself in a new light. What a vision we have ahead to explore: 'The universe as home for man!' "

For further reading:

Bergmann, Peter G., *The Riddle of Gravitation: From Newton to Einstein to Today's Exciting Theories*, Scribner, 1969.
Penrose, Roger, "Black Holes," *Scientific American*, May, 1972.
Ruffini, Remo, and Wheeler, John A., "Introducing the Black Hole," *Physics Today*, January, 1971.
Thorne, Kip S., "The Death of a Star," *Science Year*, 1968.

The Search
For a Human
Cancer Virus

By Lee Edson

**The virus, once considered unrelated to human cancer,
has become the focus of the most promising research yet**

Excitement has never run higher among cancer researchers. In the
quest for the cause of cancer, virologists, specialists on viruses and virus
diseases, have reported a flood of promising findings. But their optimism is tempered by cautious skepticism.

Typical of this new surge is a striking series of reports issued during
the last half of 1971. In July, virologists Leon Dmochowski and Elizabeth S. Priori of the University of Texas in Houston announced that
they had isolated a virus they suspected to be the cancer agent in the
tumor tissue of a child. In December, 1971, virologist Sarah E. Stewart,
of Georgetown University in Washington, D.C., declared that she, too,
had found a suspicious new virus in a human tumor. A few days later,
Robert McAllister and Murray Gardner of the University of Southern

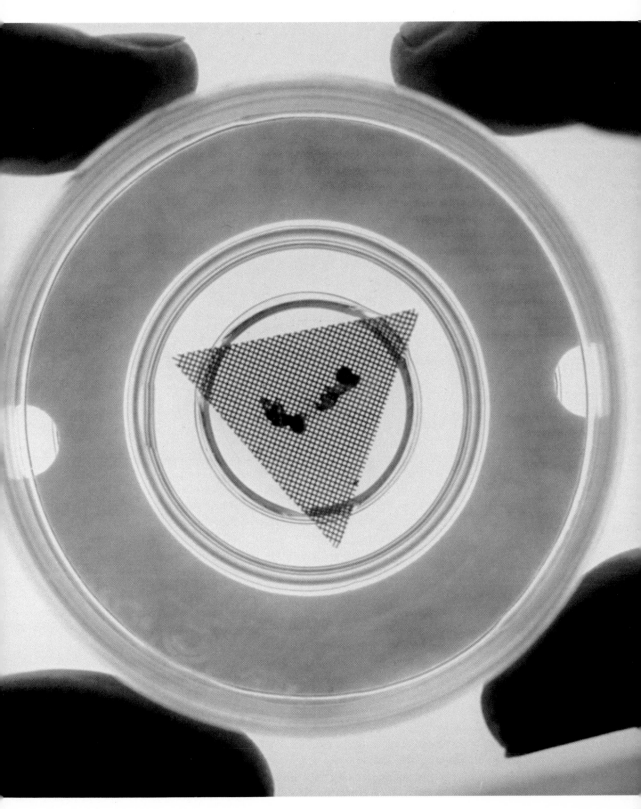

A Mini-Message Spells Disaster

Host cell

Virus

California in Los Angeles announced that they also had a new candidate for a human cancer virus.

For a fleeting moment, it looked like the long, frustrating, seemingly hopeless search for a human cancer virus had turned an important corner. But within weeks, doubts arose in the scientific community. Several scientists claimed that their tests on the virus isolated by Dmochowski and Priori showed that it was a mouse virus that had contaminated the culture. Virologists dismissed the Stewart discovery as lacking adequate experimental proof. The McAllister-Gardner finding was also attacked. However, it was not as easily discredited and, in 1972, this virus and two others held center stage as the excitement mounted.

Viruses have only recently become respectable subjects for cancer workers. Medical researchers once thought of them only as "bugs" that spread through the air or by physical contact to cause such contagious diseases as flu, measles, and polio. If viruses cause cancer, why did not doctors treating cancer patients come down with the dread disease themselves? And how can a viral cause explain why many more Japanese men living in Japan contract stomach cancer than do Japanese men who live in the United States?

The first strong link between viruses and cancer came in 1908 when two Danish biologists, V. Ellermann and O. Bang, injected chicken leukemia cells into a healthy chicken and observed that the disease was transmitted. Then, only a few years later, biologist Francis Peyton Rous of the Rockefeller Institute for Medical Research in New York City linked viruses and cancer even more closely. He injected a cell-free

The author:
Lee Edson is a science writer whose articles have appeared in many magazines and books. He wrote the article "Attempting to Understand the Brain" in the 1972 edition of *Science Year.*

92

Virologist Renato Dulbecco, *above right,* showed that one kind
of cancer virus enters a normal cell, *above left,* and attaches its
own DNA (orange) to the host cell's DNA (red). The viral DNA
has genes that cause the cell to grow and reproduce in the wild
manner characteristic of cancer. Each new cell formed in the
resulting cancerous growth has a copy of the deadly message.

extract from the tumors of cancerous chickens into healthy chickens.
The injected chickens rapidly developed tumors and died. Rous was so
afraid to suggest that the cancer-causing ingredient might be a virus
that he called it simply an "agent."

More than 50 years later, the scientific community finally realized
the importance of Rous's experiment, and he shared the 1966 Nobel
prize for physiology and medicine. What had awakened the scientific
world was the slow-but-steady increase in the number of experiments
proving that viruses could cause cancer not only in birds, but also in
other animals, including mammals. By the late 1960s, researchers had
discovered 128 different animal cancer viruses, and the eventual discovery of a cancer virus in man seemed inevitable.

However, it is much more difficult to prove that a suspect virus
causes cancer in a human being. In animals, scientists need merely to
inject the virus into healthy specimens and watch for signs of the disease. This would be unthinkable in human beings, so indirect, circumstantial evidence must suffice. For example, finding the same virus in
every case of the disease and then showing that this virus causes cancer
in test animals and in human cells grown in the laboratory would be

important evidence. But, despite one of the most intensive hunts in the history of science, a human cancer agent remains as elusive as ever.

Viruses are a far from easy quarry. Submicroscopic structures—about a trillion of them can fit on a dime—they can sometimes be detected only through complex chemical tests. They come in all shapes, from simple spheres to a configuration that resembles a miniature lunar excursion module. But they all share one characteristic. They contain either deoxyribonucleic acid (DNA) or ribonucleic acid (RNA), the chemical molecules of heredity. Thus, viruses are essentially packets of genetic information.

Viruses are often described as a bridge between the living and non-living worlds. Outside a host cell, viruses are as inert as stone. Inside, they spring to life. In many infectious diseases, they reproduce repeatedly and ultimately destroy the cells they inhabit. The case with cancer is different. Here, rather than being destroyed, cells multiply and grow abnormally. During the 1950s, Renato Dulbecco of The Salk Institute for Biological Studies in San Diego showed how an alien particle of DNA from a virus may slip into a cell's DNA, and change the cell's programming, causing it to produce the uncontrolled growth that is cancer. Each new cell that is produced in the cancer then contains a copy of the deadly viral DNA.

Dulbecco also pointed out that the viral DNA in a cancer cell is arranged in a configuration that normally prevents reproduction of the virus. This avoids the cell destruction that usually occurs in viral infection, and assures perpetuation of the malignancy.

However, nearly all naturally occurring animal cancers to which viruses have been linked involve not a DNA, but an RNA virus. Dulbecco's work did not explain how such a cancer virus could work. In 1964, Howard M. Temin, a young biochemist studying tumors at the University of Wisconsin, proposed that an RNA cancer virus could use its RNA as a template, or pattern, to produce DNA. The reverse process, DNA producing RNA, is a key step in gene expression. At first, few scientists accepted Temin's proposal. Then, in 1970, Temin and molecular biologist David Baltimore of the Massachusetts Institute of Technology announced, independently, that they had discovered in several tumor viruses an enzyme that was using RNA as a template for making DNA. The enzyme, called reverse transcriptase, has since been found in all RNA tumor viruses, and it has become a standard procedure to analyze any suspected cancer virus for the presence of the enzyme.

Thus, some of the most basic mechanics of how viruses can produce cancer became known. Yet, a few scientists still doubted that viruses do cause cancer in human beings. However, some of them have reassessed their convictions in view of studies indicating that at least some forms of cancer under certain conditions might be contagious.

For instance, in 1971, researchers Peter Greenwald and Nicholas Vianna of the New York State Department of Health and Jack Davies of the Albany (N.Y.) Medical College made a surprising discovery.

Thirteen cases of Hodgkin's disease, a cancer of the lymphatic tissue, had appeared among members of the Albany High School graduating classes of 1953, 1954, and 1955 and their outside associates. The 13 cases is well above what one might expect to find by chance in such a small group. The researchers interviewed the living victims, and traced the history of those who had died. They found that each victim had had contact with one or more of the other group members. Further investigation traced the origin of the disease to several focal points, including a girl who had dated many of the young men who eventually came down with the disease. The girl, however, has remained healthy. The investigators have since located 27 more cases of Hodgkin's disease among the approximately 300 graduates. In addition, the scientists have uncovered other Hodgkin's disease clusters in various U.S. cities.

Clusters of cancer cases have occurred a number of times before. One of the most bizarre was in South Pasadena, Calif., in 1968. Two children, a sister and brother, came down simultaneously with the same form of a cancer known as a sarcoma. Next, a playmate of the pair developed the same disease, as did three adults in the neighborhood. And, to top off the mystery, a neighbor's dog fell ill of the same sarcoma, the children's pet cat came down with leukemia, and the family parakeet also developed a tumor.

Could such cases simply be coincidences? Most statisticians would say yes. In addition, the scientists and public health officials studying the New York Hodgkin's disease clusters point out that the disease is too rare worldwide to be contagious in the straightforward way that, say, smallpox is. Certainly the bulk of the evidence does not support the idea of contagion. But, at the moment, according to such scientists as virologist F. Kingsley Sanders of the Sloan-Kettering Institute for Cancer Research in New York City, contagion may be possible under special circumstances. Sanders points out that the majority of infected people may be healthy carriers of the disease. The girl in the Albany case might be such a person.

A virus that virologist Magdalena Eisinger, one of Sanders' colleagues, isolated from the cells of Hodgkin's disease patients is one of the three candidates still running strong in the human cancer-virus sweepstakes. This suspect is one of a relatively widespread DNA-containing group of viruses called herpes viruses. Each member of the group is spherical with a 20-sided internal structure that contains its DNA and some proteins.

Some herpes viruses cause relatively mild illnesses in human beings, such as mononucleosis and cold sores. But other herpes viruses have been associated with cancer. For example, Marek's disease, a tumor of the lymphatic tissue in chickens, is caused by a herpes virus. Another herpes virus is found in the tissues of human beings with Burkitt's lymphoma, another lymphatic cancer. Because Hodgkin's disease is also a lymphatic cancer, Sanders thought that it might have a herpes virus associated with it.

The largest of the several kinds of cells in the lymph glands of Hodgkin's disease victims has long been suspected of being the malignant cell. However, nobody had ever been able to grow quantities of this cell outside the body. The Sloan-Kettering team suspected that this cell died in cell culture because short-lived cells that are isolated with it somehow prevented its growth. The scientists extracted tissue from the lymph glands of a victim of the disease. Then, using a small piece of stainless steel screen to filter out the smaller cells, they succeeded in growing the suspect cells for the first time.

Some of the large cells were tested by injecting them into hamsters. When some of the hamsters developed cancer, the experimenters felt they had found the right type of cell. Using sophisticated chemical techniques, they looked for foreign RNA and DNA molecules in some of the malignant cells. They found evidence of viral RNA, but were unable to isolate a virus containing it. Then, some of the malignant cells in the nutrient medium unexpectedly began producing new types of cells called blastoids. Similar cells appear in the lymph glands of patients at certain stages of Hodgkin's disease. Tests revealed that the blastoids also contained the viral RNA. But, in addition, they contained viral DNA. A look at the blastoids through an electron micro-

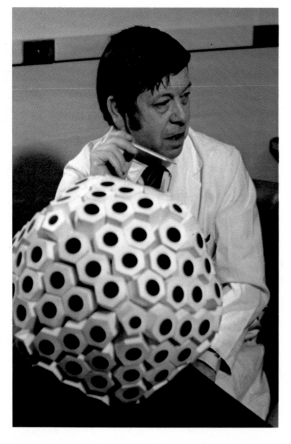

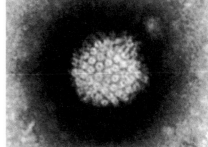

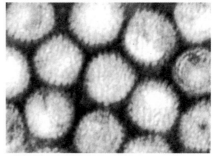

Virologist F. Kingsley Sanders, *left,* displays a model of herpes virus, one of which he has linked to human cancer. Other herpes viruses cause cold sores in humans, *top,* and cancer in chickens, *above.*

scope showed a previously unknown herpes virus. Whether this virus, the RNA virus, or any virus causes Hodgkin's disease, or is simply a harmless passenger in the cell, remains to be proven.

Another virus in the running is one of the so-called B-type, RNA viruses. Such a virus causes breast cancer in mice. Through the electron microscope, the B-type virus looks like a sphere with spikes projecting from its outer edge. It sometimes also shows a taillike section.

In February, 1971, virologist Dan H. Moore and his associates at the Institute for Medical Research in Camden, N.J., found in the breast milk of two groups of women, a B-type virus very similar to the mouse breast tumor virus. One group included some of the Parsis of India, members of an intermarrying religious sect whose women suffer a high

One Cancer—Two Suspects

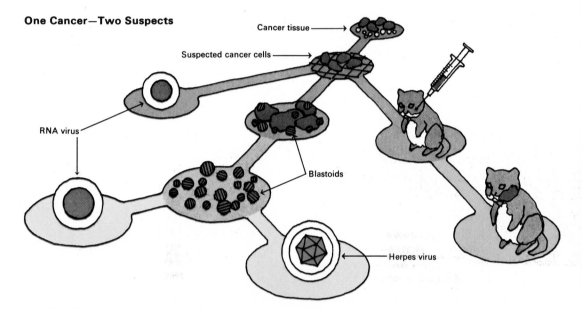

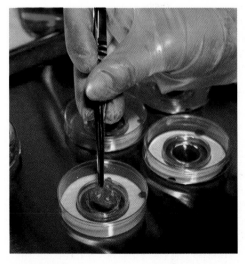

Sanders placed tissue from the glands of patients with Hodgkin's disease on a fine stainless steel screen and put it in nutrient liquid, *left*. The screen filtered out the large suspected cancer cells, *above*. Hamsters injected with some of the cells developed tumors. Tests revealed evidence of RNA virus in some of the cells. Other cells developed smaller cells called blastoids, which are also found in some stages of Hodgkin's disease. Sanders also found evidence of the RNA virus in the blastoids. In addition, he isolated a second virus, this time a herpes virus from the blastoids. Sanders says either or both of the viruses may be involved in Hodgkin's disease, and he is now doing experiments aimed at pinning down their roles.

Molecular biologist
Sol Spiegelman, *left,*
noted that a mouse
breast cancer virus,
below left, closely
resembles a suspected
human breast cancer
virus, *below.* Thus,
he felt that the mouse
virus might be used to
help prove the human
virus's role in cancer.
Each virus is magnified
over 225,000 times.

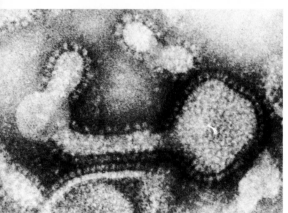

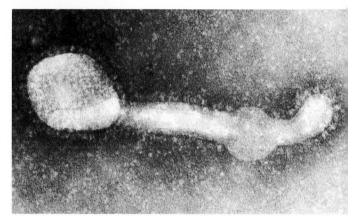

rate of breast cancer. The other group was composed of American women with a family history of breast cancer. Could it be that the DNA made by such a virus lurked within the breast cells of these women, waiting to trigger the start of cancer?

Although later work with a larger sample of women showed no such correlation between their backgrounds and incidence of the virus, there is some evidence that the virus may be a cause of human breast cancer. Molecular biologist Sol Spiegelman and his associates at Columbia University in New York City found the cancer-associated enzyme, reverse transcriptase, in the milk of women with the B-type virus. Then, in January, 1972, Spiegelman and his group reported experiments linking the B-type virus even more closely to human breast cancer. If the virus found in the human milk was being made by a portion of DNA of viral origin within the human breast cells, the scientists thought that they should be able to use the viral RNA to detect that

Mouse Tracer for Human Virus

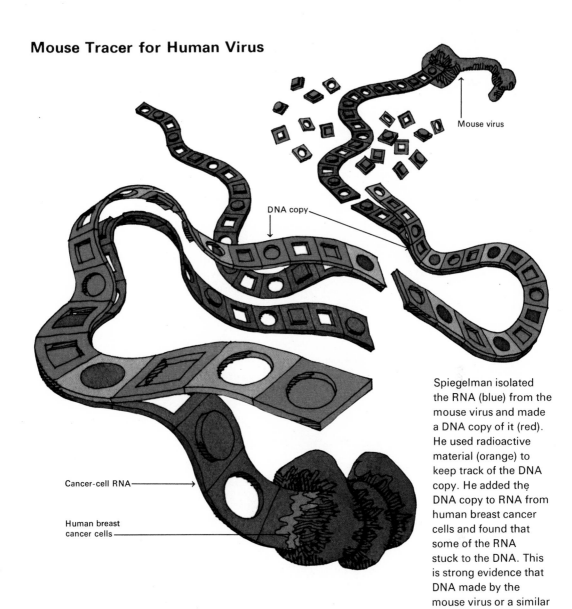

Mouse virus

DNA copy

Cancer-cell RNA

Human breast cancer cells

Spiegelman isolated the RNA (blue) from the mouse virus and made a DNA copy of it (red). He used radioactive material (orange) to keep track of the DNA copy. He added the DNA copy to RNA from human breast cancer cells and found that some of the RNA stuck to the DNA. This is strong evidence that DNA made by the mouse virus or a similar one is active in human breast cancer cells.

DNA. This is reasonable because RNA made on a DNA template will stick to that DNA on contact. The same is true of DNA made on RNA.

There was one big problem, however. The B-type virus from humans was hard to extract and could not be grown artificially by any known culture method. But the scientists had a plentiful supply of the mouse breast tumor virus. They reasoned that since the human virus strongly resembled the mouse virus, perhaps the RNA of the human virus was also similar to that of the mouse virus. So, they decided to gamble and use the RNA from the mouse virus in their experiment.

To the mouse virus RNA, they added reverse transcriptase and the raw materials needed to synthesize DNA, some of which was made radioactive. As expected, a slightly radioactive strand of DNA complementary to the viral RNA was formed, and the researchers separated this DNA from the RNA.

If there was similar DNA in human breast cancer cells, it might be producing RNA that would stick to the radioactive DNA. So the scientists isolated RNA from such cancer cells and added the radioactive DNA. The results were conclusive—a significant amount of sticking between the radioactive DNA and the tumor-cell RNA in 70 per cent of the trials. The implications were clear—the human cancer cells contained portions of DNA made by a virus similar to, or the same as, the B-type mouse tumor virus.

The third virus still under suspicion is the one discovered by McAllister and Gardner. It is one of a group called C-type viruses, which also are RNA viruses. Viewed through an electron microscope, a C-type virus appears as a sphere with a dense, dark core. Researchers have found some 30 different kinds of C-type viruses associated with cancer in animals. In 1972, tumor expert Thomas Kawakami of the University of California at Davis found C-type viruses in tumors from

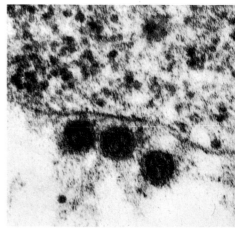

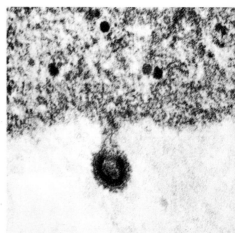

Virologist Murray Gardner, holding a virus culture, *above,* was part of a research team that discovered a C-type virus that may cause cancer in humans. Three of the viruses, *above right,* are budding from a human cancer cell. The C-type virus known to cause leukemia in cats is shown for comparison, *right.* Both viruses are magnified 100,000 times.

Human Cancer in a Cat

Gardner and virologist Robert McAllister injected human cancer cells into cat fetuses. Less than two months after birth, one of the kittens had developed a brain tumor that was composed of human cancer cells. A previously unknown C-type virus that also may be of human origin was associated with the tumor.

Human cancer cells

C-type virus

woolly monkeys and gibbons, two close relatives of man. Many scientists believe similar viruses are involved in at least some human cancers.

For 20 years, McAllister, a pediatrician-turned-virologist, doggedly searched through hundreds of human tumors for suspicious viruses. One day, he discovered that he could keep the cells from a patient's muscle tumor alive and growing indefinitely in a special nutrient liquid. This is highly important because scientists have had difficulty keeping human cancer cells alive and reproducing in the laboratory.

McAllister studied these cancer cells with all the care of a proud parent. He tried to coax hidden viruses from them with chemicals, X rays, and even with the help of another virus from a cat with leukemia. Then one day, McAllister and Gardner decided to inject some of the cells through the walls of a pregnant cat's uterus and into the developing young. This was not just a whim. Cats have proved useful as experimental animals, and very young kittens are particularly useful because they have little immunity to disease.

McAllister and Gardner injected the cancer cells into the embryos of three pregnant cats, and tumors developed in some of the kittens within 60 days after birth. Most important, however, the tumors were composed of human cells just like those that were injected. Furthermore, a C-type virus was discovered in some of the tumors.

Gardner and McAllister have run the virus through numerous tests. They have compared it to known animal viruses, weighed it, and checked it chemically and found that it contains reverse transcriptase. "So far, the virus matches no other known," says Gardner. "So we hope that it may at long last be a C-type virus of human origin."

The C-type virus is also at the center of an ingenious theory that has gradually gained some acceptance since it was first proposed in 1969. According to two virologists, Robert J. Huebner, head of the Viral Carcinogenesis Branch of the National Cancer Institute, and his co-worker George J. Todaro, the source of most cancers in man is a section of the cell's own DNA. This package of genes is normally transmitted, through inheritance, from parent to offspring like the gene for blue eyes or red hair. According to the theory, however, under some conditions, this string of genes produces a C-type virus, an RNA copy of itself that can escape the cell. Normally, the deadly genes remain within the cell in a quiescent, or "switched off," or inactive, state until "turned on" by some environmental factor such as radiation, a chemical in polluted air, or merely the internal changes of old age. Further, Huebner and Todaro say that the string of cancer genes has a Jekyll and Hyde character. It has a good purpose within an embryo where it promotes the fast growth that is needed in the early stages of development. But it turns evil if it is not turned off when the organism reaches a more developed stage. Then, the package of genes satisfies its drive for fast growth by building new unwanted tissue—a tumor.

The evidence collected so far to prove this speculative theory is intriguing, but remains circumstantial. For instance, in one type of experiment, animal cells that were free of C-type virus were grown in the laboratory for periods of time equivalent to old age in live whole animals. Experimenters carefully avoided contamination of the cells by viruses. Yet, C-type viruses suddenly popped up, as if propelled by a hidden trigger. The DNA needed to produce them was evidently in the cells all the time. This experiment has been repeated in several laboratories, in cell cultures and, also, in whole animals.

Another type of experiment, performed first at the Bronx Veterans Administration Hospital in New York City and at Stanford University, also starts with cells that have no C-type virus. However, when these cells are exposed to radiation or certain chemicals, they become cancerous, and C-type viruses appear.

If the Huebner-Todaro theory proves correct, it will be impossible to conquer cancer by using vaccines and other conventional antivirus methods. A vaccine produces antibodies that attack and destroy viruses in the blood stream. But, how can medical scientists penetrate the cell's nucleus? And, can the tiny cancer-producing package of genes ever be pinpointed in the uncharted territory of a nucleus packed with 10 million other genes? If it can, how can we change or eliminate it from the billions of cells in a human body?

Despite the growing interest in the Huebner-Todaro theory, most scientists believe that viruses in the sense we have always known them cause some forms of human cancer. Vaccines may someday be developed that are useful against these cancers. Hope for this type of weapon was kindled by the announcement in March, 1971, that scientists at a U.S. Department of Agriculture research laboratory in East Lansing, Mich., had developed a vaccine against Marek's disease.

An even more promising vaccine was reported in April, 1972. Immunochemist Jesse Charney, collaborating with Moore, developed the first vaccine against breast cancer in a mammal. He killed mouse breast cancer virus by exposing it to formaldehyde. Then he injected the killed virus into mice. These mice did not get breast cancer even when they were later injected with large amounts of the live cancer virus. Unvaccinated mice given the same dosage of the live virus developed the familiar mammary tumors. Of course, Charney points out, this has been done only in mice. However, he has also discovered that some of the antibodies from human breast cancer patients will act against mouse breast cancer virus. This suggests that there may be an immunological relationship between the virus in the mouse cancer and the suspected (but not yet proven) virus in the human cancer. Perhaps the first vaccine against a human cancer is not so far off after all.

For further reading:

Chedd, Graham, "First Steps to Malignancy," *New Scientist and Science Journal,* July 29, 1971.

"Footprints of Human Cancer Viruses?," *Nature,* Dec. 5, 1970, p. 907.

Ozer, Harvey L., "Current Concepts on Viruses and Cancer," *The Science Teacher,* September, 1971.

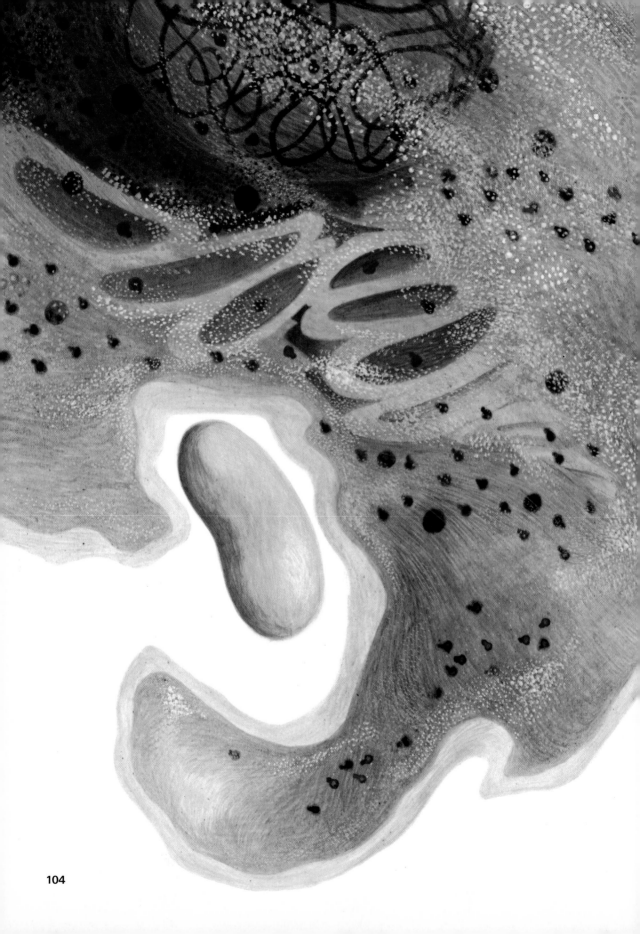

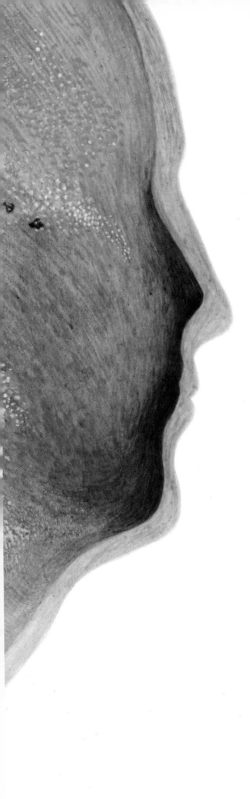

The Curious Ancestry Of Our Cells

By Jerald C. Ensign

Eons ago, primitive cells may have captured and enslaved others as a critical step in the evolution of complex plants and animals

Three American biochemists reported in October, 1971, that they had used genetic material from bacteria to repair a defect in human cells. The cells, taken from a galactosemia victim, lacked a gene that produces an enzyme needed to break down the sugar galactose. Researchers Carl Merril, Mark Geier, and John Petriociani of the National Institutes of Health in Bethesda, Md., inserted a gene from the bacteria into a laboratory culture of the cells. The cells then began to break down galactose. Apparently, the bacterial gene was providing the missing enzyme.

The medical implications of this achievement may be great, but this was almost certainly not the first time that a lowly bacterium has come to the aid of a more complex cell. Many scientists believe that every cell in the human body, and those in virtually all animals and plants, could not function without using genes that originated in bacteria. These genes were acquired at a time long before man existed, a time when life probably consisted only of billions upon billions of single cells floating about in the warm seas of an otherwise barren earth. In fact, these bacterial genes were essential for the evolution of these simple cells into the complex living organisms we know.

All life on earth today is packaged in two quite different types of cells. The smaller and simpler of the two, called procaryotic cells, includes only bacteria and blue-green algae. The larger, more complex cells, called eucaryotic, make up all other forms of life. A significant difference between the cells is seen in their deoxyribonucleic acid (DNA), the chemical substance of which their genes are composed. The genes determine the cells' inherited properties. In a eucaryotic cell, most of the DNA is in strands that are enclosed within a membrane-wrapped nucleus. During cell division through mitosis, a highly complex cell apparatus divides these strands.

The simpler procaryotic cell has no nucleus. In this cell, the DNA forms a single closed loop that is attached to the inside of the cell's membrane. Scientists have never detected any complex cell apparatus to help this loop divide during cell reproduction.

Perhaps a more important difference between the two cell types is that only eucaryotic cells contain tiny, membrane-enclosed bodies called mitochondria and chloroplasts. Both of these, unlike any other bodies in the cell, have small amounts of their own DNA. Mitochondria are present in all eucaryotic cells and are the site of the cells' energy-generating processes. Chloroplasts are found almost exclusively in green plant cells. They contain chlorophyll, which is used for photosynthesis.

Mitochondria and chloroplasts are somewhat similar in appearance, most of them being rod-shaped and approximately $1/7,000$ of an inch long and one-third as wide. Depending on its size and kind, a cell may contain from one to as many as a thousand of these tiny bodies.

That all cells are either eucaryotic or procaryotic—with no cells having intermediate properties—presents a most puzzling riddle to biologists. The evolution of living organisms has normally proceeded through a sequence of slow changes from one life form to another. Evolutionists can thus trace the history of life by placing related organisms in a continuing chain of development. Something entirely different must have occurred in evolution on the cellular level. But, what could it have been? Many biologists now accept the "capture theory." It states that some cells were enslaved by other cells early in the evolution of life. This theory was first proposed by Russian biologist Constantine S. Mereschkowsky in 1905, but it was not taken seriously until the early 1960s, when scientists made the discovery that both chloroplasts and mitochondria contain their own DNA.

Most scientists believe that life on earth began more than 3 billion years ago. Much of the earth at that time was literally bathed in vast seas containing a complex and concentrated mixture of dissolved organic materials—amino acids, nucleic acids, sugars, and proteins. This mixture is called the primordial organic soup.

The author:
Jerald C. Ensign is a professor of bacteriology at the University of Wisconsin. He also writes the Microbiology article in the Science File of *Science Year*.

The Two Cell Types

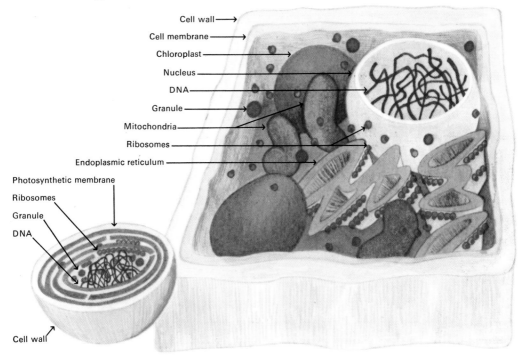

Cell wall →
Cell membrane →
Chloroplast
Nucleus
DNA
Granule
Mitochondria
Ribosomes
Endoplasmic reticulum

Photosynthetic membrane
Ribosomes
Granule
DNA
Cell wall

All cells are either procaryotic, such as blue-green alga, left, or eucaryotic, such as a modern green-plant cell, right. Only the eucaryotic cell has a nucleus, endoplasmic reticulum, chloroplasts, and mitochondria. Other differences include ribosome size and amount of DNA.

From this soup arose the first living thing, able to consume some of the soup's organic materials to grow and reproduce. Just how this occurred, and exactly what this first cell was like, is a mystery. This primitive cell must have been simple. Probably it was little more than a few genes and a few proteins. But, it had one great advantage—it had no competition for the plentiful supply of dissolved organic material. The cell multiplied rapidly.

Over millions of years, multitudes of cells were produced, some of them virtually identical to the original parent. However, mutations, or changes in genes, produced many variations. Such changes occurred relatively frequently, because the sun's intense ultraviolet rays were strong mutational agents. Numerous mutations and the addition of new genes eventually produced cells that were more complex and sophisticated than the first cell. But, they were still simple compared to all cells today. Oxygen, which is used to produce energy in many cells today, was not present in the earth's early atmosphere. To obtain energy, the early cells used the relatively inefficient process of fermentation, producing energy by the breakdown of organic foodstuffs.

Proponents of the capture theory believe that two new cell types evolved from these early cells. One type, destined to become eucaryotic, grew increasingly larger, but kept its primitive metabolism based on fermentation. Since the dissolved nutrients of the organic soup were now being quickly depleted by the exploding numbers of cells, this was

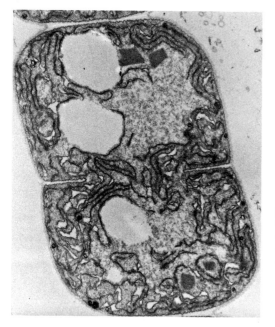

When a procaryotic cell, such as a blue-green alga divides, *above,* its DNA (in the light, grainy regions) remains more or less dispersed. When a eucaryotic cell, such as one from an onion, divides, *above right,* its DNA forms dense structures called chromosomes.

a difficult road to take. The larger the cell, the more food it requires. Thus, it became essential that the larger cells find a new food source to survive. They solved their dilemma by one of the most significant events in the evolution of life. The cells began to engulf undissolved particles of organic matter.

Survival for these cells became a matter of developing an efficient means of digesting such food. The endoplasmic reticulum, a complex internal membrane, developed at least partly in response to these needs. Food particles could be trapped and digested in pockets created by folds in the membrane. Also, as the cells became more complex, they had to develop a mechanism for distributing the genetic material during cell division. This was the beginning of reproduction by mitosis.

The other type of cell, destined to remain procaryotic, evolved in quite another fashion. It stayed relatively small and simple, but it developed increasingly sophisticated and efficient means of obtaining energy from dissolved foodstuffs. First, some of these cells discovered how to combine inorganic materials such as nitrates and sulfates with their organic foods to produce more energy than fermentation could. These cells became primitive bacteria, whose descendants live today in airless places, such as the bottoms of polluted ponds and lakes.

It was only a small biochemical step for some of these primitive bacteria to develop a simple form of photosynthesis, a process that harnesses the most plentiful supply of energy on earth—sunlight. Using this energy, cells could synthesize their own organic food from plentiful inorganic chemicals. In time, this process was refined to break water down into oxygen and hydrogen. The hydrogen was used in the cells to synthesize food, and the oxygen was released as a waste product. At this stage, the organisms that we now call blue-green algae were born. In addition, one of the most significant events for the course of later evolution was occurring—oxygen became part of the earth's atmosphere.

It is not surprising that the large cells engulfed and digested some of their smaller, metabolically more clever cousins. And it does not stretch the imagination too far to assume that occasionally the engulfed cells were not digested. Instead, they became permanent residents within the predator cell. The first time that a predator engulfed but did not digest a blue-green algae cell, the stage was set for a dramatic development that abruptly changed the course of evolution.

In effect, the host cell then enslaved the algae cell, the equivalent of a present-day chloroplast, forcing it to produce extra organic compounds by photosynthesis. This gave the larger cell a tremendous advantage over other cells. Its growth and development were suddenly no longer limited by lack of energy, for it could siphon off food in abundance from its little green slave.

Meanwhile, more and more oxygen accumulated in the atmosphere. Some of the primitive bacteria that were using inorganic substances to help produce energy began to evolve an even more efficient form of energy production. They used oxygen to obtain energy through respiration. The large predatory cells soon engulfed and enslaved the respiring cells just as they had the photosynthetic cells. The resulting combinations were the earliest eucaryotic organisms.

Respiration proved to be a great advantage. Modern eucaryotic cells contain either mitochondria alone or mitochondria and chloroplasts, but none contain only chloroplasts. Apparently any early cell that contained only chloroplasts, being unable to obtain extra energy in the dark, was soon crowded out by respiring competitors. Thus, the early eucaryotic cells that contained only respiring bacteria—premitochondria—were the ancestors of the multitude of organisms we now call animals. The early eucaryotic cells that contained both premitochondria and blue-green algae—prechloroplasts—were the forerunners of today's eucaryotic green plants.

Biologists who dare venture a guess estimate that these first eucaryotic cells enslaved their respiring bacteria and blue-green algae about a billion years ago. Since then, the cells have become completely dependent on their mitochondria and chloroplasts. And these structures, in turn, have become dependent on their hosts.

A crucial test of the capture theory of evolution would be to grow mitochondria or chloroplasts outside their host cells. This would allow scientists to make direct comparisons between these bodies and procaryotic cells. But all efforts to coax mitochondria and chloroplasts to grow in another environment have failed. The organisms have so efficiently adapted that they no longer can survive free.

There is, however, much other evidence supporting the capture theory. For example, the characteristics of all forms of life have always been controlled and determined by genes, which are composed of DNA.

Therefore, in a sense, evolution is the progressive change in DNA. The closer any two organisms are on an evolutionary chain, the more alike their DNA molecules will be.

The DNA in the nuclei of eucaryotic cells floats free in the nucleus, is an open strand, and is associated with proteins called histones. During the 1960s, the efforts of investigators from more than a dozen laboratories revealed that the DNA of both mitochondria and chloroplasts is very different. It is a closed loop bound to the outer membrane, and it is not associated with histones. It resembles the DNA found in bacteria and blue-green algae.

The close relationship between procaryotic cells and the mitochondria and chloroplasts is best shown, however, by the chemical similarity of the DNA strand itself. The genetic code expressed in DNA contains only four basic code letters—A, C, G, and T, for the four chemical bases adenine, cytosine, guanine, and thymine. The bases are in a sequence along the strand. Scientists can determine the ratio of the total amount of one base to the total amount of another base in a cell from any organism. A comparison of these ratios for any two organisms indicates how closely they are related. For instance, the ratios would be very much alike for any two birds but quite different for a bird and a mammal, and even more different for a bird and a plant.

In work done over several years and still in progress in 1972, scientists have compared the base ratios in mitochondria and chloroplasts to the ratios in the nuclei in host cells, ranging from yeasts to mammals. They have found that these ratios are markedly different. However, the ratios of the four bases are amazingly alike in all the mitochondria examined, no matter what the species of their host cell. The ratios are similarly alike in the chloroplasts from all the plant cells examined.

The implications are clear. All the mitochondria must have come from a common ancestor, and so did all the chloroplasts. Had they developed originally as parts of the varied cells in which they now reside, the base ratios in their DNA should be as different as the base ratios in the DNA in the nucleus of their host cells.

Another, more precise, comparison of the DNA was reported in March, 1972. Biochemists G. H. Pigott and N. G. Carr of the University of Liverpool, England, reported experiments that compared the sequences in which the bases appear along the DNA strand. The scientists found that the sequences in seven species of blue-green algae are very similar to the sequence in chloroplasts from the eucaryotic microorganism *Euglena gracilis*.

Further evidence that bacteria and blue-green algae are the progenitors of mitochondria and chloroplasts comes from their ribosomes, small bodies used in protein synthesis. In addition to the ribosomes in its mitochondria and chloroplasts, the host eucaryotic cell contains ribosomes within its *cytoplasm* (the substance that surrounds the nucleus). These are the bulk of the cell's ribosomes. Procaryotic cells contain only one kind of ribosome. They are smaller than those in the cytoplasm of

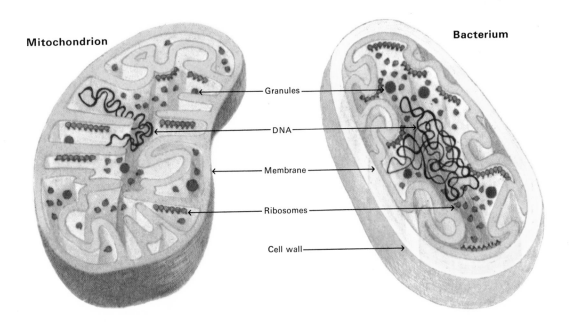

Mitochondrion

Bacterium

Granules

DNA

Membrane

Ribosomes

Cell wall

Revealing Comparisons

Mitochondria strongly resemble bacteria, *above,* and blue-green
algae resemble chloroplasts, *below.* These similarities support
the theory that mitochondria evolved from primitive bacteria
captured and held within eucaryotic cells, and chloroplasts
evolved from similarly captured primitive blue-green algae.
The shorter DNA in both mitochondria and chloroplasts indicates
fewer genes. The captured cells probably lost many genes that
they did not need when they became residents of other cells.

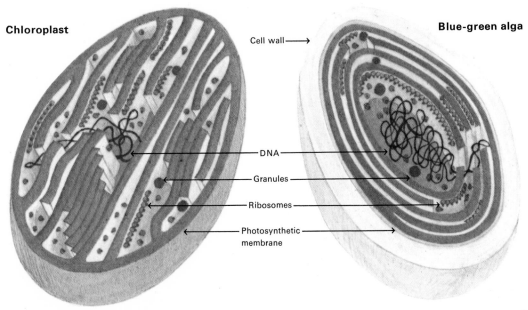

Chloroplast

Blue-green alga

Cell wall

DNA

Granules

Ribosomes

Photosynthetic
membrane

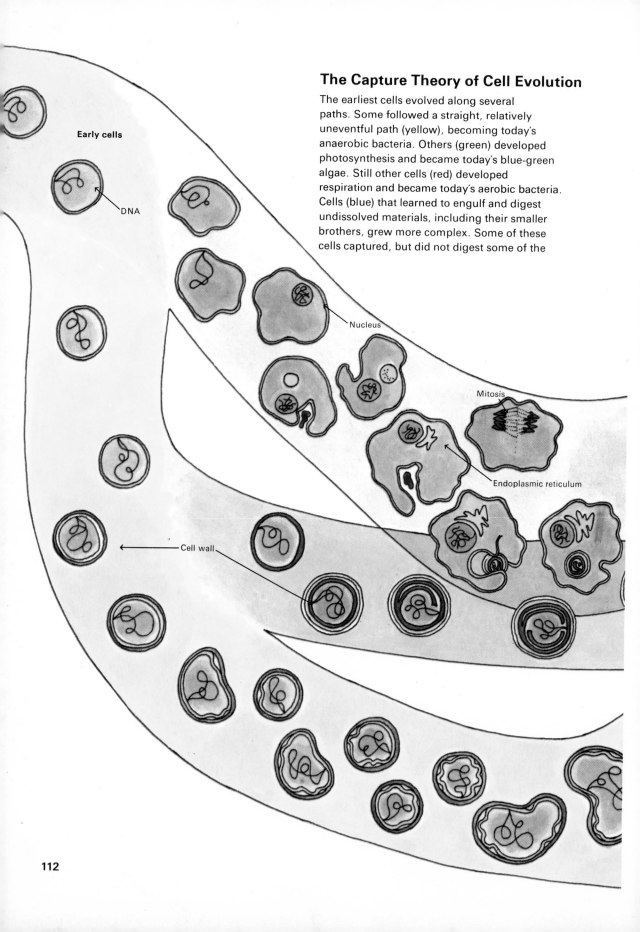

The Capture Theory of Cell Evolution

The earliest cells evolved along several paths. Some followed a straight, relatively uneventful path (yellow), becoming today's anaerobic bacteria. Others (green) developed photosynthesis and became today's blue-green algae. Still other cells (red) developed respiration and became today's aerobic bacteria. Cells (blue) that learned to engulf and digest undissolved materials, including their smaller brothers, grew more complex. Some of these cells captured, but did not digest some of the

Early cells

DNA

Nucleus

Mitosis

Endoplasmic reticulum

Cell wall

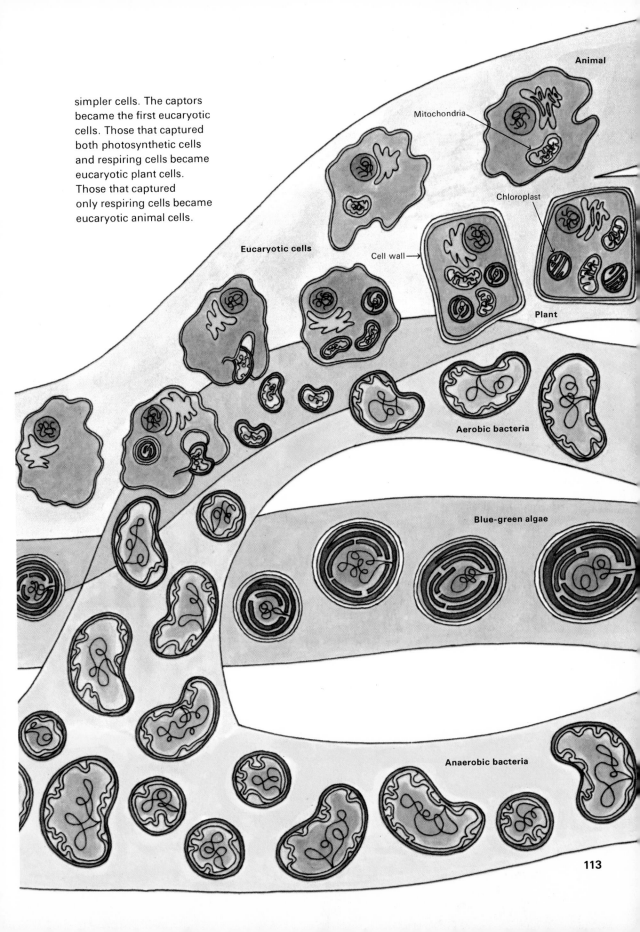

simpler cells. The captors
became the first eucaryotic
cells. Those that captured
both photosynthetic cells
and respiring cells became
eucaryotic plant cells.
Those that captured
only respiring cells became
eucaryotic animal cells.

Animal

Mitochondria

Chloroplast

Eucaryotic cells

Cell wall →

Plant

Aerobic bacteria

Blue-green algae

Anaerobic bacteria

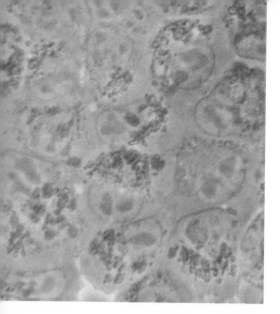

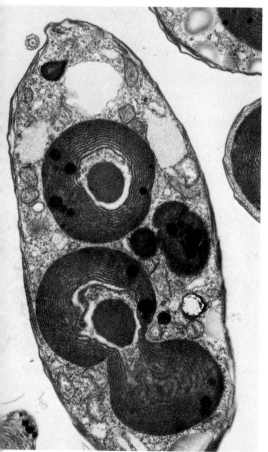

Support for the capture theory comes from experiments in which mouse cells, *top,* engulf and hold chloroplasts (green) and from blue-green algaelike chloroplasts, *above,* of eucaryotic alga, *Cyanophora.*

eucaryotic cells. They also differ in chemical composition and in their tendency to disintegrate spontaneously into subunits. In 1967, two teams of scientists, working independently, isolated ribosomes from within mitochondria and chloroplasts. One team was led by microbiologist Hans Noll at Northwestern University and the other by biologist W. Edgar Barnett at the Oak Ridge (Tenn.) National Laboratory. These ribosomes differed considerably from the ribosomes in the cytoplasm of the cells from which they were isolated. But they were about the same size and were very similar chemically to ribosomes from procaryotic cells.

Exactly the same pattern—similarities between mitochondria, chloroplasts, and procaryotic cells—exists in the mechanics of protein synthesis. In 1968, a group of molecular biologists led by James Schwartz of New York University School of Medicine in New York City and microbiologists A. E. Smith and Kjeld A. Marcker of the Medical Research Council in Cambridge, England, found that protein synthesis in mitochondria begins when a complex molecule, called N-formylmethionine-transfer-RNA, attaches to the ribosomes. Protein synthesis starts in the same way in bacteria, but not in the cytoplasm of eucaryotic cells.

Many antibiotics used to treat bacterial diseases in human beings kill the bacteria by inhibiting protein synthesis on the bacteria's ribosomes. These antibiotics, however, do not affect protein synthesis in the cytoplasm of human cells, because the ribosomes there are chemically different. In 1968, microbiologist Anthony W. Linnane and his associates at Monash University in Australia showed that these antibiotics also inhibit protein production by mitochondrial ribosomes. These same drugs likewise block protein synthesis in chloroplasts. Linnane showed, too, that the chemical cycloheximide greatly inhibits protein synthesis by most eucaryotic ribosomes, but has no effect on the ribosomes of bacteria, blue-green algae, chloroplasts, or mitochondria.

The capture theory is also supported by dozens of examples of eucaryotic cells that contain recognizable bacteria and blue-green algae. In some cases, these procaryotic cells function as mitochondria and chloroplasts. For example, the microscopic ameba *Pelmyxa palustris* ordinarily lives anaerobically—that is, without oxygen and respiration. There is evidence, however, that this ameba can replay a critical stage in cell evolution. It will take in certain respiring bacteria that then begin to function as mitochondria, conferring upon the ameba the ability to respire.

There is one example of blue-green algae that live in a fungus. These algae also function as chloroplasts, producing food for the fungus. In addition, these algae will grow independently in the laboratory. Two kinds of one-celled organisms, *Glaucocystis* and *Cyanophora*, also contain what look like independent blue-green algae. Biologist William T. Hall of Rockefeller University in New York City has shown that, in *Cyanophora*, these function as chloroplasts do in plant cells. However, *Cyanophora* can also engulf and digest food particles, as do animal cells. In addition, scientists have been unable to grow the blue-green algaelike structures of *Cyanophora* outside the host cells. This raises the question of whether the structures are actually abnormal chloroplasts rather than blue-green algae. Biologist Eberhard Schnepf of the University of Heidelberg, in Germany, feels that they are examples of an intermediate stage in the evolution of blue-green algae to chloroplasts.

Although such experiments and observations strongly support the capture theory of cellular evolution, they may ultimately be of more practical significance. For example, one intriguing possibility for easing the world's food shortage is to transplant chloroplasts or blue-green algae into animal cells, perhaps even those of human beings. Theoretically, this would allow an animal or human being to use the energy that is provided by sunlight to produce at least some of his food from gaseous carbon dioxide and water.

Such a fantastic idea might be classified as a poor example of science fiction were it not for the work of biologist Margit Nass of the University of Pennsylvania. In 1969, she demonstrated that mouse cells could engulf chloroplasts isolated from spinach and African violet leaves. These chloroplast transplants remained in the mouse cells for at least five days, and they maintained their ability to photosynthesize. Nass did not report, however, whether the ingested chloroplasts provided food to the host cells. Unfortunately, the ingested chloroplasts did not multiply in their new hosts, as they do in plants, under the stimulation of light. If this problem, along with a myriad of others, can be overcome, an offshoot may be that one day people will come back from their summer vacation proudly bearing a sungreen.

For further reading:

Goodenough, Ursula W., and Levine, R. P., "The Genetic Activity of Mitochondria and Chloroplasts," *Scientific American*, Nov. 1970.

Nass, Margit M. K., "Uptake of Isolated Chloroplasts by Mammalian Cells," *Science*, July 4, 1969.

Pigott, G. H., and Carr, N. G., "Homology Between Nucleic Acids of Blue-Green Algae and Chloroplasts of *Euglena gracilis*," *Science*, March 17, 1972.

Raven, Peter H., "A Multiple Origin for Plastids and Mitochondria," *Science*, Aug. 14, 1970.

A Matter
Of Fat

By Jean Mayer

**Research on how the brain controls appetite, plus
measurements of physical activity, prove that obesity
is caused by more than just a love of good food**

During the 20 years that my colleagues and I have studied teen-agers
in Boston, we have seen an alarming 50 per cent increase in adolescent
obesity. Today, at high school graduation, one out of every five of these
young people clearly weighs more than he should for good health and
appearance. Obviously the excess weight accumulates because he eats
more than he needs to keep his body warm, his organs functioning, and
himself moving in work and play.

How much people eat is controlled, in part, by appetite and hunger.
We all experience these feelings and realize how their differences in intensity affect our desire for food. Appetite is mainly psychological; it
depends on our pleasant memories of taste, sight, and smell. Even after
a large meal, we may still want dessert. Hunger, on the other hand, is a

less-civilized matter. It is mainly physiological, and the hungrier we become, the more appetite we have for any kind of food. A few days without food, and we are driven to devour almost any substance within reach, no matter how repulsive it normally appears.

Hunger and appetite are, regrettably, far more complex than such simple explanations would lead us to believe. In fact, they are regulated by a great many interrelated factors within our bodies.

Some facts basic to the understanding of how our intake of food is controlled were established as early as the 1770s, when Albrecht von Haller, the Swiss physiologist, suggested that the emptiness or fullness of the stomach excites nerves that cause our sensations of hunger and satiety, or fullness. It was not until 1912 that Haller's theory was proved experimentally by Walter B. Cannon, a physiologist at Harvard University. Cannon inserted a balloon into the empty stomach of a medical student and connected it to an outside pressure gauge. Whenever the student reported hunger pangs, the gauge showed his stomach was contracting vigorously on the balloon.

In 1916, Anton J. Carlson, a physiologist at the University of Chicago, reported that during the first few days of fasting, the normal tension of the empty stomach, as well as the frequency and intensity of its contractions increases steadily. He suggested that the vagus nerves, which link the brain with the stomach and other internal organs, are the main pathway for these gastric impulses and that the primary hunger center must be in the brain's medulla oblongata, from which the vagus nerves arise. However, French pathologists Camus and Řoussy in 1913 showed through autopsy that some extremely obese persons had brain lesions. These lesions were in or impinging on not the medulla, but the hypothalamus, a part of the brain located on the floor of the brain hemispheres, just above the pituitary gland.

Experiments with rats based on this pioneering observation demonstrated during the 1940s that when the ventromedial areas, two small centrally located areas of the hypothalamus, are surgically destroyed in a rat, it overeats and becomes obese. Refining this technique, I was able, by 1954, to create the same kind of obesity in mice, which normally consume four times as much food per unit weight as rats do. This was the first time such work had been done with mice, permitting us to compare the metabolic characteristics of this kind of obesity in more than one species of animal.

In 1950, psychologist Neal E. Miller and his associates at Yale University examined the feeding behavior of rats with lesions in the ventromedial areas. They showed that while these animals overeat (we call such rats hyperphagic), they will not work as hard for food as normal animals. The scientists suggested that they might, in fact, be less hungry. To test this possibility, psychologist James Anliker and I trained normal mice to press a lever to obtain food. The frequency with which they pressed the lever was a measure of their hunger. As the mice became satiated, they slowed down and stopped pressing the lever. We

The author:
Jean Mayer, professor of nutrition at Harvard University, is a noted authority on obesity and the regulation of hunger. As a special consultant to the President in 1969, he organized and then chaired the First White House Conference on Food, Nutrition and Health.

Why We Stop Eating–the Glucostatic Theory

When the stomach is full, nerves inform the cortex. But it is high levels of glucose and insulin in the blood that stop hunger. They activate the satiety centers, which inhibit the feeding centers and, via vagus nerves, stop hunger contractions. Meanwhile, muscles also use the glucose for energy; any excess is stored in the liver or fat cells. As the glucose level falls, satiety centers become inactive and hunger contractions resume.

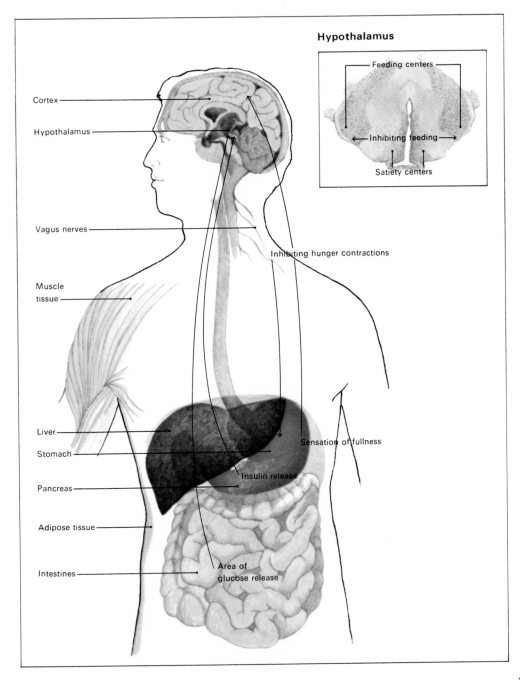

Hypothalamus

Feeding centers

Inhibiting feeding

Satiety centers

Cortex

Hypothalamus

Vagus nerves

Inhibiting hunger contractions

Muscle tissue

Liver

Sensation of fullness

Stomach

Insulin release

Pancreas

Adipose tissue

Intestines

Area of glucose release

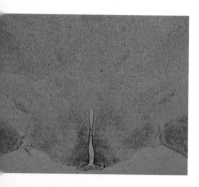

The hypothalamus in a normal rat's brain, *top,* contains satiety centers. These can be destroyed, *above,* by injecting the chemical goldthioglucose. The fatter of the two rats, *right,* was injected with the compound, which made it overeat.

found that after we gave them hypothalamic lesions, they continued to press the lever at the same rate, but over a longer period of time. This showed that while their hunger was not increased, they could not satiate themselves. The experiment confirmed the suspicion that the ventromedial areas are satiety centers, or brakes on a feeding mechanism that otherwise is constantly working.

Research by Indian physiologist Bal K. Anand indicated that another part of the hypothalamus controls the feeding mechanism. Working with Dr. John R. Brobeck at Yale during the 1940s, Anand showed that destroying a pair of small lateral areas of the hypothalamus to the right and left of the satiety centers caused animals to stop eating and drinking. One reason the animals might have stopped eating was that they stopped drinking and thus became dehydrated. However, Seoras D. Morrison and I, working at Harvard in the early 1950s, proved that this was not so. When we gave the test animals water through a tube that was inserted into their stomachs, they did not automatically resume their eating. They had stopped eating because the lateral areas, the feeding centers, had been destroyed.

Later, neuroanatomist Edward Arees and I, with a new technique, were able to actually trace the nerve fibers going from the satiety centers to the feeding centers. This gave a solid anatomical basis to the idea that the two centers of the hypothalamus are closely related and that what is regulated is not hunger but satiety.

With at least part of the brain mechanism identified, the next problem was to find out what determines satiety. We already knew that the sole fuel of the brain is blood sugar (glucose), which originates from food in the intestines and is stored in the liver as glycogen. We suspected, however, that the amount of glucose available to the cells in the satiety centers might also regulate hunger.

This central role of glucose forms the basis of the most widely accepted theory on the role of the hypothalamus in regulating food intake. Experiments in my laboratory had demonstrated that the satiety centers contain cells that are particularly sensitive to glucose and hormones, such as insulin, to which the rest of the brain does not respond. We suggested that the rate at which these cells, called glucoreceptors, take up glucose from the blood determines whether the satiety centers are active and the feeding centers quiet or, conversely, whether the satiety centers are quiet and the feeding centers active.

Many different types of experiments have supported this glucostatic theory. For example, by recording the electrical impulses in the hypothalamus of rats, cats, and monkeys while glucose is being taken up, Anand demonstrated that the satiety cells are electrically very active and the feeding centers very quiet. In man, we have learned that hunger and stomach contractions are absent as body tissues actively take up sugar, and vice versa.

Experiments in animals in 1968 showed that food intake increases after injections of substances that decrease the rate at which glucose is taken up by the glucoreceptors. We used such substances as glucose analogues, which are chemically similar to glucose but which block the enzymes involved in the glucose uptake. A second substance is phloridzin, which causes the loss of glucose through the kidney into the urine, and a third lowers blood glucose through other means, such as stimulating the pancreas to produce too much insulin.

An experiment that I reported at the meeting of the Federation of American Societies for Experimental Biology in April, 1972, also demonstrates the glucostatic mechanism. We had demonstrated in 1955 that we could destroy the glucoreceptor brain cells of test animals with an injection of a chemical compound consisting of a heavy metal and glucose (goldthioglucose). The animals subsequently overate permanently and, as would be expected, got fat. Goldthio compounds that do not contain glucose did not have that effect. This suggested that the glucose carries the gold to the glucoreceptors and the metal poisons them. By demonstrating in 1972 that mercurythioglucose acts in the same way, we again confirmed this hypothesis.

On the basis of the glucostatic mechanism, Saovane Sudsaneh, a Thai scientist, and I were able to show in 1959 how the satiety centers control a group of nerve cells called the Nucleus of the Vagus and, through the vagus nerve, the stomach contractions and the secretion of hydrochloric acid in the stomach. We showed that when the ventromedial area is destroyed, an increased use of glucose no longer inhibits stomach contractions. In other words, the same cells that control cerebral mechanisms of hunger also control gastric phenomena, harmonizing the whole organism. In 1968, Arees and I were able to follow the exact neural pathway going from the ventromedial area of the hypothalamus through the bundle of Schutz, a system of nerve fibers, to the Nucleus of the Vagus.

We are thus beginning to understand what part the brain plays in the hunger mechanism. Similarly, we have begun to learn what happens in the tissues. In particular, we had thought until the early 1960s that fat cells were passive and did not influence the energy balance of the body. We have learned, however, that the adipose tissue—the blood vessels, fat cells, and other cells that give the fat cells nourishment and structure—is very active. The amount of adipose tissue does not simply indicate whether intake is greater than output or vice versa, but the tissue actively competes for blood nutrients, glucose in particular, and consequently influences the body's food intake. We discovered this by measuring the difference in the level of glucose in an artery going into adipose tissue and in a vein leaving the tissue.

In current experiments, we are finding that the adipose tissue of obese mice is particularly adept at producing fat. Reducing the animals' weight decreases the fat content of the fat cells, but does not decrease the number of enzymes that make the fat. As a result, the adipose tissue tends to replenish itself. If we reduce the weight of these mice by giving them less food and more exercise, they regain weight more slowly than if we simply gave them less food.

The fat cells of a normal rat, *right,* are smaller and fewer per sample than those of a fat rat, *far right.*

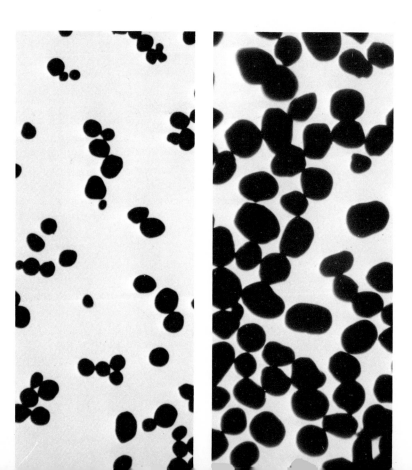

How Dieting Affects the Fat Cell

An overweight person stores excess fat, or triglycerides, top left. They break down into fatty acids in the blood vessel wall, enter a fat cell, and enlarge the lipid, or fat, droplet. Dieting reverses the process, center. A year later, he is still at his correct weight, right. Another person, who may be inherently overweight, has a larger fat cell, bottom left. He may reduce, center, but regains excess weight easily, right, perhaps because his fat cells contain more enzymes.

In Short-Term Overweight

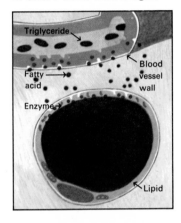
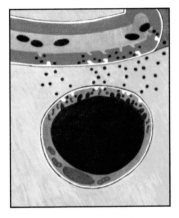
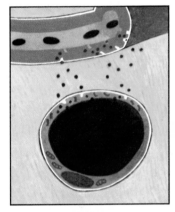

In Long-Term Overweight

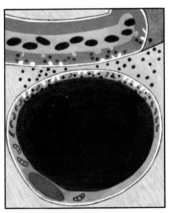
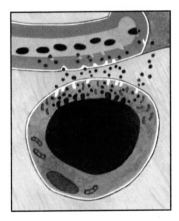
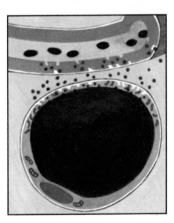

In laboratory animals, we thus find two basic kinds of obesity. One is due to something amiss in the brain mechanism that regulates food intake. I call these regulatory obesities. In the other kind, metabolic obesities, something is biochemically wrong with the animal's metabolism, so that its adipose tissue makes too much fat or does not burn it readily enough. There is a great amount of evidence that similar categories exist in human beings.

Many obesities in the laboratory animals are genetic in origin. For example, the obese-hyperglycemic mouse, weighing up to 120 grams (about 4 ounces) instead of the normal 30 grams, inherits genes that create more fat cells and higher blood glucose and blood insulin than does a normal mouse. The fat mouse also has a curious abnormality in the metabolism of its fat cells.

There is good evidence that genetics is also very important in determining obesity in man. Structural studies done in Sweden during the early 1960s and confirmed by Dr. Jules Hirsch at Rockefeller University in 1968 show that while the adipose cells of most obese individuals are filled with fat, some obese humans also have a greater number of fat cells as well. According to Hirsch, the number of fat cells seems predetermined, except perhaps for some increase during the first year of life under the influence of overfeeding. Also, obesity seems to run in families. In the Boston area, an average of 7 per cent of the children born to thin parents are obese at high school age. The rate is 40 per cent if one parent is overweight, and 80 per cent if both parents are overweight. A large-scale study in London in the mid-1960s revealed that children adopted at birth do not show this association with the weight of their adoptive parents. This indicates that family food habits may play a far less important role than many think.

If obesity is so completely determined by internal and inherited mechanisms, can we hope to find a cure or at least minimize its effects? Experimental and clinical scientists have proved that a lack of exercise

A litter of mice includes two that are extremely obese. They inherited genes that created excessive numbers of fat cells.

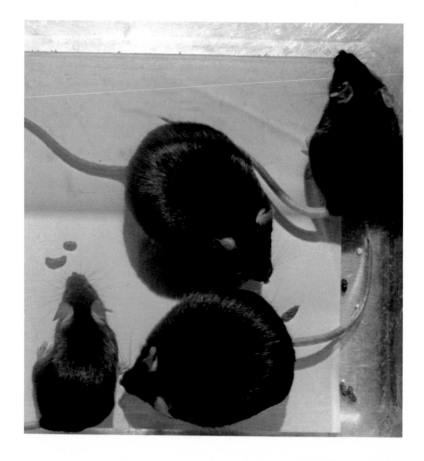

is often associated with obesity. Yet, in the past two decades, in the United States at any rate, the role of exercise in weight control has been minimized, if not ridiculed. Americans, once justly proud of being the most vigorous and active group in the Western world, are daily absorbing propaganda advocating prolonged self-enforced starvation as the only way to stay in shape. This advice seems motivated, in most cases, by a desire to promote a fad diet or diet pills. It is bolstered by two misconceptions, both of them, unfortunately, initially plausible.

The first of these is that relatively few calories are burned up during exercise. If this were true, it would follow that even a great amount of physical activity could have little effect on the balance between calories taken in and calories used up. This balance, of course, determines how much a person weighs. The second faulty notion is that at any initial level of caloric intake, an increase in physical activity automatically triggers an increase in appetite and is, therefore, self-defeating. Let us look at the evidence.

The first misconception, minimizing the cost of physical activity, is obvious to anyone who has studied a table of recommended dietary allowances or caloric requirements. For example, the daily allowances for men in the National Research Council's table vary from 2,400 calo-

As two mice from the same litter, one obese and one of normal weight, move, their cages tip, allowing a device to measure their activity. The experiment shows that normal mice are much more active than obese ones.

ries for sedentary men to 4,500 calories for very active men. Nor does the latter figure represent an upper limit. Laborers, soldiers on the battlefield, and athletes may require up to 6,000 or more calories. Surely a factor that can more than double the number of calories a person uses up is not one to be casually ruled out as of no great importance in his efforts at controlling his weight.

And yet, magazine and newspaper articles, radiobroadcasts, and pamphlets prepared by state, municipal, and private health educators seemingly illustrate the futility of exercise as a reducing aid with assertions such as this: "The caloric equivalent of a pound of fat can be matched only by walking 36 hours, splitting wood for 7 hours, or playing volleyball for 11 hours." These extremes of activity cited by the enemies of exercise picture any wearying performance as being accomplished in a single, uninterrupted stretch. Actually, the energy is used up whether the activity is performed in a day or in a year. If you chop wood for a half hour every day, it would, according to this reasoning, be equivalent to 26 pounds of body fat a year. Similarly, a half hour per day of handball or squash would use up 16 pounds of fat during the same period. Multiply this by 10 and you see that in a decade such exercise would make you disappear altogether.

It is certainly more useful to cite the actual cost in calories of different types of physical exercise. Such determinations have been carried out with great accuracy in a number of countries. The results are expressed in terms of an average man of definite physical characteristics. For a 150-pound man, the cost of walking, over and above the cost of resting, varies from 115 to 565 calories per hour, depending on how fast he walks. Swimming will consume twice as many calories. The peaks reached in some athletic competition, such as cross-country skiing, may approach 1,300 calories per hour.

Even more important, the cost of activity depends on the person's size. In exercises where no heavy object outside the body is moved—for example, jogging—the cost of exercise increases proportionately to body weight. If excess weight impairs body movement, making the person awkward and inefficient, the cost of exercise increases faster than does body weight. An overweight person thus burns up more body fat for the same amount of exercise than does a person of normal weight.

The ratio of the cost of physical activity to weight has another important result. Any increase of caloric intake above the balance level causes far less weight increase in an active individual than in a sedentary person. This is because the active person will use some of this extra intake carrying whatever fat he is storing around the tennis court, the golf course, and other places. By contrast, in the sedentary person, only the slow increase in passive heat loss, which follows the expansion of his waistline and body surface, eventually starts matching the daily calorie addition in, say, an extra dessert at supper. In effect, the active person has two brakes on weight gain: increased cost of exercise (a very potent one), and increased basal metabolism. The sedentary individual has only the less potent brake, and his weight is likely to fluctuate. When it stabilizes, it does so at a higher figure.

Let us now examine the second major argument—that exercising to control weight is a self-defeating practice for obese people. Increased activity does trigger an increase in appetite in persons who were reasonably active to start with. The mechanism that regulates appetite in active people adjusts the amount of calories consumed to those burned. As a result, if the body is suddenly called upon to do increased work, it will not consume vital tissue for energy. However, an inactive person's appetite is at its lowest level, and at this level moderate exercise increases calorie usage but not appetite. Thus, underexercising, not overeating, may well be the most important cause of obesity today.

As is so frequently the case, studies of small rodents paved the way for observations of this phenomenon in human beings. At Harvard's

Although both mice have a damaged hypothalamus, the mouse on the left stayed thin because he was made to exercise and given less food.

In sports and play, the obese child tends to watch while his companions perform.

department of nutrition in the early 1950s, my collaborators and I studied how the food intake of white rats varies when their exercise is varied. We trained a large group of rats to run on a motor-driven tread-mill. Then we divided them into a number of smaller groups, which were exercised, respectively, for 1, 2, 3, and up to 10 hours daily. We measured their food intake and weight changes for several weeks. Then we compared these rats with others that had been left unexercised in their cages as a control.

We found that rats exercised one or two hours daily ate somewhat less than did the unexercised rats and gained no weight, while the inactive animals slowly gained weight. From two hours daily onward, food intake increased as exercise increased. Although these active animals gained, their weight stabilized at a lower level than that of the sedentary rats. Making the animals swim produced the same result. Within the range of normal activity, appetite reliably adjusted their food intake to the amount of energy expended. Below that range, appetite caused the rats to eat more than they needed.

Studies of genetically obese mice in our laboratory during the past 20 years show that they are at least 10 times less active than their normal brothers and sisters, and that inactivity precedes the development of obesity. Exercising these young obese mice slows down excess weight gain considerably, and appetite does not increase correspondingly. The same effect occurs when we "breed" exercise into genetically obese mice. This can be done with a gene called, appropriately, the waltzing gene, which makes them seem to be constantly chasing their tails.

Decreased activity seems to play a particularly important part in childhood obesity. In general, psychiatrists dealing with this vexing problem believe that obese children overeat for a variety of reasons: As an escape from tensions at home, as a substitute for the affection of one parent or both, as a result of parental neglect or oversolicitousness, because of a desire for increased size and recognition, or as a protection against shyness. Child psychiatrist Dr. Hilde Bruch of New York City noted in *The Importance of Overweight* in 1957 that many obese children are inactive; but it is only recently that the basic assumption has been questioned: Do all obese children eat excessively as compared to children of normal weight, or could some of them gain weight simply because of underexercising?

A study we conducted in 1956 answered this question in an older group of youngsters. It showed that inactivity is indeed the major factor in perpetuating obesity in many, if not most, overweight adolescents. Our examination of the dietary intake and daily activities of equal groups of overweight and normal-weight high school girls, matched for age and height, showed that the obese students fell into two groups. One group, by far the larger, contained girls who ate no more than the normal-weight girls, but who exercised considerably less. They preferred activities during which they could sit rather than walk or run. They watched television four times as long as did the girls in the normal-weight group. The second group ate more than the normal girls did, but exercised normally. They were red-cheeked and cheerful and, while overweight, they appeared less fat and more muscular than the inactive group. This group shows that there is more than one cause of obesity. Other studies that we conducted in the early 1960s produced the same results with boys.

More recently, we have used a motion-picture technique to measure this phenomenon more accurately. Every three minutes, we had a camera take a three-second shot of obese and normal youngsters at play. Physiologist Beverly Bullen and I analyzed 29,000 such shots. They show that obese youngsters playing volleyball are immobile 80 to 90 per cent of the time (as compared to 50 per cent for the normal); playing singles tennis, 60 per cent (as compared to less than 20 per cent for the normal). While normal youngsters swim, expending 5 or more calories per minute above the amounts used at rest, most of the obese youngsters float, sit, or stand in the water expending no more than 1 calorie. We are studying why the obese are so inactive and whether it antedates or follows obesity. At any rate, there is little doubt in our minds that this inactivity is one of the main factors that causes the obesity to be self-perpetuating.

Since this seems to be the case, people should benefit greatly from an exercise program. From 1965 to 1970, we conducted an experimental study in the public schools of a large Boston suburb. Several hundred overweight students, from 1st grade to senior high school, exercised every day for one school period. By measuring them throughout the study with an instrument called a skin-fold caliper, we found that a large proportion substantially reduced the amount of fat in their bodies. We are now measuring many of these young persons again to see how permanent the change has been.

To maintain his proper weight a person must not only watch his diet, he must also take up daily exercise.

We are also continuing to actively explore the causes of the inactivity of overweight youngsters. Studies of babies suggest that a tendency to inactivity and overweight is present at birth. This was demonstrated by a study I reported in 1968 with Dr. Hedwig Rose, a pediatrician. Following 31 infants from birth to 15 months, we found that the fat babies tended to be inactive babies with a very moderate appetite, while the active babies ate much more but weighed less.

Other findings suggest that the obese may find it difficult to carry out automatically (without thinking, and without an external rhythm) any sort of rhythmical, patterned motion, such as running or swimming, that involves the large muscles of the body. The obese do well enough in nonrhythmical activity, such as the shot-put and archery, even when strength is required. And their rhythmical motions are greatly facilitated by highly accented rhythmical music. But they may be handicapped for more ordinary, continuous activity, an indication that I am currently testing with Harvard psychiatrist Peter Wolf. We have devised special recording systems to test the ability of obese teen-agers to carry out finger-tapping exercises. The tapping is recorded on magnetic tape and then analyzed by computer.

All of my research on the role of exercise in obesity indicates it is in many instances a "disease of civilization." Automobiles and labor-saving machines combine to decrease physical exercise. For many of us, this mode of life has so decreased our physical activity that we no longer respond to a further decrease in exercise by a further decrease in appetite. Excess calories, accordingly, accumulate as fat. If we want to avoid this, we must either exercise more or eat less and feel hungry all our lives. Discounting the value of playing nine holes of golf because it works off only the equivalent of a slice of apple pie à la mode is dangerous. For, the person who believes this not only will not play the nine holes of golf, he may also have an appetite so regulated that he will eat the pie and ice cream anyway.

It is not easy to suddenly change one's mode of life to include daily exercise. Walking is time-consuming and is often made difficult because of traffic. Facilities for adult sports are often inadequate. Many sports, although theoretically available, are expensive to cultivate. Calisthenics, perhaps the most effective method if done well, are uninspiring, if not actually boring. Yet Americans are becoming more aware of the need for exercise if fitness is to be maintained. We must find attractive and inexpensive ways to meet the need.

For further reading:

Deutsch, Ronald M., *The Family Guide to Better Food and Better Health,* Meredith Corp., 1971.

Mayer, Jean, *Human Nutrition,* Charles Thomas, 1972.

Mayer, Jean, *Overweight,* Prentice-Hall, 1968.

Vogue, April 1, 1972, "Fat Cells," an interview with Dr. Jules Hirsch.

Wyden, Barbara W., "The Fat Child Is Father of the Man," *The New York Times Magazine,* Sept. 13, 1970.

The Nutritious Trace

By Walter Mertz

The lack of only a few atoms of a chemical element can cause changes in your body that may take years to notice

Dr. K. Michael Hambidge, a Denver pediatrician, analyzed some human hair clippings in 1971 to determine what trace elements are present in normal hair. In February, 1972, he announced an unexpected finding. The hair was taken from 340 children and young adults who appeared well nourished and healthy. Yet, about 8 per cent of the children between 4 and 16 years of age had very low levels of the chemical element zinc in their hair, less than 70 parts per million. This amount was far below the average concentration of 155 parts per million. Intrigued, Dr. Hambidge examined these zinc-deficient children and discovered that as a group they were shorter and weighed less than the others. Their appetites were poor, and many of them were less sensitive to the taste of food.

Dr. Hambidge gave these children small dietary supplements of zinc for from one to three months. With only this change in their diet, the children's taste and appetite improved, and the zinc level in their hair rose toward normal. They began to eat better, and their height and weight began to increase.

Today, nutritionists, aided by analytical chemists and doctors, are determining our need for dozens of chemical elements, such as zinc, that are found only in "trace" amounts in the body. And their research suggests that all of us, like the children in the Denver area, are not getting enough of the many chemical elements that our bodies need to perform their intricate physiological processes. As we rely increasingly on highly processed and prepared foods, on crash diets, food fads, and the offerings of vending machines and drive-in restaurants, our nutrition is becoming even more precariously balanced.

Only 11 elements make up more than 99 per cent of the human body. Six of these—hydrogen, carbon, nitrogen, oxygen, phosphorus, and sulfur—are the building blocks of proteins, fats, carbohydrates, and nucleic acids. Calcium, the bulk of our bones and teeth, is, together with potassium and magnesium, an essential part of all cells. Sodium and chlorine are present as salts in the body fluids.

The rest of the human body, the less than 1 per cent, consists of the trace elements. Their concentrations vary from several hundreds parts per million to less than a few parts per billion. For instance, our blood serum contains only one part per million of zinc. Cobalt is present in a much smaller amount—0.85 part per billion.

It is not easy, even for scientists, to picture just how small such concentrations are. One part per million, for example, is one drop of an additive in 14 gallons of fuel; one part per billion is one drop in 14,000 gallons. A Boeing 707 jetliner could fly from New York City to Athens, Greece, on this amount of fuel. If the 707's engines depended on the additive, that one drop would determine whether or not the plane reached its destination.

Our bodies are just this dependent. Many chemical elements are known to be essential to the body in such minute amounts. These include fluorine, chromium, iron, copper, zinc, and iodine. They are important because they allow the chemical reactions of the other 99 per cent of the body to take place.

Recent research has shown that trace elements function in four ways. The first way is to help move common elements throughout the body. For instance, iron is a vital part of the red blood cell pigment, hemoglobin, which carries oxygen to every cell in the body.

A second function of many trace elements is in enzyme systems. Enzymes are large, extremely complex protein structures that speed up biochemical reactions. Most of the thousands of known enzymes contain one or more trace metal atoms. Biochemists have isolated enzymes in pure form from tissues and measured their activity. When they remove some of the metal, the enzyme activity decreases. When they

The author:
Walter Mertz, an M.D. and biochemist, is Chief of the Vitamin and Mineral Nutrition Laboratory of the U.S. Department of Agriculture in Beltsville, Md. He codiscovered chromium as an essential trace element.

A Nutritional Table of the Elements

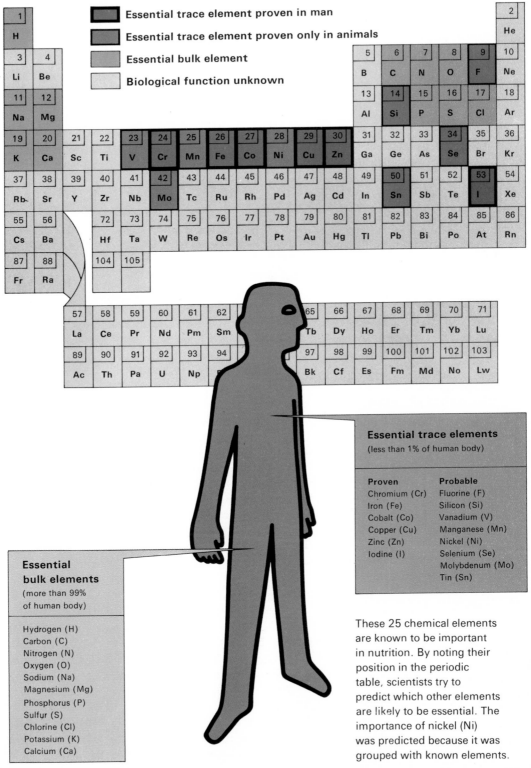

Legend:
- Essential trace element proven in man
- Essential trace element proven only in animals
- Essential bulk element
- Biological function unknown

Essential trace elements
(less than 1% of human body)

Proven	Probable
Chromium (Cr)	Fluorine (F)
Iron (Fe)	Silicon (Si)
Cobalt (Co)	Vanadium (V)
Copper (Cu)	Manganese (Mn)
Zinc (Zn)	Nickel (Ni)
Iodine (I)	Selenium (Se)
	Molybdenum (Mo)
	Tin (Sn)

Essential bulk elements
(more than 99% of human body)

Hydrogen (H)
Carbon (C)
Nitrogen (N)
Oxygen (O)
Sodium (Na)
Magnesium (Mg)
Phosphorus (P)
Sulfur (S)
Chlorine (Cl)
Potassium (K)
Calcium (Ca)

These 25 chemical elements are known to be important in nutrition. By noting their position in the periodic table, scientists try to predict which other elements are likely to be essential. The importance of nickel (Ni) was predicted because it was grouped with known elements.

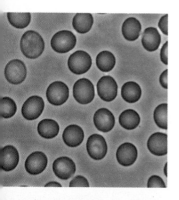

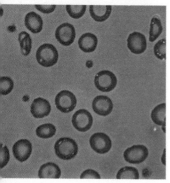

Photomicrographs show how red blood cells in a patient with iron deficiency anemia, *above,* differ from normal red cells, *top.* Insufficient iron in the diet decreases the hemoglobin in the cells, giving them a ring shape. Such cells supply inadequate oxygen to the body tissues.

replace the metal, the activity returns to normal. The trace element can hold the enzyme's protein subunits together, or it can bind the substrate—the chemical substance upon which the enzyme acts—to the active center of the enzyme. Only a small amount of metal is needed. Each minute, a single enzyme molecule can interact with thousands of substrate molecules.

A third way in which trace elements aid the body is to assist the action of hormones. Hormones are chemical substances secreted into the blood stream by our endocrine glands, such as the pancreas and the thyroid. They regulate important physiological functions and are effective in very small concentrations because they need only to interact with a few strategic sites on or in the cells. The trace element helps make the hormones so effective.

A fourth mode of trace element action is suspected, but not yet clearly understood. Nucleic acids, the carriers of our genetic information, contain rather high concentrations of many trace elements, such as vanadium, chromium, manganese, iron, cobalt, nickel, copper, and zinc. In test-tube experiments, these elements have been found to influence nucleic acid metabolism.

Some trace elements function in several of these ways. Iron, for instance, is a constituent of every living cell. It is the active site in many enzymes as well as an oxygen carrier in hemoglobin. Each unit of hemoglobin contains one atom of iron as the active center. The iron atom is so important that without it hemoglobin cannot be made and oxygen cannot be carried. As one result, the blood, and thus the complexion, becomes pale, because the hemoglobin-oxygen combination gives it its bright-red color. The round trip from the heart to the body's outermost cells takes a red blood cell up to a minute. Only after it returns to the lungs can the hemoglobin pick up a new oxygen molecule. Thus, many atoms of iron are needed to carry oxygen, whereas only a few iron atoms are required for enzyme reactions.

The body does not excrete any significant amount of iron, but about 1 milligram (mg), or about $1/30,000$ of an ounce, is lost daily when cells lining the gut are sloughed off and lost in the stool. However, loss of blood also means a loss of iron, and menstruating women lose an additional 0.5 mg per day.

Unfortunately, only 10 per cent of the iron in what we eat can be absorbed by the intestines. Therefore, the body's daily iron requirement is 10 times what the body loses—about 10 mg for males and 18 mg for menstruating females. Even a good diet may not provide this amount, so, under law in 34 states, flour and bread are enriched with iron salts. From 8 to 12.5 mg of iron is added to every pound of bread, for example, and the U.S. Food and Drug Administration is considering a proposal to increase this to 25 mg.

Persons with enough iron in their diets store this element in their tissues, mainly in the bone marrow. The first effect of an insufficient iron intake is the disappearance of this reserve. Only then do changes in the blood—and a pale complexion—occur. Although severe iron deficiency anemia is rare in the United States, nearly 25 per cent of some groups of people that have been studied have hemoglobin levels below normal. Researchers are studying the consequences of this condition. Preliminary results indicate that slightly iron deficient children have a significantly reduced attention span. This, of course, has an important effect on their ability to learn.

Iodine is an essential trace element found in thyroxine, a thyroid hormone that regulates the production of energy in the cell. This hormone is also necessary for growth and the development of intelligence and many aspects of personality. Our blood contains only 75 parts per billion of thyroxine, and each molecule contains 4 atoms of iodine. Without iodine, thyroxine molecules cannot be produced.

We need about 140 micrograms (μg) of iodine each day. When we do not get enough, the thyroid gland tries to compensate by growing bigger. This leads to the typical enlargement of the neck called goiter. Severe forms of iodine deficiency, which disturb the physical and mental development of children, are now rare, but visible goiter still occurs in areas where the iodine intake is low. In 1920, about 25 per cent of the U.S. schoolchildren in the "goiter belt," which includes the St. Lawrence River, Pacific Northwest, and Great Lakes regions, were afflicted.

An atom of iron, *black,* is the essential part in every molecule of myoglobin, a muscle protein. The iron combines with an oxygen atom (W) to store oxygen until it is needed for energy.

How the Body Uses Trace Elements

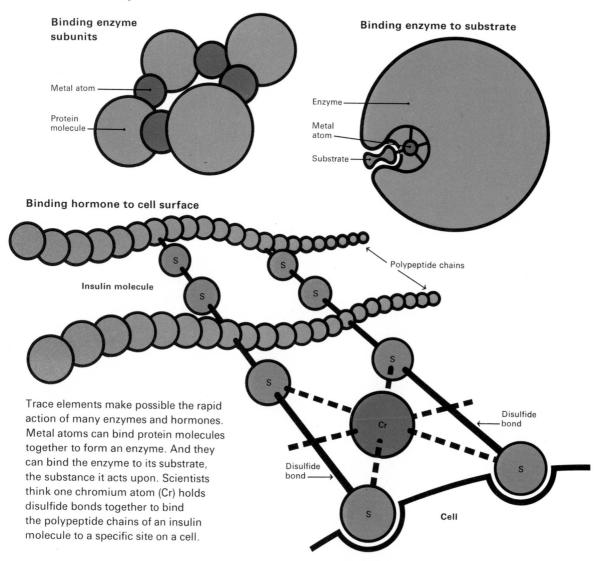

Binding enzyme subunits

Metal atom

Protein molecule

Binding enzyme to substrate

Enzyme

Metal atom

Substrate

Binding hormone to cell surface

Polypeptide chains

Insulin molecule

S

S

S

S

S

S

S

Cr

Disulfide bond

Disulfide bond

Cell

Trace elements make possible the rapid action of many enzymes and hormones. Metal atoms can bind protein molecules together to form an enzyme. And they can bind the enzyme to its substrate, the substance it acts upon. Scientists think one chromium atom (Cr) holds disulfide bonds together to bind the polypeptide chains of an insulin molecule to a specific site on a cell.

This led to the program of iodizing table salt, which reduced iodine deficiency to less than 3 per cent of that population. Iodized salt contains 1 part iodine in 10,000 parts salt, assuring users of enough iodine even in those areas where the iodine in food is low.

Zinc is an essential component of most enzymes and is important for normal growth and development. Our bodies need about 15 mg of zinc a day. A severe deficiency, which occurs in the Middle East, results in extremely retarded growth and sexual development. Milder forms of zinc deficiency occur in the United States.

Scientists accidentally discovered another role for zinc. In 1953, biochemist William H. Strain and Walter J. Pories, then a medical student at the University of Rochester School of Medicine made tests to see whether experimental wounds in rats healed faster when the animals were fed increased amounts of certain amino acids. There was a theoretical reason to suspect this effect, so these scientists were not surprised when the extra amino acids greatly improved healing. The surprise came when they repeated their experiment under almost identical conditions and got completely negative results.

But Strain and Pories were convinced that their first results were valid. They then found that the amino acid used in the second experiment had come from a different batch. That first batch, they began to suspect, may have contained an unknown contaminant that was responsible for the rapid healing. After many careful tests, they found that the effective batch contained much more zinc than the other one. Subsequent experiments showed that it was the zinc that had caused the animals' wounds to heal faster.

In 1965, Strain and Pories applied their findings to human beings. They tested young men of the U.S. Air Force who entered Wright-Patterson Air Force Base Hospital in Dayton, Ohio, for minor surgery. The men received identical treatment except that half were given a zinc supplement. The surgical incisions of the patients who were given zinc healed significantly faster.

It has since been found that only those persons with low zinc levels in their blood, tissues, or urine respond to zinc supplementation. The young airmen must have been zinc deficient, at least at the time of their operations. People with a zinc deficiency are unlikely to be aware of it, however. Usually only the stress of a disease or an operation can make the deficiency apparent.

Another seemingly unrelated development was reported in 1967 by Dr. Robert I. Henkin, a neurophysiologist at the National Institutes of Health in Bethesda, Md. While studying some rare diseases, he noticed that certain patients lost their sense of taste whenever they took a particular drug he had prescribed. They told him that everything they ate tasted like sawdust. He found that their ability to recognize, or even detect, sour, sweet, bitter, or salty taste from test solutions applied to their tongue was greatly reduced. Dr. Henkin knew that the drug he was giving interfered with the metabolism of certain trace elements and thus might also have caused a trace element deficiency. When he gave his patients a supplement of copper or zinc, their taste gradually returned to normal and they could enjoy their food again.

Dr. Henkin later found that many otherwise normal Americans have impaired taste. They range from those who cannot recognize taste differences at all, to those who use large quantities of salt and sugar because they cannot taste a normal amount. A moderate increase in their intake of zinc corrects this condition. These people, like the children Dr. Hambidge studied in Denver, had not been meeting their individual zinc requirement through their diet.

Chromium is another element important in the overall functioning of the human body. It assists the hormone insulin, which is produced

**Trace Elements
In Nucleic Acids**

Metal
atom

Nucleic acid

Nucleic acids, which carry our genetic information, contain high concentrations of many trace elements. Test-tube experiments have shown that these metal atoms influence the acids' metabolism.

by the pancreas during digestion. Insulin helps to convert glucose, the sugar absorbed into the blood from the intestines, into energy or storage forms, such as glycogen or fat. It does this by making cell membranes permeable to the blood sugar. Consequently, any excess sugar quickly disappears from the blood into the cells. When the body does not produce insulin, sugar metabolism becomes less efficient, the blood sugar level increases, and sugar begins to appear in the urine. This signals the disease diabetes mellitus.

Chromium is not part of the insulin molecule, but it interacts with insulin when the hormone attaches to the strategic sites on the body's cells. The blood stream contains only 0.4 to 4 parts per billion of insulin. If, as scientists believe, one atom of chromium interacts with one molecule of insulin, the effective concentration in the blood would be only 4 to 40 parts per trillion. Without this amount of chromium, insulin is not completely ineffective, but the body requires much more insulin in order to function normally.

The body excretes chromium every day, mainly through the urine. To compensate for this loss, an adult needs from 5 to 10 μg of this trace element. The body's absorption of chromium compounds varies greatly—from less than 1 per cent to 25 per cent of a given dose. Because the field of chromium nutrition is new, we do not yet know exactly how much chromium our bodies require or how much of this element is present in different foods.

Chromium deficiency results in the body's ineffective use of sugar. Severe protein-calorie malnutrition in the populations of underdeveloped countries is often associated with severe chromium deficiency. The resulting sugar intolerance can be corrected when chromium intake is increased. Although severe chromium deficiencies occur only in a few isolated cases in the United States, there is evidence that mild chromium deficiency exists as a result of some diseases, in women who have had several children, and in older people.

All of this recent work showing the importance of iron, iodine, zinc, and chromium has been possible because we can measure these elements at concentrations as low as one part per billion. In the past, elements were detected by being chemically isolated and weighed. This was satisfactory as long as we were concerned with milligram quantities. But the incredibly low concentrations of most trace elements cannot be detected by this method.

About 40 years ago, we discovered that certain trace elements form intensely colored compounds with certain chemicals. The intensity of the color gave the concentration of the trace element. This method greatly increased our ability to identify elements. Unfortunately, many elements form complexes of similar color with any one chemical. So this method, too, was replaced by more sensitive ones.

Today, our detection procedures begin with burning a sample of tissue, or food, which removes most of the bulk elements. To identify the remaining concentrated metal elements, we look for particular wave

lengths of light that only the elements can emit or absorb. For instance, when an ash rich in iron is applied on a carbon electrode and a strong current passed through it, an electric arc forms which vaporizes the ash and causes the iron to emit light of its own specific wave length. The intensity of this light is proportional to the amount of the element placed on the electrode. This method is called emission spectrometry.

We can also use the property of elements to absorb the same light that they emit. We shine light that has the same wave length produced by iron, for example, through a vaporized specimen. The higher the concentration of iron in the vapor, the more light it will absorb, and the less light will penetrate to the detector. This method is called atomic absorption spectroscopy. Both emission and atomic absorption spec-

To conduct ultraprecise experiments, rats on a diet deficient in a trace element must be kept in an ultraclean environment and can be given only purified food and water. Filters remove dust, which carries trace elements, from the air. Special sleeves on all-plastic isolator allow the scientist to handle a rat without touching it.

141

Before analyzing a sample of hair for its zinc content, Dr. Mertz burns it with atomic oxygen, *above,* to destroy all organic material. The sample, *right,* in machine's chamber, is being quickly reduced to ash

troscopy, in the most sophisticated apparatus, can measure quantities as small as a billionth of a gram.

Even so, the laboratory analysis of trace elements in the body is still in its infancy. There is much that we do not know. Of all the elements in the periodic table, a biological function in man or animal has so far been demonstrated for only 25.

There is only one way to prove that an element is essential: The experimenter must produce a deficiency disease because of a lack of this—and only this—element. He does this by feeding laboratory animals a diet that has all the known essential nutrients except the trace element in question. If a deficiency disease results, and then if the deficiency symptoms are removed by adding this element to the diet, the element

A whole body counter is used by scientists at Hines Veterans Administration Hospital near Chicago to find the amount of the trace element cobalt in the human body.

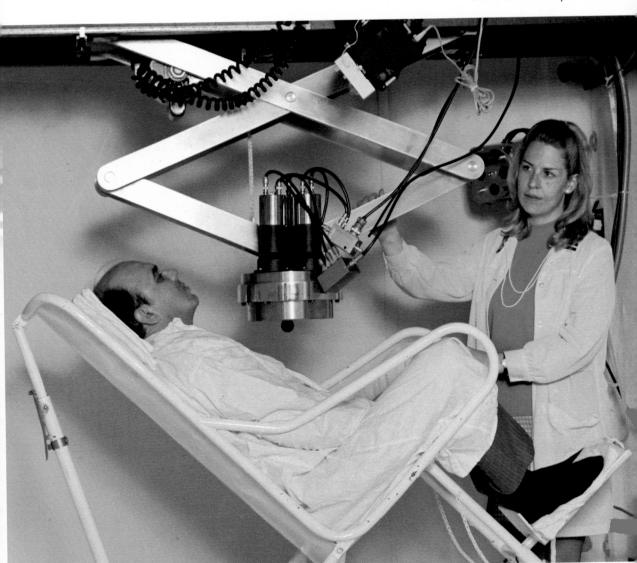

is considered essential. If other investigators show the same pattern for several different species, the evidence gains importance.

Producing a deficiency can be very difficult, however, particularly with elements that are needed only in extremely small amounts. It is often not enough to purify the food of laboratory animals. The water and all parts of the environment, including the air, must also be purified because they contain significant amounts of trace elements. To this end, researchers must keep the animals in special isolators in which the air constantly circulates through filters to remove dust particles that might contain trace elements.

A deficiency can cause many symptoms. Retarded growth is the easiest to detect, but more subtle anatomical or physiological changes, such as slight bone deformations or sugar intolerance, are just as meaningful. In human beings, symptoms of trace element deficiencies are known for only six trace elements: fluorine, chromium, iron, copper, zinc, and iodine. Although a human deficiency has not yet been observed for other elements, many more of them may be essential to human nutrition. In fact, traces of eight other elements—silicon, vanadium, manganese, cobalt, nickel, selenium, molybdenum, and tin—in addition to the above six are considered essential to other animals. Probably they are also essential for man. Silicon, the most recently detected, was reported to be essential by nutritionist Edith M. Carlisle, at the University of California, Los Angeles, in April, 1972. She had raised chickens with a diet deficient in silicon and observed retarded growth and bone abnormalities in the birds.

This small amount of chromium is the lifetime supply needed by the average person.

Some elements we can reasonably predict to be essential on the basis of their place in the periodic table. For example, the biological function of nickel was unknown until about 1969. Yet scientists predicted it was essential because of its place between two essential elements, cobalt and copper. Today, the element bromine has no known biological function. Yet bromine is a member of the halogens. Three of them—fluorine, chlorine, and iodine—are all known to be essential elements for man. Such strictly chemical considerations can guide scientists to concentrate their study on the most promising trace elements.

Despite the belief that nutritional deficiencies are rare in the United States, nutritional problems with trace elements do exist in this country, to some extent in many of us. As we have seen, parts of the population are known to be deficient in four trace elements—iron, iodine, zinc, and chromium. But these deficiencies are not insurmountable. Malnutrition, which can be deficient nutrition or overnutrition, is, in the United States, mostly a matter of poor eating habits and poor knowledge of what makes up a good diet. Mankind could not have survived if nature had not provided us with a perfect supply of all essential trace elements in foods—if our diets are balanced.

Trace element nutrition differs in one important aspect, however, from the traditional bulk nutrition. If we do not get enough calories, we become hungry, and, if we get too many, we have a stuffed feeling: Our body tells us where we are. This regulatory mechanism does not exist for the trace elements. The changes that result from too little or too much of them are subtle, and it often takes years to even notice the symptoms. Therefore, we have to rely on the findings of nutrition research and adjust our food habits accordingly.

Each day we should eat some meat, fruits, vegetables, and dairy and cereal products. We should also eat a large variety of foods, because a trace element that is lacking in one food can be found in another. It is very dangerous, however, to try to supplement our dietary copper, for example, by eating a lot of nuts because they are rich in copper. While concentrating on getting plenty of this trace element, we may forget many other requirements.

We can greatly increase the availability of several trace elements in a food by eating another at the same time. For example, eating meat or a food rich in vitamin C (ascorbic acid) with any food containing iron will make much more of the iron available to the body. Thus, the old concept of three square meals a day is also a sound basis for good trace element nutrition, and any lasting deviation from it, such as going on one-sided crash diets or limiting ourself to particular "organic" foods, can result in serious damage.

But there is another, equally important side to the concept of a balanced diet. Nature provides us with a perfect balance. But food processors have learned to divide many foods into "desirable" parts and those that they throw away. In many instances, however, the desirable portions are the calorie carriers, and the discarded ones contain the essential vitamins, minerals, and trace elements. This is certainly true for pure fats and refined sugar, as well as for such starches as white flour, which has lost the wheat kernels' iron-rich protective coat during milling. Whenever we eat too much of these processed foods, we do it at the expense of the trace-element-containing foods, and we are inviting deficiencies in the long run. The teen-ager who consumes large amounts of soft drinks and snack foods eats less during his three square meals unless he wants to gain weight. Consequently, he takes in fewer of the essential trace elements that are so important to his health.

We began by comparing the essential trace element cobalt in our blood to one drop of an additive in 14,000 gallons of jet fuel. Nobody in his right mind would fly in a large airliner that did not have the proper fuel. Should we not be as concerned with the completeness of the fuel that keeps us healthy and alive—our daily food?

For further reading:

Arehart, Joan Lynn, "Trace Elements: No Longer Good Vs. Bad," *Science News*, Aug. 14, 1971.

Mayer, Jean, "Zinc Deficiency: A Cause of Growth Retardation?" *Human Nutrition*, Charles Thomas, 1972.

Mertz, Walter, and Cornatzer, W. E., editors, *Newer Trace Elements in Nutrition*, Dekker, 1971.

The Making Of a Champion

By Daniel F. Hanley

Training, guided by the science of muscle development, diet, and drugs, is extending the dimensions of human performance

On July 9, 1971, Shane Elizabeth Gould, a 5 foot 7 inch, 126-pound Australian girl, 14 years old, smashed the world record in the women's 400-meter free-style swim. She finished the race in 4 minutes 21.2 seconds. Her time was 43 seconds faster than the time that was posted by 6 foot 3 inch, 190-pound Johnny Weissmuller, later the movies' Tarzan of the Apes, when he won the men's 400 meters at the age of 20 in the Olympic Games of 1924.

What happened during those 47 years that permitted a mere slip of a girl to outswim the mighty Tarzan? There are a lot of factors—starting to swim at an early age, learning to do it right at the very beginning, better training and coaching techniques, and, not the least of all, applying what we have learned about the human body and the limits of physical performance.

When a person trains for an athletic event, the focus is either on a particular muscle or group of muscles for a specific purpose, or on general muscular development for overall fitness. We used to emphasize developing strength and endurance with a lot of long-distance work. But Shane and other outstanding swimmers have taught us much

about training, especially the advantage of using relaxed rhythm and agility drills, as well as a technique called quality speed work. Swimmers do more than just swim. They also weight train for strength and do dry-land exercises for agility, power, and flexibility, and to develop the muscles in the upper part of the body.

Perhaps the most advanced training is required by athletes who compete in the decathlon. This contest demands a superbly conditioned human being who can excel in 10 different feats of running, jumping, and throwing. These events include: the 100-, 400-, and 1,500-meter runs; the 110-meter high hurdles; javelin throw; discus throw; shot-put; pole vault; high jump; and broad jump. In the first four, quick reflexes, speed, endurance, and power are the prime requirements. In the last six events, coordination, strength, and agility are equally important to cultivate.

The decathlon lasts two days. When there are many competitors, as in the Olympic Games, this means a long, grueling schedule that may last from 12 to 14 hours each day. During this time, the decathlete must keep his competitive desire boiling as well as his muscles warm.

Bill Toomey, an Olympic competitor, failed to quality for the United States Olympic team in 1964. But he trained regularly and worked hard at steadily developing his skills. At the 1968 Olympics in Mexico City, he scored a world record performance for the decathlon with a total of 8,193 points.

As physician for the U.S. Olympic team, I often talk with Toomey about training. "The decathlon is primarily a running event for strong individuals," he says. "The difference between those who are successful and those only moderately so is their dedication to running sprints in the speed work of their training routine."

Toomey began running in high school, and over the years he developed, by trial and error, a unique warm-up program. Instead of jogging around the track to warm up, he would practice running a few easy 100-yard runs, then walk back to his starting point. He increased his speed each time so that the last three or four runs were almost as fast as he could go. He would do about 10 of these each day. Then he moved on to the 220 and did three or four of these, the last two very fast efforts, and wound up the running section of his training with a fast 330 or two. Because Toomey felt that sprinting is more important than strength building, he did not try to run a 1,500-meter race in practice, even though it is one of the decathlon events.

Toomey calls this type of training—a few top performances each day instead of just spending a lot of time in distance running—quality training or "pyramiding the load." When he ran long distances at slower speed, he soon reached a point of diminishing returns for the effort he expended. "In college," Toomey recalls, "I had difficulty running the 440 in under 49 seconds—too much quantity and not enough quality training." For the 1968 Olympics, he did from four to six quality 220s every day instead of the 16 he did while he was in college.

The author:
Daniel F. Hanley, physician at Bowdoin College in Brunswick, Me., is chairman of the medical training committee for the U.S. Olympic team.

Rory Kenward, training for 1972 Olympic
Games decathlon, warms up by jogging, *left*.
He then runs 100-yard sprints 8 or 10
times, *above*, gradually increasing speed.
Finally, he runs an all-out 220 yards, *below*.

In warming up for the hurdles, *above,* Kenward stretches muscles of the groin area and the hamstring muscles, which move the lower leg. He runs hurdles every day, *right,* because he believes it develops the speed and flexibility he also needs for pole vaulting and high jumping.

His time improved so much that he won the 400-meter event in the 1968 Olympic decathlon in 45.6 seconds, the fastest time that had ever been recorded by a decathlete.

What Bill Toomey learned by trial and error, physiologist Otto G. Edholm reported as scientific fact in 1967 in his book *The Biology of Work*: "The most remarkable finding is that the amount of training needed to produce an improvement is very small. One maximal contraction [of a muscle] each day, maintained for less than a minute." According to Edholm, no improvement will occur unless the exercise is done at 50 per cent or more of the maximum possible. However, practice has shown that 70 per cent is a more realistic figure and one that is universally used. One way to determine the 70 per cent maximum for the biceps muscle, for example, is to see how many times a person can lift 50 pounds in 5 minutes. He should then begin his training with 70 per cent of that effort (number of contractions or amount of weight). Muscle strength and endurance adapt continuously, so he must regularly adjust the 70 per cent factor upward.

This form of training has another advantage. Athletic trainers and physicians know that microscopic tendon injuries increase at a tremendous rate when maximum contractions are repeated daily. Thus it is obvious that an athlete's most efficient training program involves load-

The California state discus champion in 1969, Kenward displays the grace and agility needed to excel in this decathlon event.

Kenward concentrates to attain a mighty contraction of his muscles as he winds up to put the shot. This 195-pound athlete can heave the 16-pound shot more than 50 feet.

ing his muscles to less than maximum and adjusting that level as his strength and endurance improve. Thereafter, they can be maintained with a minimum number of contractions three to four times a week as well as with an irreducible amount of damage to the tendons. The more the athlete works at it, within reason, the bigger, better, and more efficient his muscles become. This may well be the reason that older athletes can surpass younger ones in such endurance events as the marathon and cross-country ski races.

Besides the obvious increase in the size of an athlete's muscles, a training program produces more subtle changes in his respiratory and circulatory systems and in the basic structure of his muscles. To understand these physiological adaptations to exercise, let us begin by looking at muscles in greater detail.

Muscles are composed of bundles of slender fibers. There are only two kinds of muscles classified by their appearance under the microscope—smooth and striated, or striped. The smooth muscles are made up of plain elongated cells of even density that have a single nucleus. They generally move slowly and power the involuntary motions of such internal organs as the stomach and intestines. They also constrict or dilate blood vessels to shunt the flow of blood to those parts of the body where the demand is greatest, as to the arms of Shane Gould for her powerful swimming strokes.

There are two types of striated muscles: skeletal, which acts both under reflex and voluntary control, and cardiac, the heart. Skeletal muscles, which can contract rapidly, account for about one-fourth of the total body mass of an adult. A skeletal muscle is made up of many giant fibers or cells with many nuclei. The fibers are arranged parallel to each other and may run the entire length of the muscle. The fibers of the cardiac muscle, on the other hand, are branched and their striations are closer together than those of skeletal muscles. The heart has its own special network of nerves running throughout its muscle fibers. It also has the unique ability to relax completely between contractions and thus work effectively for years without tiring.

Regular exercise improves the size and efficiency of the heart muscle. Dr. Per-Olof Åstrand, a Swedish physiologist, showed in 1964 that the cardiac output, the maximum amount of blood the heart can pump in a minute, also increases with training. This increase is largely due to the greater amount of blood ejected during each beat, called stroke volume, and to more efficient contractions of the heart muscle. Later, studies on Olympic athletes during high-altitude training in preparation for competition at the 7,575-foot altitude of Mexico City confirmed Åstrand's findings.

The average person's heart can pump about 4.5 liters (4.7 quarts) of blood a minute while he is resting; during exercise, this may easily rise to 20 to 22 liters. After a few weeks of training, an athlete will average about 36 liters of blood a minute during violent exercise. The world class athlete who is in top physical condition has recorded an output of 42.3 liters during heavy exercise—twice as much as that of an untrained person.

If exercise, such as running or swimming, activates enough of the body's skeletal muscles, the respiratory system improves its ability to take more air into the lungs, extract more oxygen from it, and send it through the lung's thin alveolar membrane to the blood stream faster. This process is known as increasing the maximum oxygen uptake. Training also improves the efficiency with which oxygen combines with the hemoglobin of the red blood cells

For the javelin throw, Kenward works to develop flexibility of the hips and shoulders.

Kenward develops strength by lifting weights. From position *above,* he lifts 300 pounds to develop shoulders for throwing and, *right,* 270 pounds to develop legs for jumping, *below*.

and with the muscle's myoglobin, a protein that stores oxygen and gives the muscle its red color.

Scientists are currently studying a component of the red blood cell called 2,3-DPG (2,3-diphosphoglycerate). They think it may allow oxygen to combine with hemoglobin easier at the lungs and be released easier and faster to the muscles. The athlete has more than average amounts of 2,3-DPG. Thus, his circulatory system can deliver more oxygen to his muscles, remove the waste products of muscle action (such as carbon dioxide and lactic and pyruvic acid) more quickly, and carry the wastes to the lungs, kidneys, and skin, which excrete them.

The red blood cells themselves benefit from training. More cells are needed to transport oxygen. In response to this demand, the body builds more red blood cells and packs more hemoglobin into each one of them. The average young adult man has from 13.5 to 14 grams (about 0.5 ounce) of hemoglobin per 100 milliliters of whole blood. Well-conditioned athletes may have 17 to 18, and, occasionally, 19 grams. Training also makes hemoglobin carry more oxygen from the lungs and release more at the muscle. Therefore, a trained athlete can supply greater amounts of oxygen to his exercising muscles, which improves his efficiency and performance.

As a training program progresses, the athlete's breathing becomes deeper and slower, both at rest and during exercise, and he can take more air into his lungs and extract more oxygen from it. As a result, he needs less air for the same amount of exercise and less time to recover from the exertion. His heart becomes even larger and stronger, and the total amount of blood in his body increases. He can also get more energy out of his blood sugar. His pulse slows down, and his blood pressure decreases both at rest and during exercise. Finally, his coordination improves and his reflex and reaction times grow shorter.

Prolonged training increases the number of capillaries for every square centimeter of skeletal muscle. Capillaries also increase in the cardiac muscle, so the heart itself becomes stronger and able to endure more exercise. That the strain of muscle exertion is bad for the heart and that athletes die young from an "athlete's heart" are myths. Sensible exercise causes no bad effects in normal individuals of any age. In later life, the additional capillaries may even be a lifesaving factor if a heart attack occurs. There is a better chance for repair and survival if a person has additional pathways for the blood to get to the obstructed portion of the cardiac muscle.

Although the human body is capable of all sorts of complicated maneuvers, the muscles themselves can do only two things—contract and relax. The contractions and relaxations of many billions of muscle fibers are necessary to carry out a person's daily tasks.

Each muscle fiber consists of hundreds of myofibrils. A myofibril is composed mainly of actin and myosin, complex protein molecules which have the special ability to shorten and lengthen their mass by contracting and relaxing. We once thought that any increase in muscle

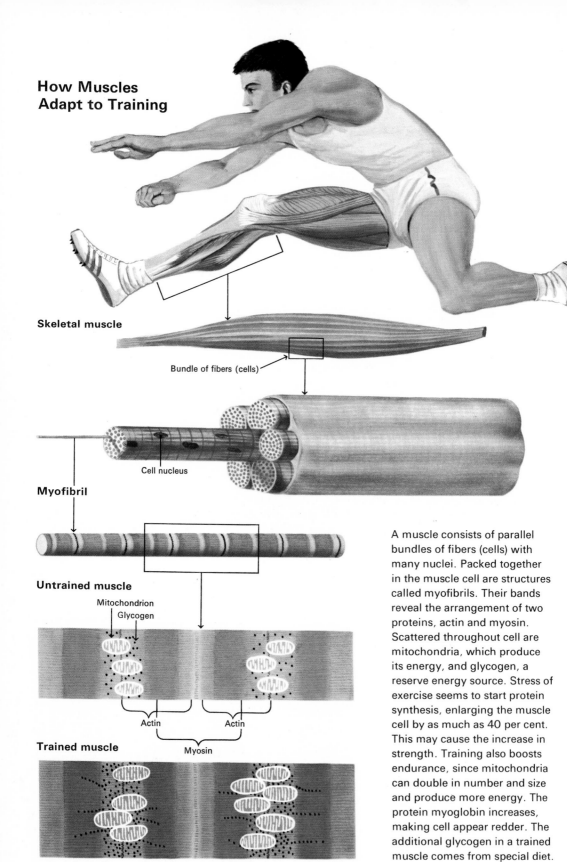

How Muscles Adapt to Training

Skeletal muscle

Bundle of fibers (cells)

Cell nucleus

Myofibril

Untrained muscle

Mitochondrion
Glycogen

Actin   Actin

Myosin

Trained muscle

A muscle consists of parallel bundles of fibers (cells) with many nuclei. Packed together in the muscle cell are structures called myofibrils. Their bands reveal the arrangement of two proteins, actin and myosin. Scattered throughout cell are mitochondria, which produce its energy, and glycogen, a reserve energy source. Stress of exercise seems to start protein synthesis, enlarging the muscle cell by as much as 40 per cent. This may cause the increase in strength. Training also boosts endurance, since mitochondria can double in number and size and produce more energy. The protein myoglobin increases, making cell appear redder. The additional glycogen in a trained muscle comes from special diet.

Generating Energy in a Muscle Cell

When muscle contraction starts, Pi (inorganic phosphate) and ADP (adenosine diphosphate) combine on cristae to produce the energy molecule ATP (adenosine triphosphate). As energy is freed, ADP and Pi form and enter mitochondrion to synthesize ATP.

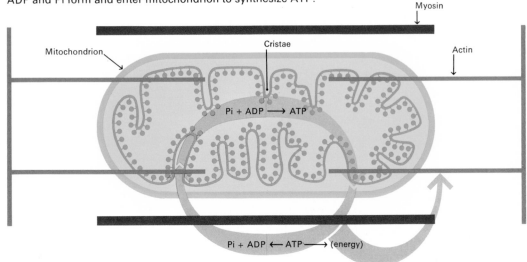

size was caused by the addition of more muscle fibers. We now know, however, that the existing fibers increase only in size and this increase comes almost entirely from additional actin and myosin.

When these protein molecules come into contact with adenosine triphosphate (ATP), an energy-rich molecule, they bind together to form a new compound, actomyosin, which contracts. ATP gives up one of its three phosphate groups, releasing energy for this contraction and changing to adenosine diphosphate (ADP). ADP then picks up a phosphate in the muscle cell to become ATP again. This quick and easy transfer accounts for much of the efficiency of the muscle as a power unit. See BIOCHEMISTRY.

Scientists do not completely understand how muscle cells convert chemical energy into mechanical energy. We do know, however, that every muscle cell contains hundreds of mitochondria, small rod-shaped structures that produce ATP. Training can double the number and size of these mitochondria, and thus greatly increase the amount of ATP available to the athlete's muscle cells.

The muscle picks up its food from the blood stream, stores it, and burns it when needed to produce ATP. Foods differ, however, in the amount of energy they can provide the muscle when they are burned with oxygen. Protein is the stingiest; it provides only 4.6 food calories for every liter of oxygen available. Fat yields 4.7 calories, while carbohydrates yield more than 4.9 calories for every liter of oxygen. Obviously a diet high in carbohydrates could give an athlete more energy per breath than could one high in either fats or proteins.

The Mechanics of Contraction

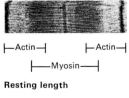

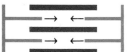

Resting length

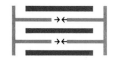

Contracting

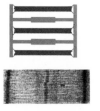

Contracted

When muscle contracts, the overlapping thick and thin filaments (actin and myosin) in a unit of a myofibril slide past each other. This causes the changes in density of band patterns seen by comparing the upper and the lower electron photomicrographs.

In spite of overwhelming scientific evidence to the contrary, many coaches and athletes still believe: "Muscle is protein; therefore, if I eat protein I'll build more muscle." As early as 1866, German scientists Max Joseph von Pettenkofer and Karl von Voit clearly demonstrated that the body burns about the same amount of protein during hard work as when it is resting. Studies reported in June, 1968, by Dr. Åstrand comparing cross-country skiers with athletes at rest confirmed this. Nevertheless, certain athletes, especially weight lifters, judo experts, and wrestlers, take great quantities of expensive protein supplements. These do nothing to help them.

In fact, physiologists Charles J. Eagan and James D. Murphy of Colorado State University reported in August, 1972, that high protein diets followed for a long period of time may actually harm performance. Protein contains nitrogen, and an excess increases the workload on the kidneys, which must then filter larger amounts of nitrogenous wastes from the blood. Strenuous exercise increases the level of uric acid in the blood and reduces the flow of blood through the kidneys. Continued high protein intake puts an additional burden on the kidneys. I have seen athletes on high protein diets at times of stress develop a significant degree of uric acid in their blood and begin to have problems with their joints.

For several years, scientists have known that fat is the major source of fuel for the muscles during light, moderate exercise. As long as the lungs and blood can provide enough oxygen, the muscles will derive their energy largely from fats. But when exercise is violent and more oxygen is needed than can be supplied, the muscles automatically begin to burn glycogen, a form of carbohydrate that is stored in the liver and in the muscles themselves. Glycogen requires less oxygen to produce the same amount of energy.

Glycogen plays a role in what is called "second wind." Tension before a contest carries over into the beginning of the event. It limits an athlete's respirations, makes his breathing more difficult, and forces him to use unnecessary muscles. As the event progresses and his body shifts to burning glycogen, he begins to relax. He can breathe easier and deeper and can take in more oxygen and use it more efficiently.

Athletic events that require endurance deplete the body's glycogen stores and force an athlete to slow down. Dr. Åstrand reported in June, 1968, experiments indicating that a high carbohydrate diet is best for endurance events. Åstrand removed small pieces of muscle fiber from the thigh muscles of 20 men at rest who had been on a normal diet. He measured the glycogen content and found 1.6 grams of glycogen per 100 grams of muscle. The men then worked their thigh muscles to exhauston by exercising on a bicycle ergometer. When Åstrand then measured a second set of muscle samples, he found that the glycogen stores had been depleted to 0.1 gram per 100 grams of muscle.

In another experiment, nine men on a normal diet and with a glycogen content of 1.75 grams continued a specific workload for 114 minutes. After three days on a diet of fat and protein, their glycogen fell to .63 gram and they could work for only 57 minutes. By following a three-day diet rich in carbohydrates, however, their glycogen levels rose to 3.51 grams and they could work for 167 minutes.

The greatest effect, however, was achieved when the men's glycogen content was depleted by heavy exercise and they were then put on a three-day diet of carbohydrates. Their glycogen stores then rose to 4.0 grams and they could work for as long as four hours. In other words, the men had increased their stamina by an astounding 300 per cent. The lesson for an athlete is simple. Eat a balanced diet regularly, but exercise to exhaustion four days before competition to deplete the glycogen stores. Then eat foods high in carbohydrates for the next three days so that the glycogen stores are filled to capacity for the event. Emphasize such foods as sugar, honey, candy bars, and other snack foods, as well as cereals, bread, pastry, spaghetti, pizza, potatoes, and other vegetables, such as rice and lima beans.

Faithfully followed, quality training and the proper diet will greatly boost an athlete's achievements. Unfortunately, some competitors

University of Nebraska scientists monitor an athlete's oxygen use, blood pressure, and heart rate while he runs until exhausted. By measuring changes in these bodily processes, they hope to find how to improve training.

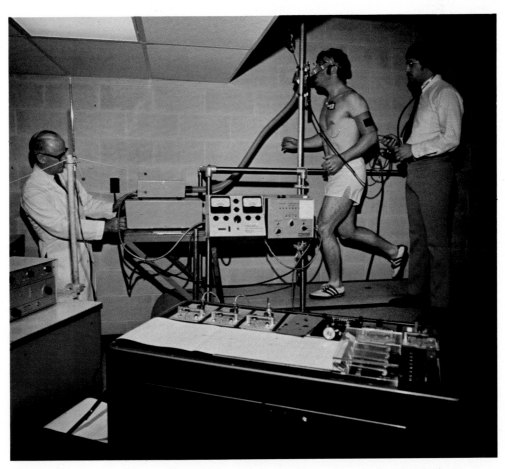

Swimmers have shown other athletes the importance of training for flexibility and power. Susie Atwood, world recordholder, *above,* pulls adjustable resistance exerciser to develop muscles for her powerful backstroke. Others in her swimming club work against each other to develop flexibility in shoulders and back, *above right,* and strength in the kicking muscles, *right.*

would like to find some short cuts. The history of sports medicine contains a long list of medications and concoctions that have been tried over the years in unsuccessful attempts to discover something that will improve performance. At the 1960 Olympic Games in Rome, for example, two of the cyclists on the U.S. team showed me some broken ampules that they had picked up from the locker room floor. The ampules had held drugs used by cyclists from other nations. "Why can't we have some of this?" they asked. My answer was simple: "It doesn't work." There is no drug that will consistently improve athletic performance in a normal, healthy individual. Nevertheless, amphetamines, caffeines, and anabolic steroids, along with huge quantities of vitamins, are popular now with the pill-popping set in athletics.

Since 1967, several programs of drug testing have been carried out at world and national competitions. Each has documented the fact that there is no wonder drug. The evidence shows, for example, that more losers than winners use amphetamines. Of the eight cyclists found to have taken drugs before their races in the 1967 Pan American Games in Winnipeg, Canada, five were among the losers and only three were among the winners. This is especially revealing when one realizes that there were 170 competitors and that winners for this study included anyone who had placed first, second, or third in any of the 87 trials up to and including the final race. In Rome the same year, the cyclists who took amphetamines finished 11th and 12th.

In a check of participants at the 1970 speed skating championships, stimulants were found in skaters who finished 9th, 12th, 15th, 18th, and 26th, but none were found in those who finished in the first five. Caffeine levels were also measured. Specimens from 23 out of 113 participants contained an excess amount. Of these, only two placed near the top. Caffeine does not seem to produce winners either.

The effects of anabolic steroids on the strength, body weight, and body composition in normal, young adult males were reported in June, 1971, by Dr. Stanley W. Casner, Jr., Ronald G. Early, and B. Robert Carlson. Anabolic steroids were given to 15 athletes at the University of Texas, Austin, three times a day for 42 days, while 14 men were given placebos as a control. Total body weight increased more in the steroid group than in the control group, but this increase was mostly water. The scientists found no significant difference in strength measurements between the two groups.

Used in sufficient quantities for long periods of time, anabolic steroids will produce sex changes in both males and females and changes in the liver's function. Other harmful effects include gastric and duodenal ulcers and even changes in personality.

Athletes are among our healthiest people and it seems ironic that any athlete would resort to practices that deliberately tear down his body. Athletes do run into one special long-range health problem, however. During the years when they are physically active, they require a large number of calories (sometimes from 5,000 to 8,000 calories a day) and they become accustomed to eating large meals. If they continue to enjoy them after their athletic careers are over and their way of life becomes relatively inactive, they will gain weight. But most good athletes understand this better than the average person because they have kept a close daily check on their weight for years. As soon as it starts to climb, they usually have the self-discipline to begin eating less so that they can maintain their correct weight.

Most athletes, however, do not stop exercising completely when they hang up their competitive equipment for the last time. Most transfer their skills to milder forms of recreational activity, such as golf, tennis, skiing, swimming, and hiking. This, plus calorie control, is enough to keep them in shape. But the big factor, as far as health and longevity goes, is that they were in top physical condition during the early years of their lives. Their circulatory and respiratory systems can continue functioning at top efficiency with little effort. Compared to non-athletes, this is a very large health plus. As a result, athletes have more years in their lives, and a lot more life in their years.

For further reading:

Edholm, Otto G., *The Biology of Work*, McGraw-Hill, 1967.

Hoyle, Graham, "How Is Muscle Turned On and Off?" *Scientific American*, April, 1970.

Margaria, Rodolfo, "The Sources of Muscular Energy," *Scientific American*, March, 1972.

Earth's Heat Engines

By Kenneth S. Deffeyes

**Massive subsurface plumes of hot rock may
be the driving force that creates earthquakes
and volcanoes, and moves the continents**

"Jason's home, and he has a new idea he wants to talk about." That neighborly message from Jason Morgan's wife was the first hint to me that the Princeton University geophysicist might have a solution to one of the most difficult problems facing earth scientists. Morgan had pioneered discoveries in the 1960s that revealed the earth was more dynamic and active than scientists had previously believed. These discoveries combined sea floor spreading and continental drift into a theory showing that the earth's crust was divided into moving plates. By 1970, the major puzzle yet to be solved was what interior mechanism caused such motions at the earth's surface.

Morgan had just returned from a symposium honoring Harvard geophysicist Francis Birch, who was about to retire. One lecture particularly interested Morgan. It described a newly revised map of the force of gravity at the earth's surface. This force varies from place to place, and can be measured by irregularities it produces in the orbit of any satellite circling the earth. The satellite dips whenever it passes over an area of high gravity. Although these gravity dips have been mapped in detail, the pattern never made much sense to earth scientists. They believed that a gravity high should be caused by something extra heavy, and they could never seem to find any such heavy matter at the locations where the satellites dipped.

Morgan had earlier made mathematical calculations that showed that light material rising inside the earth could lift the earth's surface slightly, creating a bulge that would not be visible but might attract a satellite. As the latest and slightly more detailed map of the satellite variations was flashed on a screen at the meeting, Morgan suddenly noticed that several places with an unusually high force of gravity, such as Hawaii, Iceland, and the Galapagos Islands, also had clusters of active volcanoes. This, added to his mathematical interpretations of gravity variations, formed the basis for his new idea.

The double clue—a gravity pattern that could be caused by material rising and clusters of active volcanoes—led Morgan to guess that hot plastic rock might be coming up from deep inside the earth at the places where the satellites had dipped. If this guess was correct, Morgan believed he might have found not only the cause of gravity irregularities but also the long-sought mechanism that causes the motions in the earth's crust. He did not mention his idea at the meeting because he needed many months to study and test it.

We know that virtually any substance will deform slowly when hot, so Morgan's proposal of the flow of material deep within the earth came as no surprise. Rock is red-hot at great depths in the earth, but the immense pressure at those depths keeps it from melting. Instead, the rock "creeps," or slowly changes shape. Creep is well known to metallurgists, who have tried every combination of metals to make turbine blades for jet aircraft engines that will not develop some creep at the high temperatures at which they operate.

Morgan quietly explained his idea to me and the rest of his colleagues and friends at Princeton. He said that material from deep within the earth was rising under the volcanoes at the points of gravity highs. Further, he proposed that the earth's entire mantle, extending down below the crust almost 1,800 miles, was creeping a few inches each year. He described the motion of that material as upward under the volcanoes, horizontally across under the crust, and then down again into the deep earth. He called the rising areas of hotter rock "plumes," and estimated, from the gravitational dips of the satellite, that each plume was about a hundred miles in diameter.

Rocks in the earth's interior, like the turbine blades of a jet engine, can slowly change shape only when they are heated at least halfway to their melting points. The outer 50 miles of the earth is too cold to flow

The author:
Kenneth S. Deffeyes is associate professor of geological and geophysical sciences at Princeton University.

Satellites dip as they pass over the earth's plumes. Such dips, exaggerated in diagram, were the first clue that plumes exist.

The San Andreas Fault near San Francisco is at the boundary between the Pacific and the American plates. Their relative movements, indicated by arrows, are the cause of all the area's earthquakes.

Locations of Plumes and Plate Boundaries

○ Plume ——— Plate boundary ▨ Locations of earthquakes

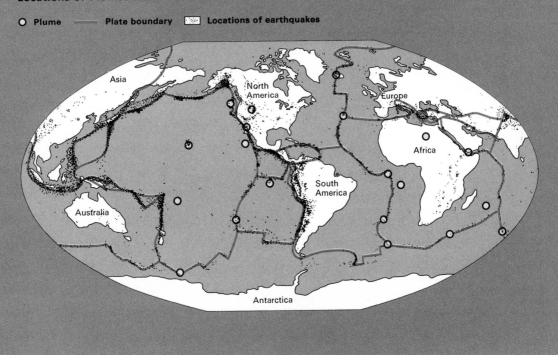

Scientists have located 20 plumes (yellow) beneath the crust of the earth. They are the driving force that moves the 20 plates that form the crust.

or creep. Material in this zone can change shape only by breaking or bending. This fracturing of the earth's outer crust causes earthquakes.

Back in 1967, Morgan discovered that the crust is divided into segments. He noticed that the world's earthquakes, which occur only in specific areas, were organized in a special way, as were the directions of fractures within earthquake areas. He found that these earthquake zones formed lines that divided the earth's crust into about 20 segments. He also showed, to everyone's surprise, that each of these segments acted as a rigid block, most of them thousands of miles across and between 20 and 50 miles deep. In time, scientists began to refer to the 20 segments as "plates," and the earth movements related to them as "plate tectonics." Morgan's ideas of plate tectonics were rapidly accepted by most earth scientists, in part because the theory accounted for a large number of observations that previously had seemed unrelated to one another. These included sea floor spreading, the upthrust of mountains, and why earthquakes occur only in certain places.

Morgan's proposal of plumes and the active behavior of most of the earth's interior suggested that the plates were merely the cool, and therefore rigid, outer boundary of a fluid and active earth. The plates simply ride on top of the deeper, hotter rock that continuously creeps radially away from the plumes.

This new theory could not be tested directly. After all, Morgan was talking about what happens 100 or more miles inside the earth, and the

167

Plate Movement over Plumes

The four dots in the Pacific Ocean, *below*, locate active volcanoes. Each is at the end of a chain of older, dead volcanoes that were formed as the Pacific Plate moved to the northwest. Until 20 million years ago, the plate apparently moved directly north.

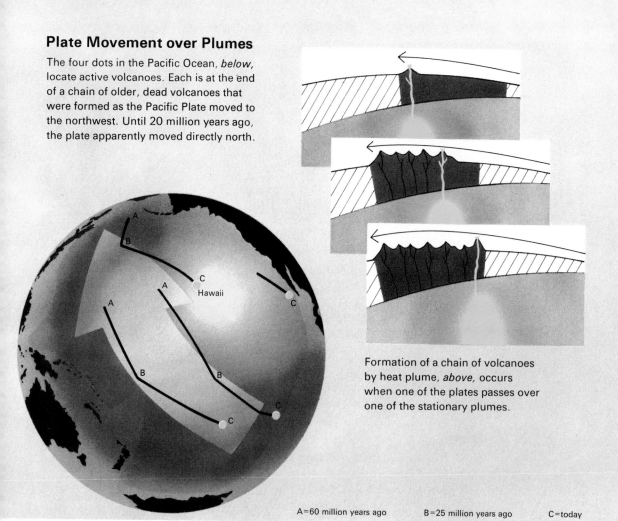

Formation of a chain of volcanoes by heat plume, *above*, occurs when one of the plates passes over one of the stationary plumes.

A=60 million years ago   B=25 million years ago   C=today

deepest wells ever drilled are only about 5 miles deep. We started making lists of indirect tests we could use to prove or disprove the theory. If a general idea such as Morgan's is true, it often has a number of observable consequences. Geology students and faculty members at Princeton enthusiastically began suggesting indirect tests. In 1972 we were still making tests, and many of us were convinced that Morgan's idea was passing them all. Convincing the wider audience of scientists throughout the world demands still more tests and a great deal of very carefully detailed work.

According to Morgan's theory, a small amount of the rock melts as it rises in the plume because there is less pressure near the surface of the earth. This melted rock appears at the surface as volcanic lava. Our first pleasant surprise in testing Morgan's hypothesis came when we discovered that volcanic centers over the plumes have lavas that are chemically distinct from the lavas of volcanoes that are not over plumes. Ever since the development of plate tectonics, geologists have known that some volcanoes occur where one plate is pushed under another. The edge of the underthrusting plate melts as it goes down, producing the lava that erupts from such volcanoes. Lava also oozes through the cracks formed when two plates move apart. But neither of these sources produces lava that is chemically the same as the lava that is from plume volcanoes.

In addition, we were delighted to learn, from studies of the minerals in the lava flows, that the lavas from the plume volcanoes showed evidence of coming from a greater depth than other lavas. This supported Morgan's original guess that we were dealing with deep-seated movements in the earth.

The interaction of the volcanoes atop the deep-seated plumes and the movement of surface plates led to another set of observations that supported the plume theory. Morgan assumed that each plume stays in one place and is continuously active for a long time. This is the simplest assumption, and it is a long-standing tradition in science that the simplest idea is worth examining first. Morgan started this test with the largest plate, the Pacific, which appeared to have four plumes. If this plate, essentially a thin, rigid shell, moves over four fixed plumes, there should be evidence of the effect of the plumes on the various parts of the plate that have passed over them. A line of extinct volcanoes would be such evidence. Morgan found four parallel trails of extinct volcanoes leading away from the presently active volcanoes over each of the plumes. These four chains of volcanic islands in the Pacific Ocean can be explained by a rigid plate sliding over four plumes. When the plate movement shifted direction, the parallel lines of extinct volcanoes changed their course accordingly.

A core of rock from Midway Island provided another piece of evidence that the plume remained fixed, and the plate moved. Midway was created by a now extinct volcano, one of those forming a trail from Kilauea, an active plume volcano on the island of Hawaii. Polarity tests of the magnetic minerals in the Midway rock showed that its lava had cooled at a latitude of about 19° north, not at its present-day latitude of 28° north.

When the theory of plate tectonics became widely accepted, scientists in several countries measured rather exactly the absolute motion of the plates over the volcanic sources. To check his plume theory, Morgan insisted on repeating these calculations himself. He wanted to make certain that the worldwide plate motions over his proposed plumes accounted accurately for the volcanic trails. Although final refinements are still to be completed, it is already clear that the plume theory holds up under these tests.

After examining the records of all known volcanoes, coupled with the data from satellites, we were certain of the location of about 20 plumes. The question of cause and effect, however, was yet to be resolved. Were the plumes stationary volcanic sources that just happened to exist beneath the plates, or were they the driving force that moved

The author, *left,* gives explanation of some of the equations that went into his mathematical model of a plume. Richard Hey, *right,* a Ph.D. candidate in the field of geology at Princeton, demonstrates the motion of a plate on the surface of a sphere.

the plates? It is usually easier to establish the association of two things than it is to prove conclusively that there is a cause-and-effect relationship between the two of them.

Several lines of evidence suggested that the plumes were driving the plates apart. Most significant, the plumes were usually found where plates moved away from one another. Altogether, 17 of the 20 plumes were located at or near the boundaries of two or more plates. Further, at several places around the world, plumes are at, or near, places where three plates come together to form a corner. An example is the Afar Triangle, where the African, Arabian, and Somali plates meet. At this three-way junction, a well-studied set of volcanoes have produced the chemical types of lava that we associate with plumes. Also at this place, the three plates are moving away from each other.

One of our students, Dick Hey, discovered a third relationship that also suggests deep plumes are driving the surface plates. On a flat earth, plates might slide around like so many barn doors. On a spherical earth, however, the motions of rigid plates follow a set of rules worked out some 200 years ago by the Swiss mathematician Leonhard Euler. Euler showed that any arbitrary motion on a sphere's surface can be described as a rotation about an axis. The earth's rotation is one such motion. It has an axis running from the South Pole to the North Pole, and its surface velocity is greatest at the equator and virtually zero at the poles. Plate motions also follow Euler's rule, only the rate of motion is 30 billion times slower. Hey noticed that Morgan's plumes could all be described as being midway between "poles" that defined an axis required by Euler's rule. He reasoned that if the plumes were driving the plates, the plates would arrange themselves so that their fastest motion was at the plumes. We have been jokingly calling this "Hey's Law." But as evidence grows that the plates are, indeed, moving fastest at their plumes, we are taking Hey's idea more seriously.

My own involvement with Morgan's idea was largely in attempts to construct a series of mathematical equations that would describe known facts about movements in the earth. Through these equations, we hoped we might discover the rates of internal movements related to the plumes. In broadest terms, such a mathematical model must consider the flow of material and the flow of energy. We must estimate the amount of material that rises in a plume each year, and find a route that takes an equal amount of material back down into the earth's interior. In addition, we must know where all required energy comes from, how it is transferred, and where it ends up. Models of this kind are fearsomely complex. Some mathematical models of the oceans and the atmosphere, for example, require hundreds of hours of computation on the very largest and fastest computers.

My mathematical model of deep earth movements also had to include the rate at which heat escapes from the earth's interior, the properties of some of its minerals, measurements of gravity variations, the time that it takes earthquake waves to travel through the earth, and measurements of the viscosity of the earth's interior. I also had to list the relevant physical laws for fluid flow, for thermodynamic behavior of the minerals, and for change of viscosity with temperature and pressure. My model thus became a list of mathematical equations that described the physical laws. If there are at least as many equations as there are unknowns, it is possible to obtain mathematical answers for each of the unknowns in the equations.

In principle, the procedure is straightforward. In reality, we must do a lot of searching for the correct relationships and try many models. After I had what I thought was a reasonably complete and accurate model, some of the answers obtained by solving the equations seemed ridiculous. In particular, one of the mineral properties, called a volume of activation, turned out to be a negative number. Volumes should be positive numbers, so something

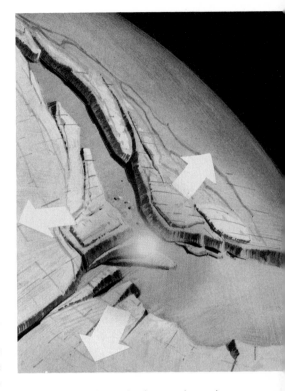

The Afar Triangle marks the meeting point of the Arabian, Somalian, and African plates. A powerful plume drives up at the intersection of the plates and forces them to move apart, *above*. The triangle is located where the Red Sea meets the Gulf of Aden, *below*.

was obviously wrong with the model. After several days and nights of puzzled testing, I began to suspect that the difficulty lay in the temperatures estimated for the earth's interior. Like everyone before me, I had assumed that temperatures become hotter and hotter as you go deeper into the earth. But if I changed to the assumption that the layer where the plume material spreads horizontally was hotter than the layer beneath it, all those troublesome negative volumes suddenly became positive.

Everything about the model now made reasonable sense. A change in temperature, where it gets colder instead of hotter as you go down, is known as a temperature inversion. Temperature inversions occur occasionally in the earth's atmosphere, where they often play a sinister role in keeping air pollution near the ground. In addition, the base of the stratosphere, about 5 to 10 miles above the earth's surface, forms a permanent temperature inversion. I had discovered a similar permanent temperature inversion about 125 miles below the earth's surface, just beneath the hot material that is spreading radially from the plumes.

To my knowledge, the only previous hint that a temperature inversion might exist in the earth's mantle came from Birch, the geophysicist whose retirement symposium triggered Morgan's thoughts about plumes. While studying the movement of earthquake waves through the earth some years before, Birch showed that a temperature inversion could explain some peculiarities in the data. At the time, however, his suggestion seemed to be only one of many possible explanations for the behavior of earthquake wave velocities.

The model that emerged from my calculations showed that every year, each of the 20 plumes brings about 7 cubic miles of rock up from the earth's interior, roughly equal to the volume of Mount Rainier. The rock rises in the plume at a rate of about 5 to 10 inches a year. It cools off by about 600° F. as it spreads out underneath the plates, and then goes down again. At the mid-ocean ridges, where plates separate from one another because of the pressure of this moving material, about 1 square mile of new sea floor is created each year. To com-

Lava from the active, plume volcano on the island of Hawaii flows into the ocean where it will become part of the land mass forming the volcanic chain.

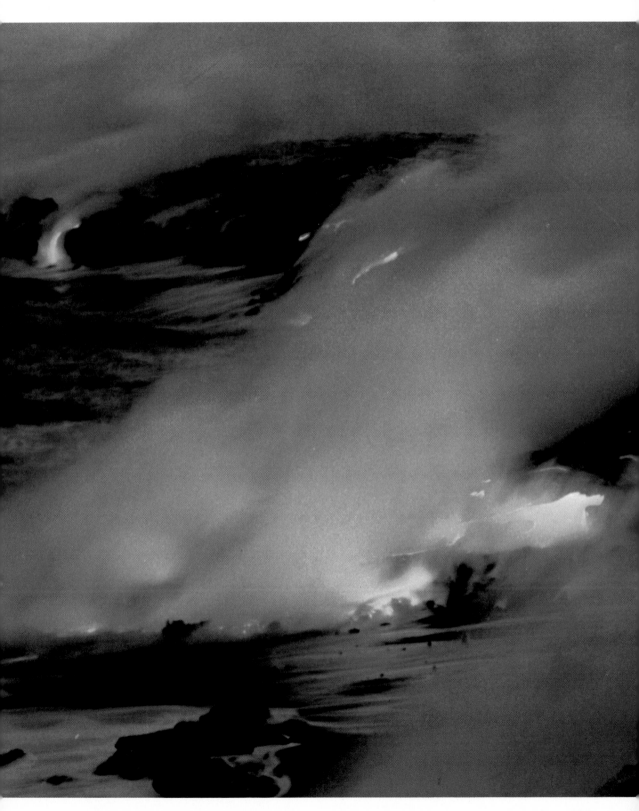

The plume under Yellowstone National Park heats the ground water until steam forms. This forces the water to spout up in a geyser.

Mexico's Pinacate volcanic field, *left*, marked with an arrow on this photo, may be over a yet unidentified plume.

The Kilauea volcano, *above*, on Hawaii is directly over a plume and presently ejects more lava than any other volcano.

Volcanoes that are not over plumes occur where the downgoing edge of a plate melts, a process that created this island near New Zealand.

pensate, a square mile of sea floor is pushed down into the deep oceanic trenches each year and returned to the earth's interior.

Now, with a mathematical model that fits the physical evidence reasonably well, many of us have become even more convinced that Morgan's idea about plumes is correct. And, every few weeks, another scrap of information falls into place. And although still not all earth scientists are convinced that plumes exist, Morgan's idea seems to be winning new champions regularly.

While supporting the existence of the plumes and the motions they cause, the model also shows us that the earth is more dynamic than we previously believed. Perhaps we should have suspected this was so. If there were no forces in the earth's interior strong enough to move the surface plates, erosion would long ago have leveled whatever land showed above the sea. There would be no freshly formed mountains or young surface rocks, because these are the results of plates pushing against each other.

The motions of plate tectonics thrust up several cubic miles of new rock each year. These motions have formed huge mountain chains such as the Himalayas and the Swiss Alps, which are relatively new by geological standards. The older chains, including the Appalachians and the Urals, were once just as high, but have been worn down over hundreds of thousands of years by wind, rain, and rivers.

On an earth leveled by erosion, there could be no additional supply of nutrients, such as phosphorus, which are essential for life, to the ocean. So even if life originally were to exist on a planet that had oceans but no plate movements, weathering would eventually eliminate all nutrients, and life would cease.

These new geological theories not only are supplying additional perspective on how the earth renews itself, they may also offer new clues about mineral resources, a new understanding of earthquake hazards, and an extension of our meager knowledge about the ocean floors. The excitement and even the controversy that enlivened geology during the 1960s, after theories of sea floor spreading and plate tectonics were first introduced, shows every sign of extending through the 1970s as the plume theory develops.

For further reading:

Calder, Nigel, *Restless Earth, a Report on the New Geology,* British Broadcasting Corporation, 1972.

Dewey, John F., "Plate Tectonics," *Scientific American,* May, 1972.

Hey, Richard N., and Deffeyes, Kenneth S.,"The Galapagos Triplet Junction and Plate Motions in the East Pacific," *Nature,* May 5, 1972.

Takeuchi, H., S. Uyeda, and H. Kanamori, *Debate About the Earth,* Freeman, Cooper, 1970.

Wilson, J. Tuzo, editor, *Continents Adrift: Readings from Scientific American,* W. H. Freeman, 1972.

The Force That Moves Continents

The Dynamic Earth

Ever since the earth was formed, its physical features have slowly but continuously changed. Continents have moved around the globe and sometimes split apart to form ocean basins such as the Atlantic. Sometimes continents collided, one crumpling and pushing up the shore of another until it folded into a series of high mountains. Geologists seeking to understand these massive movements, and the mechanism that causes them, have uncovered much information about the dynamic interior of our planet earth.

The continents and sea floor are the uppermost parts of some of the 20 giant plates that form the earth's outer shell. These plates vary in size and shape. The largest of them forms most of the floor of the Pacific Ocean. The smallest is covered by Greece and a portion of the floor of the Aegean Sea.

The rigid plates are much like the hard shell of a walnut. Made of basalt, granite, and other relatively light materials, they actually float on the much denser, hot, plastic rock, composed mostly of olivine, that lies below. Olivine is the main ingredient of the earth. It is a compound of iron, magnesium, silicon, and oxygen, mixed with lesser amounts of other elements. At great depths, pressure keeps it highly compact, but when it occasionally finds its way to the surface, its crystals "relax," and change into less dense forms.

The picture of rigid plates sliding about over a more plastic inner earth explains continental movement, earthquakes, and sea floor spreading. But the final piece in the puzzle, the driving mechanism, has long eluded geologists. Some of them now believe that heat deep in the earth thrusts up molten material in pipelike plumes, 100 miles in diameter. These drive the plates by creating currents in the hot, plastic rock beneath them. About 20 of these plumes are scattered around the earth.

Most of the earth's interior heat comes from the decay of radioactive elements (primarily uranium and potassium) in its rocks. Such radioactive decay occurs in most rocks all over the earth, but those on the continents and ocean bottoms rapidly lose their heat to the atmosphere and the water. The heat given off by rocks deep in the earth, however, is trapped, building up the tremendous forces that create the plumes.

How to Use This Unit

The roughly rectangular pit superimposed on the photograph on the opposite page marks the ½-million-square-mile area "excavated" in the Trans-Vision® that follows. To remove huge layers of the earth and see its inner mechanisms, first turn back the opposite page to expose the acetate overlays and the accompanying text. The text for the first overlay is below. For each succeeding overlay, the text is on the far right-hand page. Examine the first overlay and read the text accompanying it. Then lift the overlay to reveal the one below, examine it, and read its text. Continue until you have examined the base page and read its text.

The Concealing Surface

At the surface of the earth there is little to indicate the tremendous activity going on below. One clue comes from earthquakes, which usually occur at plate boundaries, and are caused by the rubbing and pushing of the plates against one another. The diamondlike symbols indicate earthquakes recorded during eight years, 1961 to 1969.

The San Andreas Fault along the Pacific Coast of North America was formed from this movement. Here, one plate – the Pacific Plate – grinds and scrapes past another – the American Plate – at a rate of about 2 inches a year. In the process, it carries a narrow slice of the North American continent along with it. If these plates continue at this speed for 20 million more years (not very long on the geological time scale), Los Angeles will eventually lie as far north of San Francisco as it now lies south.

One of the plumes that moves the Pacific and American plates lies under Cobb Seamount. This 9,000-foot-tall submerged mountain 310 miles off the coast of the state of Washington is composed almost entirely of solidified lava.

The giant plumes that power plate movement sometimes cause volcanoes and geysers to erupt on the earth's surface above them. Plumes can also cause volcanic activity indirectly where their force drives the edge of one plate beneath another. This type of action created Mount Rainier and other peaks in the Cascade Range. They were once active volcanoes, and still have measurable heat in their interiors.

Science for the People of China

By John Barbour

American scientists discovered that the fundamental change in China's social structure has caused a dramatic shift in the directions and uses of its science

The doors to mainland China, closed to Americans since 1949, have been reopened. In mid-1971, visitors from the United States began to get a firsthand look at modern China. In the vanguard were seven scientists who brought back exciting reports.

The scientists included Chen Ning Yang, a physicist from the Stony Brook campus of the State University of New York who shared the Nobel prize in physics in 1957. Yang was born in Ho-fei, China, in 1922, and came to the United States in 1940. Professor Arthur W. Galston, a plant physiologist at Yale University, and Professor Ethan Signer, a Massachusetts Institute of Technology (M.I.T.) biologist, made the trip together. Four American physicians traveled as a group–Paul Dudley White, the noted cardiologist from Boston; Dr. Victor Sidel,

Ancient statue and modern chart show where acupuncture needles should pierce the body. This healing art, many thousand years old, is widely used as an anesthetic in China's new medical science.

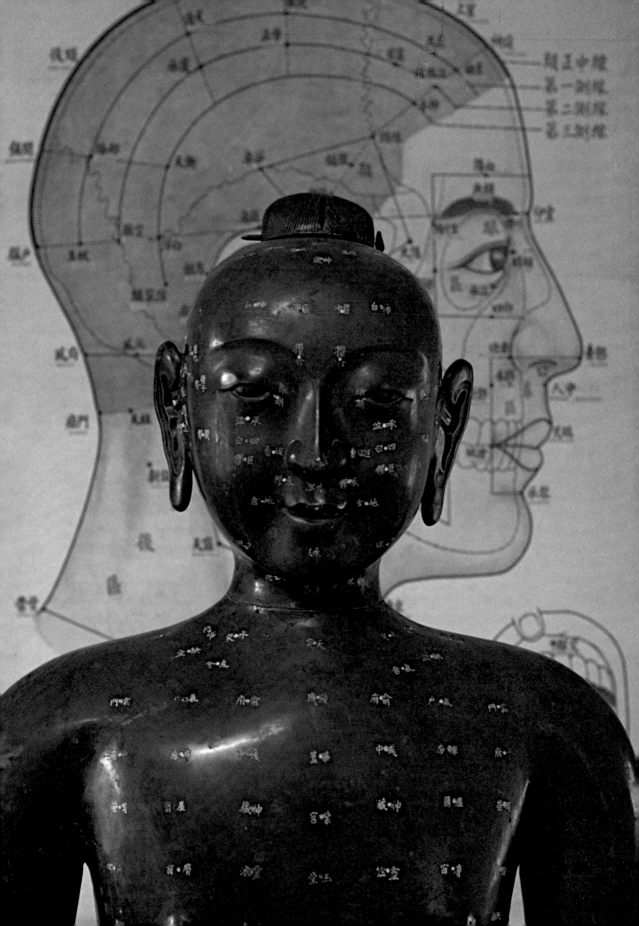

Chinese students sing as they ride through the streets of Peking. U.S. scientists found such enthusiasm widespread among Chinese people.

chief of the social medicine department at New York City's Montefiore Hospital; Dr. Samuel Rosen, emeritus professor of clinical otology at the Mount Sinai School of Medicine in New York City; and Dr. E. Grey Dimond, who had, since 1969, been provost for the health sciences at the University of Missouri, Kansas City.

Despite the Bamboo Curtain, Americans had known that science and technology were alive in China. The giant nation had become the first to synthesize insulin, had produced a hydrogen bomb, and was working on rockets and space satellites. However, the role that science plays in the daily lives of about 800 million Chinese people had been hidden from inquisitive American eyes.

Although the first scientist visitors represented several disciplines, their reactions to their trips were remarkably similar. They found an energetic, enthusiastic cadre of Chinese scientists, and a body of scientific work that, if not impressive in its quantity, was impressive in the new directions it is taking.

They talked of how China has blended modern science with relics of a deeper heritage. Plant hormones and human manure nurture the same crops. Ancient acupuncture and modern open-heart surgery are practiced in the same hospital. Scientists dirty their hands working in factories and fields, and peasants help to determine the uses of technology in communal council meetings. Yang explained: "Mao has a new proverb—that a man must walk upon two legs." It means many things: that the best of two opposing worlds, for instance, or two opposing ideas complement each other. The past must walk with the present.

So China blends the new with the old, and uses both. "I don't pretend to be an expert on this after 15 days of studying the secrets of the mystic East," said Galston, "but if there is anything that characterizes Chinese science, it is realism."

The author:
John Barbour is a science writer for the Associated Press. His article, "Probing the Pawnee Grassland," in the 1972 edition of *Science Year,* won the AAAS-Westinghouse science writing award.

The visitors saw Chinese science wed to the day-by-day needs of the people. They found science accommodated to ancient traditions, forged and bent into a useful tool, shaped to the practical needs of farm and factory. After seeing it all, the visitors could not help wondering whether there might not be some value in bringing science in America closer to the American people.

The Americans found medical science to be particularly active in China. Its direct application to the needs of the people was obvious. China has carefully preserved traditional Chinese medicine out of respect for the past, and because it is close to the people. "They are trying very hard to combine what they call Chinese medicine with what they call Western medicine," Dr. Sidel reported. "The traditional includes herb treatment. It includes acupuncture, the treatment of illness by piercing the skin with long needles. And it includes moxibustion, which is the technique of applying localized heat, usually by burning a root near the skin. They're also trying to preserve the traditional doctor's methods of dealing with people—his long and patient explanations—and the trust that people, especially in rural areas, have always had in him." The program marries old methods to modern ones.

"Training the traditional doctors in modern medicine gives them 200,000 or 300,000 more doctors for this huge country," Dr. White noted. "We visited a commune hospital. There were a number of beds available, an outpatient department, and a good deal of surgery of the simplest kind, even appendectomies. The difficult surgery was referred to hospitals in Peking and other cities."

The most profound change in China's medical program came with the Cultural Revolution of 1966, a violent drive for moral, political, and social reform. Thousands of city doctors were sent into the countryside. They taught modern methods to the country doctors, brought modern medical treatment to remote masses, and trained "barefoot doctors" and other paramedical personnel. The barefoot doctors—often villagers who never had had formal medical training—were taught basic public health standards and such skills as how to administer first aid and inoculations.

"In America, we don't need M.D.s to do 90 per cent of what M.D.s do," said Galston. "We could use paraprofessionals, and we're starting to train them. The Chinese are far ahead of us. For setting a bone in a flexible splint, which the Chinese use—they don't believe in rigid casts—you can use a technician. Midwives are frequently just as good as obstetricians. To give an inoculation you don't need an M.D. I think we are too sophisticated in that sense."

The training of barefoot doctors produced some surprises, not only for the rural patient, but also for the doctor who left a comfortable life in the city to be "re-educated" in the villages. Dr. Sidel remembers a story told by Dr. Hsu Chia-yu, now a deputy director of the internal medicine department in Shanghai's largest hospital: "Part of what was terribly moving in Hsu's story was how he trained his barefoot doctors.

He would go out and live with them in the villages in which they worked. And living with a barefoot doctor means literally that. Life is still quite hard in rural China. A house has no spare bed—nor even a spare pillow. So you literally sleep with the barefoot doctor you're training. Hsu told us how, before they went to sleep at night in that one bed, they would discuss the patients they had seen during the day. And they would talk together about their dreams for a better China. It was sort of a mutual education, and something Hsu had never experienced before as a city boy in Shanghai. In the old China, he would never have had a chance for contact with the peasants."

The American scientists all agreed that the most startling thing they saw was acupuncture anesthesia. According to legend, acupuncture originated 3,600 or more years ago, when a Chinese emperor noted that soldiers wounded by arrows sometimes recovered from chronic illnesses. The oldest known text on acupuncture dates to about 2500 B.C. According to this text, two competing forces of energy, yin and yang, in theory, flow through the body along 12 meridians. Illness results when an imbalance develops in yin and yang. Acupuncture charts of the body show 1,000 points where needles can be inserted into the body, to right the balance and restore health. Often these are remote from the site of the illness. These same points have remained virtually unchanged over the centuries.

Dr. White saw the modern application firsthand—the long, slender needles inserted into various parts of the patient's body and twirled, the only anesthetic during a brain operation: "The patient was wide awake and I was talking to her while they opened the skull and exposed the tumor. The patient may have tea or fruit during the operation. After the operation, the patient came right out of it [the anesthesia].

"Acupuncture anesthesia is relatively new—maybe 15 years old," Dr. White said. "Now, I suppose a third of their major operations are done with it. They still use ordinary anesthesia, and a modern anesthetist is always in reserve should acupuncture fail."

Another witness to the technique, Dr. Rosen, saw the acupuncture anesthesia method in use several times. He reported: "I have seen one of the most venerable arts of Chinese traditional medicine applied in the most modern of contexts in today's China. I have seen it used successfully, not once, but 15 times—in brain operations, thyroid adenomas, gastrectomies, laryngectomies, and tonsillectomies—performed in Canton, Peking, and Shanghai. Each time, the stand-by Western-style anesthetist's skills were not needed. This relatively new application of an ancient form of medical practice has not only replaced anesthetics but has permitted the patients, in every case, to leave the operating room or the dentist's chair alert and smiling—some on a stretcher, others even walking."

"At first, when Signer and I came back to the United States and started to talk about this," Galston said, "the medical profession greeted it with disbelief. They reminded us we were not competent

A "barefoot doctor,"
a peasant with some
basic medical training,
ministers to an injured
worker in the field.

medical observers, and suggested that what we saw resulted either from hypnotism or the surreptitious injection of drugs, that acupuncture had a long and undistinguished history. But it turns out there's something to it." Even so, the apparent success seemed mystifying and American scientists sought some explanation they could understand.

One of the visiting doctors asked whether the thoughts of Communist Party Chairman Mao Tse-tung might not have had a hypnotic or suggestive effect on the patients. "Perhaps," a Chinese surgeon replied simply. "But we have been producing the same effect in rabbits and in cats, and as far as we know they have not been influenced by the thoughts of Chairman Mao."

"Certainly the evidence of our own eyes was that it worked," said Dr. Sidel. "That, of course, doesn't tell us why it worked, and the Chinese were very frank in telling us that they didn't have any clear idea why it worked. They indicated to us they would like to see it studied everywhere. They have their own research teams working on it."

Since his return, Dr. White has conferred with interested U.S. anesthetists and surgeons who are prepared to study acupuncture.

Under acupuncture anesthesia, a patient underwent an ovarian cystectomy at the Third Teaching Hospital at Peking Medical College. Doctors used an array of slender acupuncture needles, *below left,* on the patient. They inserted one at the left of the spine, *below right.* Another was inserted in the patient's ear, *bottom left.* Three more were inserted in each leg and foot, *bottom right.* The pulsator between the patient's feet supplied a weak direct current to the needles. In most other respects, the operation, *opposite page,* is similar to surgery in the West.

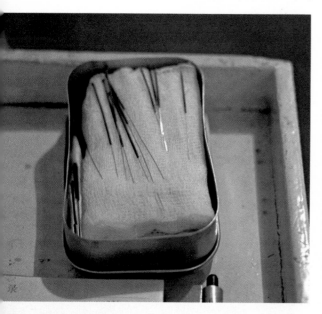

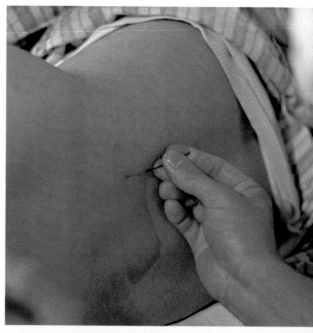

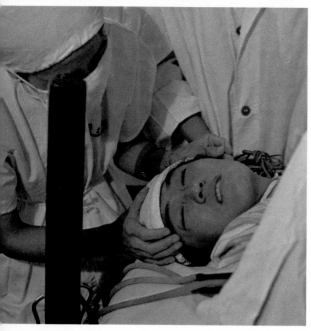

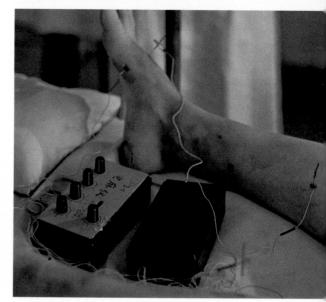

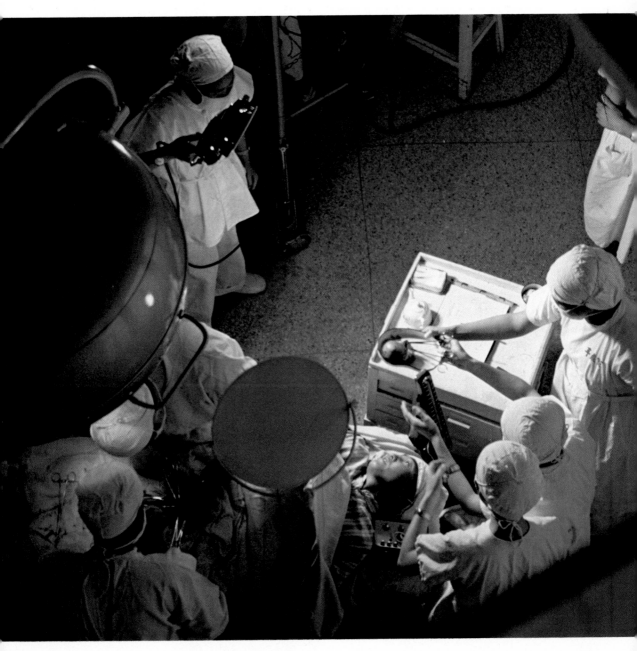

Perhaps China's most striking scientific progress, the visitors found, has been in public health. Venereal disease (VD)—historically rampant among the poverty-stricken Chinese—and prostitution have been wiped out, thanks largely to a program led by an American-born doctor, George Hatem, who has lived in China since 1933. Born in Buffalo, N.Y., he is Chinese now, with a Chinese family and name—Dr. Ma Hai-teh. The Chinese call him Dr. Ma (Horse) in what may be a reference to his long, shaggy hair.

The first target in the VD drive was the Peking-Tientsin area, where teams of local women explained to the prostitutes how they could lead a different kind of life. "When everything was prepared, we closed down every brothel in Peking in one night," Dr. Ma said. An intensive medical treatment campaign followed, with a single medical team treating and curing as many as 1,200 VD cases every two weeks. So confident are the Chinese of their conquest of VD that they have given up premarital tests and tests of pregnant women, the prime methods of VD control. Too few cases were found. Now, when a VD case is reported anywhere in China, Dr. Ma, as a senior official in the government health program, attends to it personally.

China is also battling schistosomiasis, a debilitating, sometimes fatal disease caused by tiny worms. These parasites, which are transmitted between human beings and the snails that infest China's waterways, settle in the human intestines. They attack the small intestine, colon, and rectum. The old practice of dumping human waste in the waterways fostered the disease. Scientists and peasants now work together to

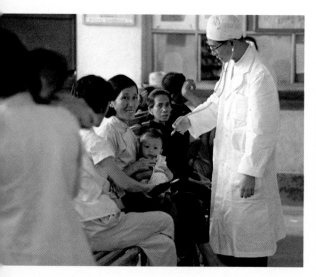

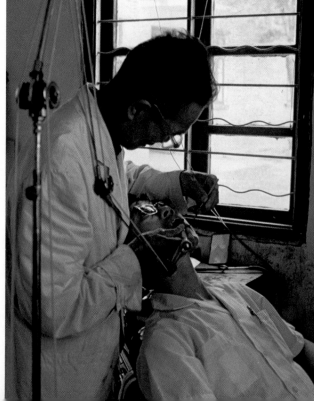

A doctor in a rural commune clinic checks over waiting patients, *above.* The new hospital at Hsing Hua commune, *right,* also provides dental services for members of the commune.

reroute watercourses, destroy the snails, and develop better waste-disposal techniques. Each makes his own contribution.

Activity in another field of science—practical science of the sort the Chinese appreciate most—also impressed the American visitors. This was the work in agriculture and horticulture, important sciences because of their influence on the Chinese food supply.

Galston, a plant physiologist, was struck more with what he found on the farm than in the laboratory. "There are several examples in which the very latest stuff, hot from the laboratory, was going right out into the commune. Take the plant growth hormone gibberellin. You make it by growing a mold and skimming the mold off, just as you make penicillin."

Applied to seedlings, gibberellin stimulates the growth of plants such as barley, but often does not improve their yield. Galston said, "It's been known in the West for 20 years now, and before that in Japan. In America, we purify the gibberellin to its crystalline form, which makes it quite expensive. One of my friends who was working on gibberellin at a Shanghai institute of the *Academia Sinica* (Chinese Academy of Sciences) was also working at the Malu commune near Shanghai.

"With the peasants, he worked out a way of making cheap gibberellin," Galston said, "by simply growing the mold on agricultural wastes at the commune, skimming the *mycelium* (the vegetative part) of the mold off the wastes, drying it in the sun, reducing it to a powder, and mixing it with a buffer. Then he filtered it through clay to remove the colored matter and toxins and it was ready to spray on the crops." At the peasants' suggestion, the scientist applied the gibberellin to the plants when they were in flower instead of when they were seedlings. They got the results they sought. Yields improved.

"There's an example of doing something very modern, very practical, very economical, and getting increased yields," Galston said. "They claim 20 per cent increased yields of barley, for example."

The necessity for pest control forces the Chinese to use chemical pesticides such as DDT, Galston said. But they realize the ecological dangers and are seeking other remedies. "We ran into one insect physiologist who was developing parasites to control the noxious gall wasp that is attacking fruit crops. They know all about these modern techniques and they have a very competent scientific corps." Signer and Galston jointly reported that 40 communes are now raising insect parasites to fight the wasp. The method was developed by entomologists at Chungsan University near Canton.

The Chinese are most innovative and free of Western prejudices in applied science—what some U.S. researchers might scorn as "mere technology." Professor Galston cited as an illustration the use and disposal of human waste. "In America, we all think we are very enlightened about this. But what do we do? We have primary and secondary sewage-treatment plants, if we're lucky. In some cases, we treat our waste, then load it on barges and dump it into the ocean. Well, the

A farmer lifts a plastic covering to check rice seedlings being grown in a modern Chinese agricultural experiment.

Chinese figure that if horse manure and cow manure are good for crops, why not human manure? They look at it quite practically.

"And they've discovered ways to ferment human wastes anaerobically [without atmospheric oxygen], which automatically brings it to a very high temperature. This kills the cysts and spores that cause disease and removes the odors," Galston said. "You then have a material you can apply to the field, which keeps the soil in good condition, grows good crops, and decreases dependence on synthetic fertilizers that cause some problems in waterways. In America," he added, "we certainly have all the technology to do the same thing, but we're not supposed to touch human waste because it's dirty and polite people wouldn't do a thing like that."

While China's progress in practical science seemed exciting, the Americans found the lack of basic research—the struggle to expand the frontiers of knowledge—was serious. Yang, Rosen, Galston, White, Sidel, and Signer all acknowledge that a nation cannot go on forever without such research, even if it taps the work done in other nations. Sooner or later, more basic research will be needed in China, even though it may not be immediately useful. Otherwise the data well will run dry; technology will stagnate.

The Chinese provide for basic research, where they consider it economical and practical. Yang found the Chinese questioning where the emphasis should be in physical research. "Physics is one of the cutting edges of technological development and the Chinese realize it," he said. "But they also understand it is a most expensive area of inquiry. In the future, the Chinese probably won't put as much strength in high energy physics because of the cost, and because high energy physics would not yield as many society-oriented results." But Yang believes they will pursue low energy physics, solid state physics, and theoretical physics, because these research lines will provide a base for industry.

China's focus on practical science does not mean, however, that it encourages only work that will generate better production or tools. Indeed, it is difficult to identify any direct benefit from a field such as archaeology. Yet the Chinese respect for the past means that archaeology is important to virtually every citizen. Archaeologists from the Sinkiang region in the west to Shantung Province in the east have recovered gold and silver, art objects, pottery, porcelain, and manuscripts. Some of the items dated from the 1500s B.C. Announcing the discoveries, a Chinese news service said: "These finds show how the feudal ruling class wallowed in luxury and dissipation...and also show the tools and skills of the Chinese laboring people of that period."

Signer and Galston reported that science education, as much as science itself, has changed radically. While American colleges want students with a demonstrated ability to learn and a wide range of non-academic interests, China takes almost the opposite approach. The prime requirement for a university student in China is a demonstrated concern for one's fellows. Candidates for a secondary education are nominated by their "production unit"—a commune, factory, or army unit. Once in school, they work hard. At Peking University, a student's schedule runs from 6 A.M. to 10 P.M. six days a week.

Dr. Sidel capsulized his feelings: "I'm not sure that kids in China are psychologically conditioned, to the extent that American kids are, to

Chinese blend the old and new in agriculture, using both tractors and oxen to till the soil.

the goal of being bigger and better than someone else. The Chinese attempt, in fact, to inculcate a very different set of goals: that you get your satisfaction, not from being bigger and better, but from how well you fit into the patterns of other people.

"The issue that keeps coming up is altruism—how to serve the people. It's very hard to convey the feeling that one gets from the Chinese, but when they say 'serve the people,' they really mean it. And that is the basis on which they are trying to build a society. That is not to predict how long it will last or where it will go."

The very flavor of the nation excited Sidel. He remarked on "the incredible hospitality, the incredible warmth, the fervor, the enthusiasm we saw on every hand," adding: "Another thing that was startling to a New Yorker like myself was the total absence of cynicism. Everyone with whom we spoke was perfectly straightforward and free of a cynical attitude of himself or society. You can't spend one hour in New York City and not come out a cynic."

The visitors saw lessons for other societies, Russian as well as American, during their visits. "It's not a bad idea to expect a year or two years of national service from everybody," Galston said. "I don't mean mili-

Science students, as in this Peking University biochemistry laboratory, told visiting American scientists that their chief ambition is to serve the people.

Students in a "cadre school" lunch in the school's courtyard. Chinese in authority attend the schools to labor and listen to criticisms of others.

tary service. I mean the kind of service that would teach young kids what this country is all about by putting them in contact with some of the problems—ghetto problems or conservation-ecology problems. I think there are many young people who would welcome an opportunity to do this between high school and university. And if they did, we would have a much better student population because they would be more mature and have had some contact with life. You know, many kids who grow up in upper middle-class homes have never done anything but accept, accept, accept, and demand, demand, demand. In that sense, we can learn something from the Chinese."

The wholesale enthusiasm, apparently genuine, among Chinese scientists surprised the Americans. "After all," Galston noted, "a professor had a pretty good deal going before the Cultural Revolution of 1966, and suddenly he was told to pack up and go live with a bunch of peasants. It was not exactly his idea of comfort or a summer vacation."

Yang saw the insistence on grass-roots work in the communes by scientists as an attempt by the Communists to break down the élitism—the sense of superiority in certain groups—that had extended all through Chinese history. Galston saw it as an attempt to break down the idea of a scientific élite that followed the pattern in Russia.

"I think this is part of the Chinese concept of a classless society," Galston added. "The Soviets, although they don't like to admit it, are one of the most class-stratified societies in the world. You know, the scientist is an absolute god in the Soviet Union. His salary is way up in the stratosphere compared with the worker's, and he has all sorts of privileges. Now, that's good in some respects. You attract the best men. But it's also bad because these guys get pretty cocky and start to think they're supermen and remove themselves from the mainstream of life. In a society like China's, they didn't want that to happen. To avoid a system

201

The head of the Chinese Academy of Sciences greeted four visiting U.S. physicians in Peking. From left are Victor Sidel, Samuel Rosen, Paul Dudley White, Kjo Mo-jo, and E. Grey Dimond.

in which professors do nothing but profess, and in a while run out of things to profess about, they are enforcing a system in which there has to be contact between classes."

"If such changes were imposed at M.I.T.," Signer said, "you'd have a small revolt. People would go nuts." "At Yale, it would be even worse," said Galston. "I think to say you had to connect your activities at Yale to some activity in the outside world would be very repulsive to a vast majority of Yale professors."

What happens to recalcitrant scientists, the ones who will not bend to the will of the government and the people? If they were purged, the American visitors could find no evidence of it. Dr. White said, "I thought that many of the doctors who had been trained abroad—men now in their 50s or 60s—might have been scrapped during the Cultural Revolution, but it certainly didn't appear to be so." He told of meeting Western-trained Chinese he had known years before. Other travelers to China told of "schools of rectification" in the countryside where scientists and other intellectuals were retrained in line with the purposes of the nation. If there was physical duress, none was reported.

Dr. White also looked for, but did not find, a single department of psychiatry in any of Peking's big hospitals. "We asked why," White said. "They answered, 'We work too hard to become psychotic.'"

The Chinese have widespread day-care centers for children to free women to work. And on the job, the Chinese woman finds, as Dr. Sidel noted, "equal opportunity for equal skills, equal pay for equal work."

Almost all the Americans found cases where scientist husbands and wives lived apart for extended periods, each pursuing his own work. Their lives did not seem damaged by the experience.

"I think family life has been transformed, but not destroyed, by what is going on," Galston reports. "In Peking, I visited a chap named Lee Cheng-li, a plant anatomist-morphologist who used to be at Yale. His wife is a pediatrician who worked at St. Raphael's Hospital in New Haven. They went back because Lee got a job at the university in Peking. When the Cultural Revolution of 1966 began, Chairman Mao Tse-tung said, 'You know, you physicians are all clustered in the cities, but 80 per cent of the people of China are out in the countryside, and don't have adequate medical care.' Lee's wife thought about this. She consulted with her husband and her children [her youngest was 16]. She said, 'I'd like to go out into the countryside with a medical brigade.' Her family said they could care for themselves," Galston said, "so she went for 18 months—inoculating, treating, establishing medical clinics, training midwives, training paraprofessionals and barefoot doctors. She came back just glowing.

"When we met her in Peking, she had just returned from her second or third trip. This is now her life. She believes in it. There's an example of somebody whose life was comfortable. There was no coercion, but she changed her life because of demonstrated national needs and because of exhortation of a very effective type."

Yang noted, "There's no doubt that in China there is still a tremendous need for many material things. But I found a spirit of enthusiasm. I look at the United States today and I wonder if there is not too much of the material things and not enough of the spirit."

"The people look healthy," Galston said. "They look well fed even in rural areas where life is hard. They look as if they have enough food and medical care, and when you compare that with the China of 22 years ago, this is a fantastic achievement."

Reflecting on his visit, Dr. White said the Chinese doctors themselves found great satisfaction in "what they've done to rescue these hundreds of millions of rural people from disease and hunger. And some of the engineers who helped put up dams and dig irrigation canals through the mountains have also helped rid China of hunger as well as floods. So that the great devastation we used to hear about—thousands dying in a famine, thousands being drowned in a flood, thousands dying of cholera—that doesn't happen anymore. This is a great achievement. This is the greatest achievement."

For further reading:

Dimond, E. Grey, "Acupuncture Anesthesia," *Journal of the American Medical Association,* Dec. 6, 1971.

Galston, Arthur W., and Signer, Ethan, "Science and Education in China," *Science,* Jan. 7, 1972.

Terrill, Ross, "The 800,000,000—Report from China," *Atlantic,* November and December, 1971.

New Peril To Our Oceans

By John A. McGowan

Worldwide fallout of pollutants from the atmosphere may be more damaging to our oceans than spilling oil or dumping wastes

On a single day in the summer of 1971, 10 to 15-million menhaden, a commercial fish widely used for fertilizer and bait, died in Escambia Bay near Pensacola, Fla. The kill was eventually traced to a low level of oxygen in the water, resulting from direct dumping of industrial wastes.

On Sept. 16, 1969, an oil barge was grounded off Fassets Point, West Falmouth, Mass. Between 650 and 700 tons of fuel oil ran into the coastal waters. Eight months later, scientists who examined the sea bottom, marshes, and tidal rivers were surprised to discover that not only was the pollution still present, but it also had spread over an area many times larger than that first affected by the spill.

These are typical of the incidents that people connect with pollution of the oceans. But, however disastrous these accidents can be to a local area, they may be only a relatively small part of the ocean pollution problem. Scientists are beginning to understand a second, far more predominant type—dif-

Industrial waste, sewage

Plant and animal decay

Ground soil

fused pollution. All but ignored until about three years ago and vastly more complex in its patterns, diffused pollution originates on the ground but wafts its way into the atmosphere as vapor or as tiny, light particles. Later, perhaps hundreds or even thousands of miles from its source, it falls or washes out of the atmosphere. Most of it ends up in the ocean. For the first time, the fragile 140 million square miles of ocean and the life within it are truly in danger.

Until less than 40 years ago, the history of pollution on this planet had been one of local point-source pollution. All animals and plants have natural waste products. This biological waste contains a considerable amount of phosphorus- and nitrogen-containing compounds and various other components. Many of these are deposited in soil or water, where they are broken down by microorganisms.

Over the course of evolution, a dynamic balance developed between the rate of waste production and the rate of degradation. As part of this process, many waste materials dissolve in the ground water that eventually seeps into streams and rivers. The streams and rivers carry the materials into the ocean. Eventually they find their way into sea plants and then into grazing fish and other animals, or they become trapped in the sediment on the ocean bottom. This was the natural cycle of waste disposal before man entered the picture. There was no pollution because the wastes never became so abundant that the disposal system could not cope with them.

The natural cycle was altered for the first time about 4500 B.C., when man first began to farm and raise domestic animals. This resulted in people clustering together in villages. The concentration of people and

The author:
John A. McGowan is a marine biologist at the Scripps Institution of Oceanography of the University of California, San Diego.

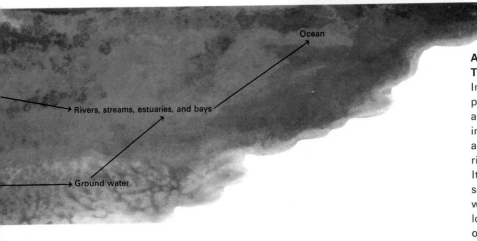

Ocean

Rivers, streams, estuaries, and bays

Ground water

A Familiar Route To Ocean Pollution
In point-source pollution, plant and animal decay seeps into ground water and is carried into rivers and streams. It is joined there by sewage and industrial wastes. The whole load is carried to the ocean. Pollutants deliberately dumped and oil spills and other accidental discharges enter the ocean directly.

animals caused more heavily concentrated quantities of waste, and point-source pollution was born. Archaeologists tell us that in this early stage of man's development he frequently moved his village sites, probably to get away from his accumulating waste. Then, man began to allow his animals to overgraze the land and he started cutting down trees to build more permanent structures. This destroyed most of the local ground cover, allowing the rain to wash away the rich topsoil. Water that formerly seeped slowly into the topsoil now ran off the barren surface, washing waste materials that would once have been broken down by soil microorganisms directly into streams.

About a thousand years later, man learned to smelt metals, and with this process, released toxic fumes into the air. Thus point-source industrial pollution was born. With increasing technological development, along with the preservation and storage of food, man's waste continued to grow. There still was no great problem because man still had plenty of space. But, the equilibrium between waste disposal and waste breakdown had begun to change rapidly.

With technological advances in agriculture, food processing, and toolmaking, man's population grew, and, along with this growth, came cities. Cities brought an even greater increase in local point-source waste production and a correspondingly greater shift in the balance between waste disposal and waste breakdown. To handle the increased burden, sewage-disposal systems were built. Ancient Pompeii, for example, had an elaborate sewer system about 2,000 years ago. Most of these early sewage-disposal systems emptied directly into rivers, streams, lakes, and the oceans.

Direct Ocean Dumping

Nearly 50 million tons of waste are dumped at some 250 known sites along U.S. coasts each year. About 80 per cent of all ocean dumping is dredge spoils—sediments dug from the floors of inland bodies of water. More than 30 per cent of these are polluted.

During the Industrial Revolution the growth of population accelerated and cities grew even larger. Most of them were located near rivers, bays, and the oceans for reasons of transportation and trade. By now, both population and technology were increasing at a geometric rate. So much waste was produced that natural breakdown processes were fast being hopelessly overwhelmed. The result was first a gradual and then a rapidly increasing rate of pollution in many bodies of fresh water. Eventually, the pollutants reached the ocean.

This growing point-source pollution has been intensified by the addition of some new substances born of the industrial age. In less than four decades —a few minutes on the evolutionary time scale— man has introduced into the environment a wide range of totally new, highly toxic chemical compounds. They include organic compounds, such as pesticides and polychlorinated biphenyls (PCBs). These compounds, along with petroleum products from oil spills and dumping and from nondegradable plastics and other solids, constitute a new type of waste material. They are so new to the environment that natural breakdown and detoxification processes for them do not even exist.

Because the amount and rates of this and all local point-source pollution are closely linked to industrialization and crowding, a Belgian and an American scientist have been able to devise a useful index of the potential of a nation to pollute itself. The index is calculated by comparing a nation's gross national product (GNP) to its area. The GNP is related to industrial, agricultural, and domestic activities. These in turn relate to a nation's ability to pollute. According to this index, countries with the highest pollution potential include Singapore, The Netherlands, Belgium, and Japan. Those with the lowest are Southern Yemen, Peru, Mongolia, and the Central African Republic. The United States ranks 21st in its ability to foul its own nest, just behind Taiwan (Formosa). Russia, at 52nd, is far down on the list.

Although the approach has been largely piecemeal, much can and is being done to reduce or eliminate point-source pollution. Many communities are beginning to apply more sophisticated treatments to both their domestic and industrial sewage. Intensive efforts are being made to control the input of heavy metals and pesticides in sewage systems. Many solids, such as bottles, cans, and newspapers, are now being recycled. Recently created state and governmental agencies are being given the legal power to ban or severely restrict the dumping of heavy metals and pesticides in the oceans. The United Nations has formed an ocean pollution study group, and scientists all over the world are studying the effects and consequences of local point-source pollution. In February, 1972, 10 European countries signed an international agreement to end dumping from ships and aircraft.

Determining the effects of the various forms of point-source pollution is a reasonably straightforward scientific problem. Eliminating the effects becomes an engineering, social, political, and legal problem. But this is not true of diffuse-source pollution. Just to understand it better will require a great deal of basic research in many disciplines—among them chemistry, meteorology, oceanography, ecology, and physiology. Many of the most dangerous and toxic substances such as DDT, PCBs, methyl mercury, gasoline, and lead and copper compounds are highly volatile—that is, they evaporate very quickly, allowing large quantities to get into the atmosphere. For example, although the annual world production of PCBs is only about 100,000 tons, 25,000 tons of PCBs evaporate into the atmosphere each year. All these pollutants are joined in the air by those from burning coal, gasoline, and oil.

Once all these substances get into the atmosphere, they may be scattered around the world by the global wind system. However, they do not remain in the atmosphere forever. Many are absorbed onto dust particles that eventually fall out of the atmosphere or are washed out in raindrops and snowflakes. Measurements show that the lead concentration in snow that fell on glaciers in Greenland in the mid-1960s was 500 times greater than in the snow found deeper in the glacier that fell in 800 B.C. The measurements showed fairly low levels from 800 B.C. to 1952, but an increase of nearly 250 per cent from 1952 to the mid-1960s.

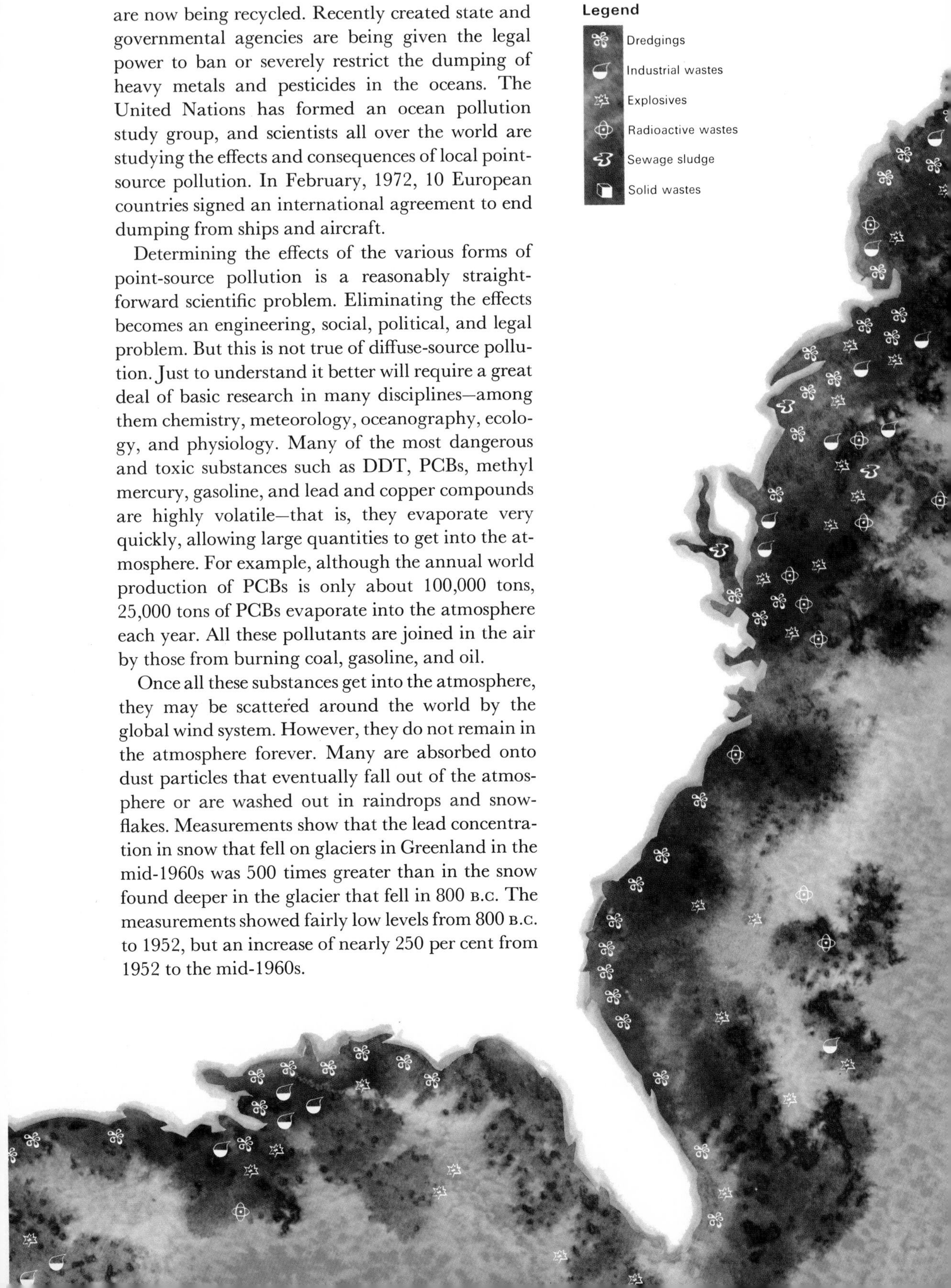

Legend

- Dredgings
- Industrial wastes
- Explosives
- Radioactive wastes
- Sewage sludge
- Solid wastes

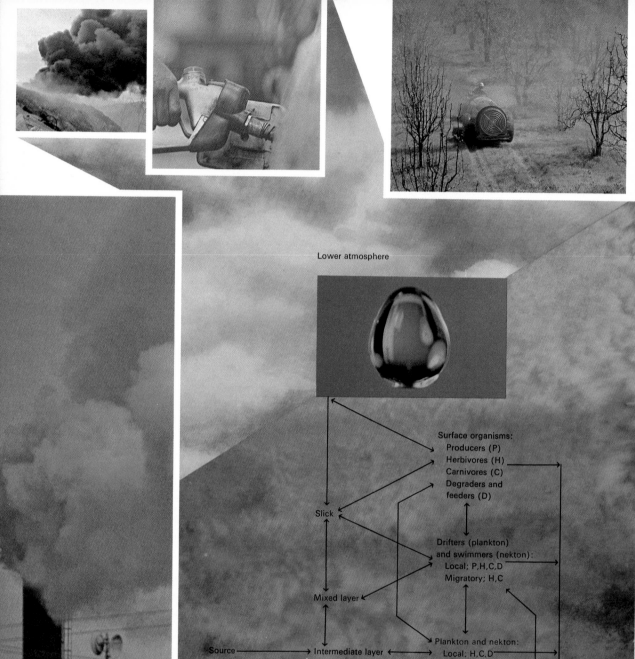

Lower atmosphere

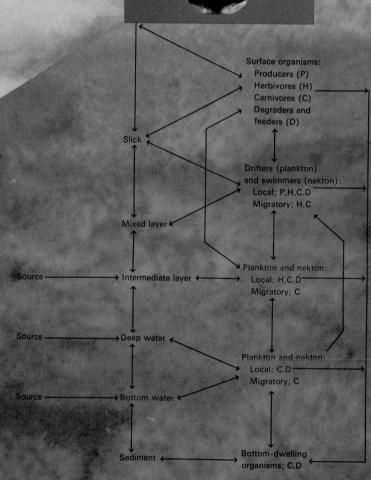

Surface organisms:
  Producers (P)
  Herbivores (H)
  Carnivores (C)
  Degraders and
    feeders (D)

Slick

Drifters (plankton)
and swimmers (nekton):
  Local; P,H,C,D
  Migratory; H,C

Mixed layer

Plankton and nekton:
  Local; H,C,D
  Migratory; C

Source ⟶ Intermediate layer

Source ⟶ Deep water

Plankton and nekton:
  Local; C,D
  Migratory; C

Source ⟶ Bottom water

Sediment

Bottom-dwelling
organisms; C,D

An Unfamiliar Route to Ocean Pollution

Pollutants from many sources enter the atmosphere and fall or
wash out in rain or snow. The journey of the pollutants through
the ocean's layers and through the plant and animal life that
make up the ocean's food chain is very complex. The arrows show
some likely flows and transfers within and between the two routes.

Although many scientists were at least partly aware of this aspect of
the global problem, it was not entirely clear how important it really
was until the spring of 1971. Then, the National Academy of Sciences
conducted a workshop to examine, in detail, the scientific problems related to its marine environmental quality program. Five major problem areas were selected:

- Identifying major recognized and unrecognized pollutants, their
sources, and their rates of input.
- Describing how these pollutants are dispersed.
- Understanding the geochemical and biological transfer of pollu-
tants in the ocean.
- Determining how pollutants affect man and other organisms.
- Finding out where specific substances in the ocean end up.

About 50 scientists, including physical oceanographers, marine
chemists, physiologists, marine ecologists, and fisheries biologists
formed task groups to study each of these problem areas. The workshop
convinced participants of what had been suspected by some at the out-
set—that atmospheric fallout of many of the most serious pollutants is
often greater than transport of pollutants to the ocean by rivers or sew-
ers. For example, of the 1,820-million-ton world annual production of
oil, it is estimated that 2.6 million tons enter the marine environment
through seepage, spills, deliberate dumping, and other "injections."
However, about 35 times that amount—more than 90 million tons—are
estimated to escape into the atmosphere.

Little more than a start has been made in understanding the extent
and complexity of diffuse pollution, but it is a start that is taking place
on several fronts. Several groups of scientists are investigating the entire system of pollutant input, transfer, and final deposition.

The group I belong to at the Scripps Institution of Oceanography is
developing a mathematical model to show what happens with such
pollutants. Our model began on land with world production figures for
pollutants including lead, copper, chromium, cadmium, mercury,
DDT, aldrin, PCBs, and petroleum. Then we estimated how much of
each is lost or spilled while it is being handled or used. A fraction of this
loss gets into the food chain on land and tends to be concentrated up
the food chain from levels of plants to *herbivores* (plant-eaters) and to
carnivores (meat-eaters). However, at each of these steps, death and
decay of some of the organisms release a certain amount that is returned to the local land. Portions of these materials also evaporate into
the local air. Once in the local air, they mix and circulate around the

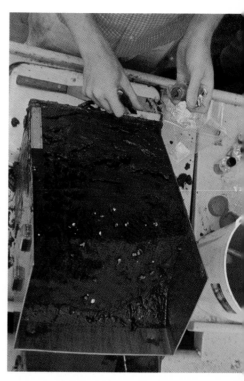

Scientists tracking ocean pollutants collect specimens, *right,* to determine numbers and types of species present. The diversity of species in the ocean provides a clue to changes in the ocean environment. Ocean sediment, *far right,* will undergo chemical analysis.

world. They then may fall out or be washed out by rain and snow anywhere. Ultimately, most of these substances reach the oceans, where many get into the upper, biologically rich, mixed layers and are taken up by plants and animals just as on land.

This uptake process may include a significant amount of concentration. This is true especially in the case of materials such as DDT or PCBs, which are not particularly soluble in seawater but are highly soluble in the fats and oils of living organisms. This principle of differential solubility is primarily responsible for the concentration of these particular pollutants up the food chain to the higher carnivore levels. In the upper, mixed layer of the ocean, just as on land, animals and plants die and are broken down by bacteria and other degraders. Hence, some of the pollutants are returned to the water and are available for uptake once again by other organisms.

A certain amount of the pollutants settle to the deeper zones of the ocean where there are also animals capable of concentrating them. As on land, these animals go through a process of further concentration up the food chain, death and decay, and an eventual recycling of the pollutants once again back to the water. The pollutants may then continue their downward drift to the *abyss* (lowest depth), and go through another recycling process through the bodies of *benthic* (bottom-dwelling) animals. Finally, the pollutants probably are buried in the sediments on the ocean bottom.

However, an additional problem—migration—complicates the process. Along with the force of gravity, which moves pollutants ever deeper, some of the organisms living in the upper, mixed layer and the intermediate depths migrate up and down daily, carrying pollutants from shallower to deeper levels and vice versa. For example, some small crustaceans and fish migrate at night from depths of 200 yards or more to feed near the surface. There they could pick up and carry back down not only a meal, but also a load of fallout pollutants. Then, when some of the migrators are eaten by permanent residents of the deeper zones, the predator acquires the pollutants along with his meal. This mechanism of active transport may move more pollutants from surface layers to the depths more rapidly than sinking, but we need more experimental evidence to prove this hypothesis.

To fully understand diffuse pollution we must regularly measure the concentration of each pollutant in various environments around the world. This means collecting an immense amount of data. We need to know the rates at which the pollutants move into the ocean. This means that we must monitor their concentrations in the air; on many different parts of the land; in land plants and animals; in the upper, mixed layer of the ocean; in the plants and animals living in the ocean; in concentrations at middepths; in the abyss; and in the sediment. We also need to continuously study the feeding rates of the important herbivores and carnivores. Equally important, we must measure rates of

A 65-foot-high tower in Hawaii, *right,* detected unusually high concentrations of lead and vanadium in air over that portion of the Pacific Ocean. A scientist samples annual ice layers in Greenland, *far right.* Tests of this ice and ice from Antarctica show a sharp increase in lead since 1940.

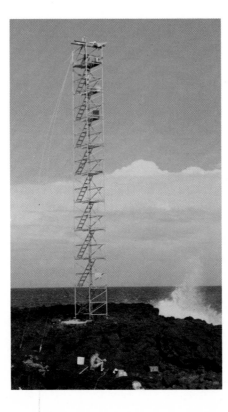

transfer between the air and the water and from the water to organisms. Above all, we need to know the rate of fallout from the atmosphere to the surface of the ocean.

Hopefully, if these studies are done over a long enough period, over a large enough area, and on a variety of the basic members of the food chain, some systematic pattern will emerge. We can then translate this pattern to our mathematical model to arrive at some estimate of the concentrations of pollutants in various parts of the system, the time they spend in each part, the rates at which they move between the parts, and the effects of man's changing usage. The goal, of course, is to use this model to set usage rates at amounts that maintain pollution levels that we think we can live with.

The alternative to this is to ban completely or severely restrict the use of toxic heavy metals, petroleum products, insecticides, and fertilizers. However, a complete ban of these materials would probably result in mass starvation and disease. For example, persistent pesticides—those that cannot be easily broken down—are essential to control malaria mosquitoes in large parts of South America, Asia, and Africa. Lead probably can be eliminated from gasoline without too much disruption, but it has extremely important industrial uses. Lead and mercury are found in coal and some petroleum products, and both metals are released into the air when the fuel burns. But these fuels are essential for generating power in many parts of the world.

It is evident that we need to use many of these materials until safer substitutes are developed. In the meantime, scientists must continue their attempts to determine what usage levels are relatively harmless to life. In other words, we will have to learn to live with a certain amount of pollution, and reliable models of diffuse-source pollution are needed to help determine how much.

Determining relatively harmless levels of pollutants in seawater will be difficult. At the present time, we have little information on most of the more important animals and plants. Some acute toxicity studies have been done in laboratories to determine the concentration at which a pollutant kills 50 per cent of the experimental organisms in 96 hours. These doses are called LD 50s (for lethal doses of 50 per cent). However, researchers noted in many of these studies that the growth of the surviving 50 per cent was affected. For example, the LD 50 of mercury for kelp lies between 3 and 5 parts per million (ppm). But, after 24 hours in seawater containing only 1 ppm, the kelp stopped growing.

Further studies have shown that the growth of several other kinds of marine plants and animals has been slowed by concentrations of mercury, copper, zinc, pesticides, and detergents far below the LD 50 levels. Because of the delicate balances between the living creatures of the sea, slowing growth rates in even a few species can seriously disrupt the entire system. Because of this, basic studies are being made on how food chains form and operate and the effects various pollutants have on them, but much more work must be done.

A 1971 study of DDT and PCBs in marine organisms by scientists at the Woods Hole (Mass.) Oceanographic Institution, under a grant from the National Science Foundation, indicated "a real cause for concern." In a variety of marine organisms collected in the North Atlantic Ocean, the Woods Hole researchers found that "PCBs are readily demonstrable in all, and DDT in most."

The scientists found a relatively direct relationship between the amount of DDT in a local environment and its levels in, and effects on, local organisms. When DDT use is controlled or cut back in a given area, levels decrease dramatically in the organisms living there. However, the scientists confirmed that the major pathway for DDT is not local use, but diffusion "from the land, through the atmosphere into the oceans, and into the oceanic abyss." Once in the ocean, DDT and PCBs may take different physical and biological routes under certain conditions. For example, in areas where there are residues of petroleum spills, the scientists found higher concentrations of DDT than PCBs in certain phytoplankton.

In addition to this type of research, studies are underway on the concentration of heavy metals, pesticides, petroleum, and other pollutants in open ocean areas far from any local point-source of contamination. Scientists are measuring the pollutant "load" in the bodies of plants and animals now so that we may have some future basis for determining if these concentrations are increasing, decreasing, or remaining constant. Researchers are collecting phytoplankton, zooplankton, deep-sea fish, squid, flying fish, sharks, dolphins, and tuna, and freezing them for shipment to laboratories where analysts check how much heavy metal, pesticides, PCBs, and petroleum they contain.

It is also evident that we need to know much more about the normal population variations in all plants and animals living in the open ocean as a result of predation, competition, starvation, aging, unusual climatic changes, and disease.

We now have a new type of disease—chemical disease—as a result of pollution. Sorting out the effects of chemical disease from all of the other factors is one of the most challenging problems in basic ecological research in the ocean. We hope our model will guide us to the vitally needed answers about the entire system of pollutant input, transfer, and final deposition. Only then will we be able to gauge the destructive effects of diffuse-source pollution and learn how much our ocean—the earth's final sink—can bear.

For further reading:

Marine Environmental Quality, National Academy of Sciences, Washington, D.C., 1971.

Ocean Dumping: A National Policy, Council on Environmental Quality, U.S. Government Printing Office, Washington, D.C., 1970.

Walsh, John J., "Implications of a Systems Approach to Oceanography," *Science*, June 2, 1972.

Supertanker

By Wade Tilleux

**As mammoth oil carriers roam the seas
in evergrowing numbers, they pose new
challenges to marine technology and safety**

The Japanese tanker *Nisseki Maru* takes on oil from a storage complex on the Persian Gulf at a rate of more than 20,000 gallons a minute. As the oil flows in, seawater ballast is pumped from another of the ship's 15 tanks to keep the ship on an even keel. A computer flashes the ship's internal stresses and balance to the control room.

After four days, the nonstop pumping ends. The tanker's hull—almost 1,200 feet long, 180 feet wide, and 105 feet high—is full. The ship holds 372,698 tons of thick, black crude oil. When the loading began, the empty tanker looked like a giant floating cork, its deck high in the air. Now, settled lower in the water, its draft, or depth below the water line, is 45 feet greater than before.

Oil lines disconnected, the sluggish tanker, assisted by tugboats, creeps toward the open ocean. It takes almost an hour for the loaded supertanker to reach its 15-knot cruising speed, when its broad bow pushes the sea ahead into a 12-foot-high bow wave.

Clothed in scaffolding in a Japanese shipyard, the immense, bulbous bow of a new supertanker is sanded and polished prior to painting.

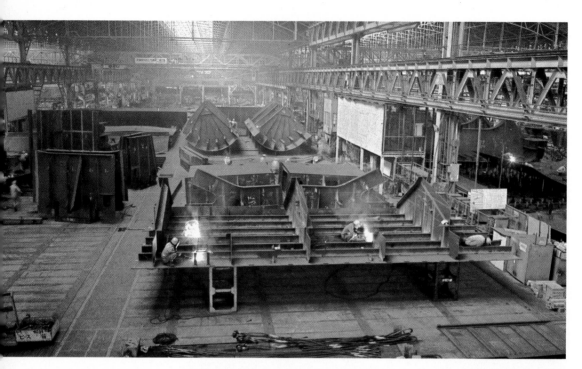

Supertankers are born
in the shop, *above,* as
prefabricated sections
are welded together.
Assembly takes place
at building dock, *right.*
The growing colossus
is honeycombed with
storage tanks, *opposite
page, top.* Workmen
join up oil transport
pipes on main deck,
opposite page, bottom.

The 11,200-mile voyage from the Persian Gulf to Japan takes about 20 days. The crew of about 40—some of them women—have private, air-conditioned rooms, a gymnasium-recreation room, and more than 4 acres of deck space on which to wander. Supertankers, because of their draft, length, and weight, are exceptionally stable. Virtually no rocking or pitching occurs, even during severe storms.

The *Nisseki Maru* is a majestic ship, an engineering triumph. Yet it is little more than a huge barge, a mammoth egg carton of oil compartments. More than half of the 2 billion tons of petroleum removed from the earth in 1971 was transported by tankers. The *Nisseki Maru* is one of 20 supertankers now in service, the largest ships in the world. Each one carries more than 200,000 tons, which is a quart of oil for every person in the United States. These ships transport oil from the oil-rich Arab states of the Middle East around Africa to Europe or the United States, or around Asia through Indonesian straits to Japan.

The tremendous size of these ships has dramatically reduced the cost of shipping petroleum and has revolutionized shipbuilding. But their size also presents a hazard—catastrophic oil pollution. The staggering consequences of a tanker wreck were dramatically demonstrated in

219

1967, when the *Torrey Canyon* ran aground on a reef in the English Channel. More than 100,000 tons of its crude oil cargo spilled, killing fish and birds and contaminating 140 miles of English coast. As tankers increase in size, such accidents would be even more catastrophic.

Ship designers, as well as ecologists, worry about a supertanker's lack of power and maneuverability. Engine horsepower—about 40,000 for the largest tanker—has been increased only slightly relative to the vessel's increased size. Since a ship's drag increases at a much slower rate than its weight, only slightly more power is needed to maintain top speed. But turning and stopping distances increase directly with weight. In a full-load test, with its propeller thrusting full astern, the *Nisseki Maru* needed 23 minutes and more than 3 miles of ocean to come to a dead stop from its 15-knot cruising speed.

These problems are somewhat compensated for by the improved hull design of a tanker. First, a computer model of the structure is thoroughly tested for potential weak spots. Then an instrumented scale model of the hull, from 30 to 40 feet long, is pulled through a towing tank in a test similar to the wind-tunnel tests used in airplane design. This test can help to predict maneuverability, various types of drag, and the effects of waves on the full-sized ship.

With bow and bridge still to be added, a new supertanker is eased into a new dock for completion, *right.* A captain's view from the bridge of the *Nisseki Maru, opposite page,* reveals the ship's thousand-foot length of freshly painted deck.

Even experienced captains go back to school before piloting supertankers, because of the ship's vastly different handling characteristics. They receive two weeks of classroom instruction, *above,* accompanied by practice on scaled-down models, *top and right,* that mimic the big tankers' power and maneuverability.

Their captains must have special training. Mammoth tankers respond very slowly to rudder signals, and special electronic devices simulate this delay in small training boats. They also simulate both the on-board anticollision radar that sets off an alarm whenever a ship approaches at sea, and the sonar docking system that detects even the slightest motion of the tanker while it is in port.

During the trip back to the Middle East, oil tanks are cleaned with sprays of seawater. In 1970, the 207,000-ton tanker *Marpessa* exploded and sank when this spraying process built up a static electric charge that ignited oil vapor in its tanks. To prevent similar explosions, tanks being cleaned are now flooded with an inert gas, such as cooled exhaust from the ship's boilers. To reduce pollution, the cleaning water drains into smaller "slop tanks" where heat separates the oil from the water before the water is dumped back into the sea.

Petroleum is civilization's major source of energy, and the need for it will double each decade. Most of the petroleum will be transported by tankers. As of 1972, the *Nisseki Maru* was the biggest supertanker afloat. However, two tankers now being built in Japan will carry 470,000 tons of oil each. And within a decade, super-supertankers may carry as much as 1 million tons.

Her tanks brimful with 372,698 tons of Arabian oil, the new *Nisseki Maru* rides low in the water on her first voyage from the Middle East.

Our Earliest Ancestors

By F. Clark Howell

**Newly dated fossils climax a decade of discoveries
that have added millions of years to the record of man**

A fragment of jawbone, uncovered in eastern Africa after rain and wind had swept over volcanic sediments, has astounded scientists because of its great age. This fossil bone is the oldest evidence of manlike creatures yet found. It proves that he roamed and gathered food on this planet more than 5 million years ago. As a result, anthropologists, who only a few years ago had little understanding of the antiquity of mankind, are taking a fresh look at man's evolution.

The fragment was discovered in 1967 at Lothagam Hill in Kenya. However, it took more than three years of careful study of this specimen and its location in the surrounding geologic deposits, as well as of elephant bones found there, to determine exactly how far in the past this manlike creature lived.

The Lothagam Hill find has climaxed more than a decade of anthropological discovery in eastern Africa, particularly around the Great Rift Valley, which extends from the Red Sea almost due south through Ethiopia, Kenya, Tanzania, and Mozambique. These east African

The *Australopithecus africanus* species, in East Africa, probably made and used tools to kill and skin animals, such as baboons.

The Great Rift Valley in eastern Africa, where the recent discoveries of ancient man were made, has a hot, dry climate that is ideal for preserving fossils.

The author:
F. Clark Howell, professor of anthropology at the University of California, Berkeley, has led many expeditions in search of early man in both East Africa and Spain.

studies, in which many students of early man, including myself, participated, have involved intensive and prolonged field work by scientists in such diverse fields as archaeology, botany, and geochemistry.

Before the last decade of discoveries, there were many unanswered questions about human beginnings. For instance, the posture, gait, diet, and mental capabilities of the earlier members of the family of man were very poorly known. Also, it was uncertain whether one or more hominid species—that includes all living people and their extinct relatives—were represented in the early fossil record. Without fossil and stone artifact evidence, there was no way to determine when and how men began to make stone tools, what they used as shelters, and whether they lived in large groups or family units.

It was unlikely, however, that many of these bones or teeth would ever be discovered. After all, early men probably lived only in places that had suitable climates and food resources. Almost all their skeletal remains would have been destroyed by scavenging animals and insects or bleached and decomposed by the sun and rain. The only bones that might survive as fossils would be those that were quickly buried by silt and sand along rivers and lakes, by volcanic ash, or within tar pits and quicksand bogs. And before these fossils could be found, disturbances in the earth's crust must have lifted the fossil-bearing deposits to the surface, so that erosion could expose them. Fortunately, just this series of events occurred along the Great Rift Valley of eastern Africa.

On July 17, 1959, anthropologist Mary D. Leakey made the first of the discoveries that have revised our knowledge of man. She found some manlike teeth that had been exposed by erosion in the lowest and earliest geologic deposits of Olduvai Gorge, in northern Tanzania. The teeth were later shown to be nearly 2 million years old. On that same day, exactly a century before, Charles Darwin had been busily engaged in correcting the proofs of his monumental *On the Origin of the Species*, the book that established modern evolutionary theory.

Olduvai Gorge is uniquely suited to preserving fossils of animals as well as men. The area is now a dry plain without permanent water sources. But several million years ago this location was part of a shallow lake. A series of active volcanoes rose to the east. The ashes of their eruptions were recognizable in the deeper layers of soil within the gorge where Mrs. Leakey worked.

Africa's Valley of Revelations

The Great Rift Valley in eastern Africa has both a climatic and a geological history that has made it a rich repository of early man's fossilized remains. Teams of anthropologists are digging at many sites in the valley in Ethiopia, Kenya, and Tanzania.

East Africa

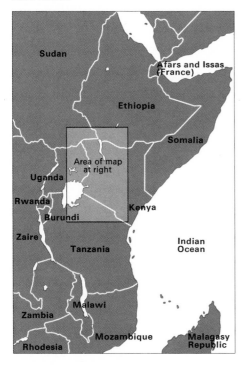

Great Rift Valley Excavation Sites

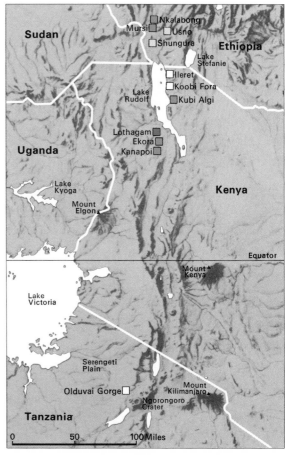

Age of geologic formations:

■ 5 million years and older

■ 4 million years and older

□ 1.8 to 4 million years old

□ 1 to 2.5 million years old

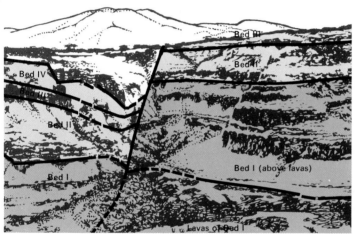

Fossils of early man can be dated by the sediment layer where they are found. The earliest unearthed in Olduvai Gorge, *above,* were found in Bed I, *left.* The gorge is about 200 feet deep there.

Using delicate camel's-hair brushes and dental picks to remove bits of the surrounding rock, she and her husband, Louis S. B. Leakey, carefully uncovered almost 400 fragments of the skull of a nearly adult individual. Her discovery was a great triumph. Both Leakeys had patiently explored the steep walls of this 300-foot-deep gorge for more than 20 years, searching for the buried evidence of early man.

When the bone fragments were fitted together, only the lower jaw was missing. From the shape of the skull and upper teeth it was clear that this primitive hominid represented an extinct being, called *Australopithecus.* It was the first time that remnants of this creature had ever been found in eastern Africa. Further digging revealed the bones of animals that had been broken open to expose the marrow, a sure indication that these animals had been used for food. What made the discovery most exciting were many stone tools found near the skull.

Further, this was the first time that an *Australopithecus* could be accurately dated. Previously, all specimens had been found in South Africa.

Part of the skull of a juvenile *Australopithecus* with all its milk teeth and its first permanent molars was discovered there in 1924. Parts of a limb and skull fragments of other adults and juveniles were found later. But all these fossils came from limestone caves or fissures that were quarried commercially for their valuable lime deposits. Blasting and mechanical digging had in each case distorted and destroyed the geological details of the area before the bone fragments were found. There was no way to determine accurately how old they were.

It had been tentatively established that the genus was hominid, or manlike, and that it had at least two species. These were named *Australopithecus africanus* and *Australopithecus robustus*. The Olduvai skull most resembled the *robustus* species.

Three different methods of dating confirm that the Olduvai specimen is about 1.75 million years old. The volcanic ashes above and below the position of the hominid bone fragments were first dated by the potassium-argon method. This method measures the amount of argon gas that has accumulated, through the slow decay of radioactive potassium, since the molten rock solidified. Fission-track analyses then measured the number of microscopic marks left in the rocks as a result

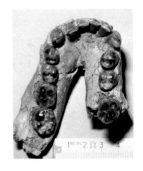

The teeth and part of a jaw were among remains of an early *Homo habilis* found at Olduvai by Louis and Mary Leakey.

The remains, *above,* in a lower level of sediment at Olduvai Gorge are thought to be a primitive dwelling. An archaeologist's detailed chart of the site, *right,* shows the location of each of the rocks that may have formed the base of this ancient structure.

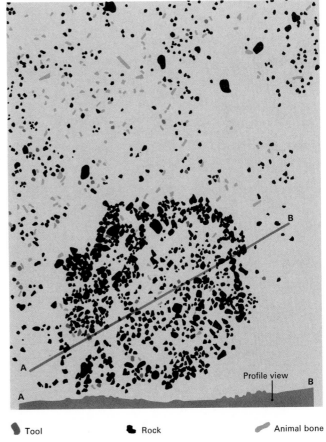

Profile view

 Tool   Rock   Animal bone

of the decay of thorium and uranium. The dates found by this process agreed with the potassium-argon dates. Finally, the deposits at Olduvai, including that in which the bone fragments were found, were fitted into the record of changes in the earth's magnetic field during the past few million years of geologic time.

Hominid remains have since been found at about a dozen places within the gorge. The ancient lake alternately flooded this site and then dried up. It had no outlet and apparently expanded and contracted with the seasons and as the climate changed. This process covered human remains with layers of mud and silt.

The Leakeys found many chipped stones of a type termed Oldowan after the Olduvai site. These tools were made by various chipping procedures. Some of them must have been brought to the site from many miles away, because they were made of a rock unlike any found nearby.

The Leakeys discovered the remains of many different species of fish, amphibians, reptiles, birds, and mammals. The mammals ranged from tiny shrews and rodents through monkeys, baboons, antelopes, and pigs, to several kinds of elephants. The animal remains were found in distinctive scatters and clusters. Sometimes, the bones of early men were found in the midst of these animal bones and stone tools, indicating that some sites served as home bases. At one site, the Leakeys found an arrangement of rocks, part of a shelter made by these hominids.

These finds demonstrated behavior that anthropologists had not suspected among such early hominids. The animal bones must have been food remains, indicating hunting and possibly scavenging. These early creatures apparently made weapons with which they hunted for food, which they shared as a group at their home base. Most of the remains are nothing more than primitive garbage heaps. As anthropologist

Glynn Isaac, *below,* has a paintbrush to remove dirt from a fossil he found near Lake Rudolf in Kenya. Stone tools, *below right,* were also found at this location.

Angular fragments 32·8

Flake fragments 17·7

Flakes 41·5

Chopper/cores 6·6

Glynn Isaac of the University of California, Berkeley, has put it, one important result of the Olduvai Gorge discoveries was their demonstration that "the habit of creating concentrated patches of food, refuse, and abandoned artifacts is among the basic features of behavior that distinguish the human animal from other primates."

Late in 1963, the Leakeys found bone fragments of a second extinct hominid species in the earliest geologic deposits of Olduvai Gorge. This was named *Homo habilis*, meaning skillful man. The remains of this species include skull, jaw, and limb parts. It was assigned to the genus *Homo*—that of *Homo sapiens*, today's man—because of the size and form of its teeth and because its brain case was noticeably larger than that of any known species of *Australopithecus*. The large cheek teeth and extensive wear on their crowns suggest that the *Australopithecus* probably lived on a coarse vegetable diet. *Homo habilis*, with smaller and less worn teeth, may have had a different diet, possibly one including meat.

Some authorities now suggest that *Homo habilis* was directly related to the older *Australopithecus africanus*, on an evolutionary line that eventually led to modern man. They also suggest that *Australopithecus robustus*, which lived at the same time as *Homo habilis*, became extinct. Other authorities, however, challenge this interpretation. They believe that this species, despite its larger brain, was only an advanced species of the genus *Australopithecus*. They argue that some characteristics of the arm and leg bones and front teeth of the *Homo habilis* species are quite primitive and unlike those of other members of the genus *Homo*. In any case, the Olduvai diggings show that both species continued to exist as late as perhaps a million years ago.

In the layers of earth above the one in which Mrs. Leakey found the 1.75-million-year-old *Australopithecus robustus* skull, she found further remains of early men and their stone tools. Some of these were of the extinct species *Homo erectus*. The highest layer of earth containing fossils of this species is about 700,000 years old. But Olduvai Gorge has an even longer record of early man from about 1.8 million to just a few thousand years ago.

Until as recently as 1967, there was nothing anywhere in the world to compare with the Olduvai discoveries. However, intensive field work at other locations in eastern Africa has since uncovered comparable specimens, as well as the fossil remains of hominids that lived millions of years before those found at Olduvai.

One location is a previously unexplored part of northern Kenya, east of Lake Rudolf. It is in the Great Rift Valley and lies more than 600 miles north of Olduvai Gorge. This area was also periodically layered by silt deposits from Lake Rudolf, and occasionally volcanic ash fell on the lakeshore. The deposits are rich in fossils that have yet to be radiometrically dated, but scientists assume that volcanic ash will eventually be found close enough for reliable potassium-argon studies to be made. After comparing the fossil mammals there with those of known age found elsewhere in eastern Africa, scientists believe that the Lake

Fossil beds in the Omo River Basin, *left,* yielded 3-million-year-old skull fragments, *below,* as well as many teeth and other bone fragments. Workers carefully sift the sandy soil at this site, *bottom,* for the smallest bits of evidence of early man.

Rudolf deposits are from 1 to 2 million years old, about the same as the deposits at Olduvai Gorge.

The Rudolf project was initiated by Richard E. Leakey, the son of Louis and Mary Leakey and the director of the National Museum in Nairobi, Kenya. He and the geologists and archaeologists of his research team have explored two principal areas with hominid fossils near the lake. The Ileret sector to the north is younger and has more fossils than the Koobi Fora sector to the south.

Most of these hominid remains are australopithecines, similar to the robust skull specimen found in 1959 at Olduvai. Because there are slight differences between these and the *Australopithecus robustus* specimens found in the limestone deposits of South Africa, some anthropologists now think that the east African specimens, including the ones found at Olduvai Gorge, are a different species. They have named it *Australopithecus boisei.*

Altogether, over 30 body parts of this species have been found. They include parts of many jaws, one complete jaw with most of its teeth intact, one nearly complete skull and most of two others, and several arm and leg bones. The limbs will enable specialists to determine the posture and gait of this species. The jaw parts have helped to establish variations in size of the different individuals. The two incomplete skulls probably came from females, because they are smaller than the massive complete skulls. This has enabled experts to determine size and other differences between the sexes in this species.

When the fragments found at the Omo River site were fitted together, they formed a nearly complete skull.

As at Olduvai, anthropologists also found skeletal parts of another species of early man at the Rudolf sites. They are less common, but over a dozen have been identified, including a few teeth, parts of upper and lower jaws, and some limb bones. The oldest of these, a lower jaw, resembles the Olduvai *Homo habilis.* All were found in levels that also had the remains of the more common *Australopithecus boisei.*

Two other fossil-bearing places in the Lake Rudolf area of Kenya have been dated by the potassium-argon method, but have not yet yielded any human fossils. The oldest is at Kubi Algi, north of Sibilot mountain. Elephant and pig remains have been found there in layers of volcanic ash and soil that are more than 4 million years old. Another, some distance inland from the present lakeshore and the Koobi Fora Peninsula, has volcanic ashes that are about 2.6 million years old. Isaac, who has directed the excavations there and at the Kubi Algi, found bones of antelopes, pigs, and hippos, some tools, and from 150 to 200 stone chips and flakes from tools that were made there by early man. The abundance of waste stone chips indicates that this probably was a primitive toolmaking factory that produced tools of the Oldowan culture. The 2.6-million-year date added nearly a million years to our knowledge of early man's toolmaking abilities.

North of Lake Rudolf, in the valley of the Omo River in southern Ethiopia, we have found fossil bones of hominids that lived more than 2 million years before those found at Olduvai or Lake Rudolf, and

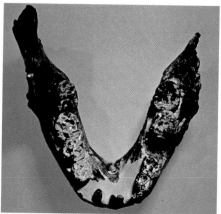

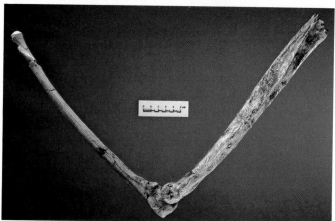

This skull of a female
*Australopithecus boisei,
above left,* was found
by Richard Leakey near
Lake Rudolf. Leakey
also found the lower
jaw of a *boisei, above,*
and the upper arm bone,
left, that fit a lower
arm bone the author
found in the Omo Basin.

more than 1 million years earlier than those who occupied the tool-making site at Koobi Fora. The Omo Basin has the longest known fossil-bearing, radiometrically dated succession in Africa, and probably in the world. Its fossil-bearing sediments are over 2,000 feet thick, and scattered throughout them are a dozen layers of volcanic ash. As a consequence, scientists have dated these sediments using the potassium-argon dating method. One set of these deposits has proved to be older than 4 million years, and others are from 1.5 to 4 million years old. The Omo complex has come to serve as an invaluable dated scale against which other fossil discoveries in eastern Africa can be compared, because its sediments contain a wide variety of fossil vertebrates.

I made my first visit to the Omo area in July, 1969, the month in which Mary Leakey discovered the now-famous Olduvai skull. I was soon convinced that the area had great potential and that its geologic history and fossil deposits should be studied extensively. Such studies were authorized in 1966, when Emperor Haile Selassie took a personal interest in the area after hearing of the work done by the Leakeys at Olduvai Gorge. Since then, the Omo Research Expedition, an interna-

How *Australopithecus africanus* appeared has been generally determined by careful study of his bones and tools he fashioned.

tional group of scientists coordinated by Yves Copens of the Musée de l'Homme in Paris and myself, has worked there each year with a staff that includes about a dozen scientists and 30 to 40 field assistants.

The Omo River has been the principal river flowing into Lake Rudolf for more than 4 million years. At times, the lake and its sediments covered much of the present Omo Basin, while at other times the river meandered back and forth across the valley depositing silt. The river also brought in deposits of ashes from distant volcanoes. In these layers, hominid fossil remains have been discovered at some 70 places in the Omo Basin. Some of these fossils are nearly 4 million years old, and the most recent ones are about as old as similar finds at Olduvai Gorge. Well over 150 teeth, 8 parts of lower jaws, part of a child's skull, and 4 limb bones have been found. The latter, coupled with the discovery of other limb parts of this species in the Lake Rudolf deposits, will undoubtedly give scientists a better understanding of the structure, proportions, and functions of the upper limb of this species. There is increasing evidence that the anatomy and functional capabilities differ in important ways from the genus *Homo*.

As elsewhere, there is evidence at Omo of at least one other hominid species. Teeth have been found at the 2.7-million and 3-million-year levels that, because of their size and shape, are clearly not those of *boisei*, but could be from *Australopithecus africanus*. The oldest occurrence is dated at about 3.5 million years. Between 2.5 and 1.9 million years this hominid is almost unknown, perhaps because it was more rare, or not even present. We surely need more and better-preserved specimens from this time range before we can determine the nature of this other hominid. Stone tools have also been found, but, as elsewhere, it is not clear which hominid made them.

The fragmentary fossil record of man has been extended even farther backwards by the discovery at Lothagam Hill, southwest of Lake Ru-

A Family Tree

Evidence of fossil bones, tools, and other artifacts suggests
this tentative chart of man's evolution over 4 million years.

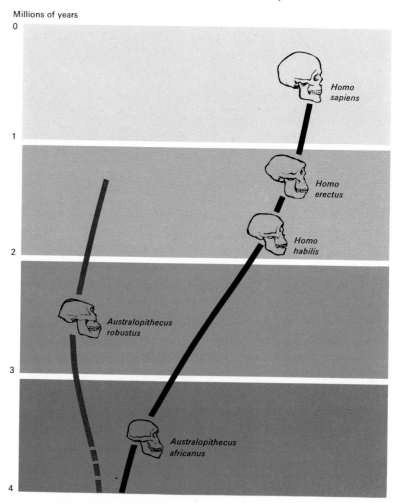

Millions of years

Lothagam Hill jawbone fragment, *left,* 5.5 million years old, is the most ancient hominid fossil yet found. Elbow bone, *above,* more than 4 million years old, was also found at the site.

dolf, where the oldest known fragment of a hominid bone has been found. The Harvard University expedition that made the discovery was headed by paleontologist Bryan Patterson, who also found an important specimen at nearby Kanapoi in 1965. His Kanapoi find, the elbow part of an upper arm bone, is more than 4 million years old. The jawbone fragment found in 1967 at Lothagam Hill is about 5.5 million years old. The fragments are too incomplete to allow scientists to clearly determine what species they came from, though specialists are sure that they are not parts of *Australopithecus robustus.* However, the parts may be from *Australopithecus africanus* individuals or from an unknown species allied to *Australopithecus africanus.* According to Patterson, the jawbone is probably that of a relatively mature female.

These discoveries have substantially expanded our conception of the time scale for man's evolution. They have filled in some of the gaps in the fossil record. And they have shown how complex is the early evolution of the family of man. In the remote past there were times when several species of men existed at the same time, even in the same area, though only one survived and continued to evolve toward modern man. Hopefully, the recovery of more fossil remains will eventually clarify the origins of man and give us an even fuller understanding of his development, a goal that scientists have pursued since the time of Darwin and a subject of interest to every human being.

For further reading:

Campbell, Bernard G., *Human Evolution: An Introduction to Man's Adaptations,* Aldine, 1966.
Howell, F. Clark, *Early Man,* Time-Life, 1965.
Pilbeam, David, *The Ascent of Man,* Macmillan, 1972.

CERN: Experiments
In Physics and People

By Robert H. March

**At this European laboratory, scientists from many
nations find they can work together in their search
for sense among the smallest particles of matter**

The CERN laboratory straddles the border between Switzerland and France, which crosses the laboratory between the proton synchrotron (PS) ring at the right and the intersecting storage rings (ISR) at the left.

Georges Charpak's job would be a smuggler's dream. He crosses an international frontier several times in the course of a day's work, and never stops for a customs check. The border he crosses is unpatrolled and even unmarked.

Charpak is a physicist on the staff of the European Organization for Nuclear Research (CERN), which lies in rolling pastureland at the foot of the Jura Mountains, about 5 miles from downtown Geneva, Switzerland. A half-mile stretch of the border between France and Switzerland cuts through the laboratory grounds.

It has been said that science knows no frontiers. At CERN, this is not merely a figure of speech. But the border is only a hint of CERN's international nature. Even the fact that the laboratory is supported by 12 European nations and affiliated with 3 others does not begin to express its character. A laboratory, after all, is made up of people, and on this level, CERN encompasses the entire world.

Countries Participating in CERN

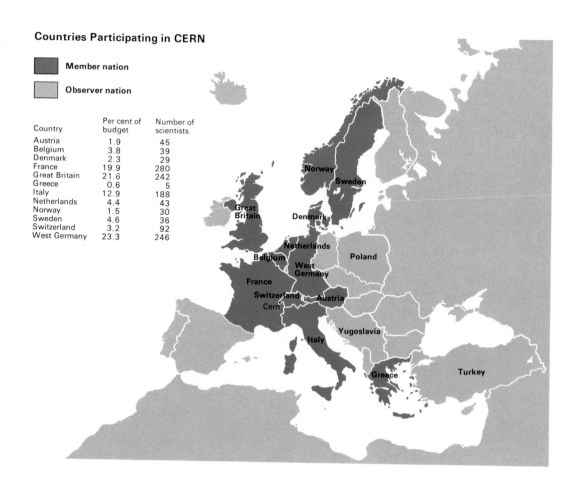

■ Member nation

□ Observer nation

Country	Per cent of budget	Number of scientists
Austria	1.9	45
Belgium	3.8	39
Denmark	2.3	29
France	19.9	280
Great Britain	21.6	242
Greece	0.6	5
Italy	12.9	188
Netherlands	4.4	43
Norway	1.5	30
Sweden	4.6	36
Switzerland	3.2	92
West Germany	23.3	246

The author:
Robert H. March is a professor of physics at the University of Wisconsin and has participated in several experiments at CERN.

The story of CERN is the story of the rebirth of European science, and of a new generation of European scientists. When Georges Charpak began his university studies in Paris right after World War II, most of the laboratories still intact in Europe were neglected and obsolete. Many of the great names that had made Europe the cradle of 20th-century physics graced the faculties of American universities. The shattered economies of European countries could scarcely support science at the prewar level. America had the laboratories, the money, and the people. The Bohemian student life of Paris was exciting, but it was not the best way to become a scientist. Charpak, however, was no stranger to struggle. During the German occupation of France, he was forced to attend high school under the name "Charpentier" to conceal his Polish origins. He was himself a member of the underground unit that prepared his false papers. This clandestine activity eventually led to his imprisonment, and, after two escape attempts, he was sent to the Dachau concentration camp. He was one of the few survivors at war's end.

Charpak then resumed his education and upon graduation from engineering school landed an assistantship in the laboratory of Professor Frédéric Joliot-Curie, then France's most distinguished nuclear physicist. As a result, physics became his field.

Then came CERN. In 1954, urged on by the United Nations Educational, Scientific, and Cultural Organization (UNESCO), the nations of Europe agreed to pool their resources in a bid for leadership in the most glamorous—and expensive—field in physics, the study of elementary particles. By 1956, the laboratory was operating on a modest scale.

Today, CERN is a world center in its field, thanks to the work of people like Georges Charpak, who joined the staff in 1959. Among his recent contributions, the best known is a vastly improved device that serves as the "eyes" a computer uses to track the motions of fast-moving particles. Charpak's device is deceptively simple—a miniature harp strung with wires that are finer than human hair. When a particle passes between two of the wires, it generates a faint electrical signal on the nearest one. The signal, amplified and fed into the computer, records the particle's position to within a few hundredths of an inch. For almost five years, Charpak used much of his time struggling to make this device work reliably.

Such a feat, however, required more than a good idea and hard work. Charpak is the first to admit the task would have been impossible without the help he received from the superb CERN staff. High pay, prestige, and a sense of adventure attract the cream of Europe's engineers and technicians to CERN. Their pride of craft is sometimes annoying to physicists who need a simple device that is just good enough to do the job. "At this laboratory," one American observed, "you ask for a paperweight and they make you a Swiss watch!" But there is no contesting the fact that such intense professional pride has helped make CERN the world leader in developing new hardware for elementary particle research.

For a physicist, a gadget, however sophisticated, is simply a means to solving the oldest mysteries in physics, the nature of matter, space, and time. The quest began more than four centuries ago in the universities of some of the same nations that now support the CERN laboratory. Today, the search requires more than any one university, or even one of these nations, can afford, in money and talent. Thus CERN, out of a

Noon hour allows some of CERN's resident and visiting physicists, *left,* to swap ideas in the cafeteria lounge. For machinists, *below,* it is a chance to play *pétanque,* the French version of a British game called lawn bowls.

Georges Charpak, a CERN physicist, lives in an international setting. At work, *above,* he leads scientists from several countries. At home, *below,* with wife Dominique, sons Yves (seated, second from left), Sarge (standing), and daughter Natalie (second from right), he chats with classmates of Natalie's from countries as near as France or as far away as Cameroon.

practical need, has become a lesson for the world in international understanding. Charpak, for example, shares with German physicist Adolf Minten the leadership of a research team consisting of 20 physicists, engineers, and technicians from 8 nations.

This international flavor carries over into the Charpak home, an ancient frame house gamely clinging to survival amid an advancing army of steel-and-glass high-rise buildings near the center of Geneva. It is not always a quiet place; the three teen-age Charpaks have made it a headquarters for lonely foreign classmates. But a guest feels the warmth of this house from the moment he is greeted at the door by Georges and his artist wife, Dominique.

Until 1971, the Charpaks lived on a mountain road overlooking Geneva, in a strikingly modern house that was designed by Dominique. "We had moved into town so the kids could go to the high school in Annemasse [a French suburb of Geneva]," Charpak explains. "Then we made a discovery. Our lives had been centered around the house rather than around people. This life is more human."

In the cosmopolitan world of particle physics, the CERN cafeteria is what the Deux Magots café in Paris once was for French writers—sooner or later, almost everyone turns up there. It is the main trunk of the

Theoretical physicist Mary Gaillard, *left,*
in CERN's library, can study reports
from every laboratory in the world.
Above, Mary and husband Jean-Marc
teach daughter Dominique to play cards.

grapevine that keeps researchers throughout the world in touch with the latest developments. A visitor from the University of California's Berkeley campus once grumbled that he could find out more about what was going on in his own laboratory by having a cup of coffee at CERN than he could by attending a dozen seminars back home.

The leisurely European lunch "hour" guarantees plenty of time for lively conversation, which picks up again at the close of the day over a cocktail. This convivial life is especially important to theorists, the paper-and-pencil physicists who trade in ideas. An idea born in a London university may pass through the CERN cafeteria to help solve a problem that is bothering a physicist in Bombay. Needless to say, not all the gossip is high-powered shoptalk. Physicists enjoy talking about such topics as politics, music, and sports just as much as anyone else. And since the support staff at CERN is as broadly international as the scientific staff, one may find a Swedish secretary (who speaks English, French, and German as well as her native tongue) arguing heatedly with a French machinist about the relative merits of cross-country and downhill skiing.

The full extent of CERN's internationalism is described by the visiting scientists, who expand the laboratory's horizons around the globe.

Approximately 200 of these visitors add to CERN's permanent professional and supporting staff of about 3,000. Some visiting researchers stay only for a few weeks, while others remain for years. Some visitors work at CERN through formal agreements, such as the one that links the laboratory to its Soviet counterpart at Serpukhov, near Moscow. Others collaborate with physicists from member nations. Any particle physicist on paid leave from his home institution can usually wangle an invitation to work at CERN. A few staff positions are reserved for scientists from nonmember nations.

CERN has kept a tight rein on the number of Americans in long-term positions, however. In the late 1950s, the United States had the lead in elementary particle research, and the Europeans feared that top-level Americans might make the laboratory a U.S. satellite. Today, the shoe is on the other foot. A growing number of young American physicists are beginning their research careers in Geneva. A tour of duty at CERN carries sufficient prestige to boost their chances in the tight U.S. job market. The "brain drain" that threatened European science during the 1950s has been reversed.

One American with deep roots at CERN is Mary Gaillard. She came to the lab in 1962, soon after her marriage to a French physicist, Jean-Marc Gaillard. The Gaillards met when she was a graduate student at Columbia University in New York City and he was a visiting scientist at CERN's American "sister" laboratory, Brookhaven National Laboratory on nearby Long Island.

A Proton Race Track

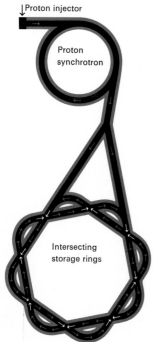

↓Proton injector

Proton synchrotron

Intersecting storage rings

Mary Gaillard has continued her career as a theoretical physicist while rearing three children. That would be a challenge anywhere in the world, but Mary believes that the physical and psychological burdens are lighter in Europe than in the United States. "Europeans don't look down on a woman for leaving young children in someone else's care most of the day," she explains. "Even if they don't have a career, middle-class women here have help with the kids."

Mary is convinced that her marriage to an experimental physicist has made her a better theorist. But the Gaillards have rarely worked together professionally. In 1971, however, they produced their first joint paper. In it, they analyzed the significance of a particularly difficult series of experiments in which Jean-Marc had been involved.

Physicists such as the Gaillards devote their working lives to the study of exotic states of matter, particles created when fast-moving protons collide and convert part of their energy of motion into mass. Few of these particles last as long as a billionth of a second. Yet, they are the key that may someday unlock the secret of how matter is created and held together. The prime tool the physicists use in this effort is CERN's

Protons from the PS, *left,* can be fed to the ISR, where they circle for days. At each intersection of the rings, *opposite page,* a small fraction of the protons collide head-on and produce new particles.

The PS control room, *above,* is CERN's nerve center. Physicists
in an experimental area hundreds of yards away, *above right,* check
their apparatus as it receives particles from the PS. In many cases,
an experiment's end result is thousands of photographs of particle
tracks that must be patiently searched, *bottom right,* for collisions.

proton synchrotron (PS), located on the Swiss side of the laboratory grounds. Its powerful electromagnets bend protons around a circular course, and a device similar to a radio transmitter speeds them up every time around. The protons reach energies as high as 30 billion electron volts (GeV). The PS and one much like it at Brookhaven have dominated particle physics research for more than 10 years.

At CERN, protons from the PS are the principal "merchandise" that crosses the French border. These tiny particles would pose quite a challenge to a customs inspector. Traveling at 99.93 per cent of the speed of light, they can be steered into a pair of overlapping race track courses called the intersecting storage rings (ISR), which lie on the French side of the laboratory. Once in the ISR, the protons engage in a submicroscopic "demolition derby." Where the rings cross, protons collide head-on. At the speeds produced by the PS, a head-on collision between two

A complex network of electronic equipment is required to make an experiment at CERN's ISR produce results.

moving protons is more than seven times as violent, in terms of energy available to create new matter, as when a moving proton hits a stationary one. The ISR, which went into operation in 1971, is CERN's latest triumph. It guarantees that CERN will continue to be a leader in particle physics research, because there is no comparable device anywhere else in the world.

The very fact that devices such as the PS and ISR have no immediate practical value is one reason that CERN could be established. International cooperation would have been out of the question in a field that might have rapid economic or military impact, favoring one country or another. Furthermore, if there was to be any particle physics research in Europe, an international laboratory was the only way. No single European nation could match the enormous sums spent by the United States in this field. Europe had known scientific leadership before, and its scientists were sure that to neglect such a fundamental field would mean accepting permanent second-class status. Research in this area also seemed the ideal choice for a "people experiment," a model for future joint efforts on more practical matters.

CERN may have been a grand idea, but making it a reality proved to be no ordinary feat. National rivalries turned the choice of a site for the laboratory into a diplomatic chess game. The neutral Swiss site was a compromise common in European diplomacy. The problem of money was finally solved by a brilliant arrangement: Each nation contributes in proportion to its national income. This system eliminated all haggling between the supporting nations, and helped to ensure that CERN would grow as Europe prospered.

Some critics gloomily forecast that the birth of CERN, attracting the most talented physicists and engineers, would mean the death of physical science in European universities. However, the reverse has actually taken place. A particle physics experiment is like a scientific expedition. Much raw data can be gathered in a few weeks in the field—at the CERN laboratory—but the real work usually begins later, when the data is analyzed back home. Physicists gathering data at CERN pressured their governments to provide better facilities at their home universities. They argued that otherwise the money invested in CERN would merely finance some other nation's glory. The scramble by universities to be the best equipped to handle CERN data has led to a level of support for particle physics in Europe that has surpassed that in the United States. And the creation of new facilities, particularly computer centers, has enabled this support to spill over into other fields of science.

Not that everything runs smoothly at CERN. Scientists are notoriously competitive, and the added international dimension makes the internal politics there fearsome. Clever experimenters can always come up with more ideas than any one laboratory could possibly test. Any physicist, whether he is at CERN or some other institution, can request time on the machines. Committees of physicists screen these proposals, picking the most promising and those that seem to dovetail with cur-

CERN's research would
be impossible without
secretaries like Norrey
McKenzie, *below,* who
handle paperwork in
several languages, and
skilled technicians such
as Erich Albrecht, *right,*
shown wiring a harplike
particle detector. After
hours, Erich, Norrey,
and friends see Geneva
night life, *bottom.*

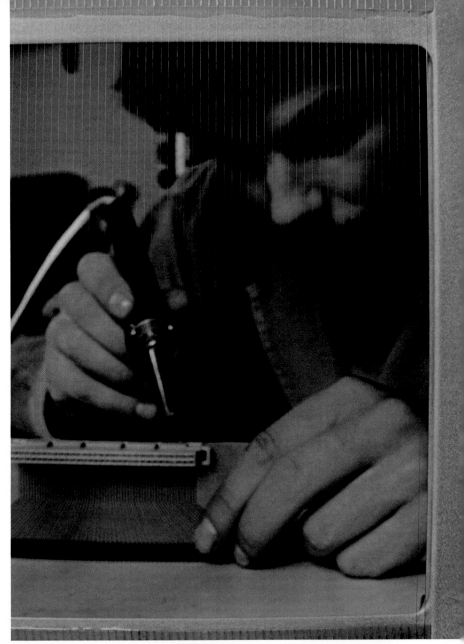

rent experiments. Budget and laboratory space limit the experiments to about 50 a year. The committees must also try to give each member nation a reasonable share of experiments. Because of this, the proposals are sometimes monstrous. Huge international teams have been formed for some experiments, more for reasons of political muscle than out of any sensible estimate of the manpower required.

All of these problems make the job of Willibald K. Jentschke, CERN's director-general, a ticklish one. A German physicist, he is the fifth man to hold the post in the laboratory's 18-year history. Most laboratory directors have to satisfy only one government. Jentschke must convince CERN's 12 that the $100 million they contribute to the laboratory each year is spent well. Furthermore, he must deal with national rivalries that sometimes surface. There are whispers that a certain division of the laboratory is run by a "French mafia," or that Italians get more than their share of the jobs in another division. Jentschke must see to it that, as far as is possible, these rumors have no basis in fact. Finally, he must cope with European professors, who by long-standing tradition are absolute lords in their own bailiwicks and are not inclined to defer to a mere director-general.

Given this array of potential pitfalls, it is a marvel that CERN has worked so well. Part of the credit should go to its host city. "Being international" is Geneva's major local industry. Geneva's 200,000 citizens regularly play host to nearly 100,000 foreigners. The city houses organizations as diverse as the World Council of Churches and the International Labor Organization. Film stars and oil-rich Arab sheiks use it as a haven from taxes and coup d'états. Corporations with international interests enjoy the flexibility provided by Swiss banks and the city's multilingual labor force. Outcasts—ranging from Protestant reformer John Calvin to Russia's V. I. Lenin—have taken refuge there.

Come nightfall, The Duke of Wellington, an English-style pub in Geneva's old town, is one good place to look for some of the younger members of the CERN staff. To Norrey McKenzie, 22 years old and a new secretary at CERN from England, "It's a bit like home, but the ceiling is far too clean." Norrey has studied both French and German, and specializes in translating technical jargon.

On the CERN staff, such talents as Norrey's are not regarded as exceptional. She is typical of many young people who flock to Geneva. Some are drawn by the charm of the city itself, small enough to walk across, but more cosmopolitan than many great world capitals. Others are attracted by the natural setting, where the swift-flowing Rhône River leaves Lake Geneva and begins its descent to the Mediterranean Sea. Summers are cool and winters mild. Some of the finest ski slopes in Europe are close enough for a day's outing. In fact, the CERN personnel bureau reports that ski injuries are a major cause of absenteeism at the lab. Mont Blanc, Europe's highest peak, can be seen on clear days and provides a popular joke about Geneva's weather: "If you can see Mont Blanc, it's about to rain. If you can't, it's already raining."

Erich Albrecht is another "foreigner" in Geneva, even though he is Swiss. He comes from German-speaking Interlaken, where he began working eight years ago as a 16-year-old apprentice machinist. In one year as a technician on the Charpak-Minten team, he feels he has discovered his craft. "Before, I would spend weeks making over and over some part of a machine I would never see. Here, I can make the whole thing, and if it doesn't work, well, it's my fault. If it does, I really have something to be proud of." Erich's pride is in the best CERN tradition, and so is his skill. While he was being photographed for this article, Erich noticed the photographer struggling with an awkward frame that holds gelatin filters for the camera. By the next day, Erich had come up with an improved frame design.

Restlessness led Erich first to a gardener's job in Surrey, south of London, where he learned English, and now to Geneva, where he is polishing up his French. Experience at CERN opens doors throughout the world, so it is likely that Erich will one day move on.

Although the people at CERN come and go, it is clear that CERN itself is in Geneva to stay. In fact, the laboratory is undergoing the greatest expansion in its history. Work has begun on a giant $310-million accelerator that will have a ring of magnets more than 5 miles in circumference. It will boost protons from the proton synchrotron up to 300 GeV. In Geneva's narrow, crowded valley it would be out of the question to acquire enough land to completely enclose this project. The U.S. Atomic Energy Commission had to acquire almost 7,000 acres of land west of Chicago for the National Accelerator Laboratory, where a similar ring began operating in 1972. The new CERN accelerator will be entirely underground, leaving most of the roads, farms, and woods intact. The project is so vast that it required a new treaty among the member nations. The giant new machine has its own director-general, Englishman John Adams, who supervised the building of the proton synchrotron over a decade ago.

So, CERN will continue its quest with new projects and new faces. What its scientists will turn up cannot be foretold. Particle physics may one day enrich the lives of all mankind, not by increasing man's mastery over nature, but by providing profound and beautiful insights into nature. Then again, it may not; science has followed blind alleys before. For the time being, the giant machines near Geneva are little more than monuments to man's curiosity. But CERN, in its human dimensions—blending different skills, different outlooks, and different tongues—can already claim success. The Georges Charpaks, Mary Gaillards, Norrey McKenzies, Erich Albrechts, and thousands of others like them, are living testimony to that achievement.

For further reading:

Casper, Barry N., and Noer, Richard J., *Revolutions in Physics* (Chapter 16), W. W. Norton and Co., 1972.
Jungk, Robert, *The Big Machine*, Scribner, 1970.
Ne'eman, Yuval, "The Strongest Force," *Science Year,* 1968.

Science File

Science Year contributors report on the year's major developments in their respective fields. The articles in this section are arranged alphabetically by subject matter.

Agriculture

Anthropology

Archaeology
Old World
New World

Astronomy
Planetary
Stellar
High Energy
Cosmology

Biochemistry

Books of Science

Botany

Chemical Technology

Chemistry
Dynamics
Structural
Synthesis

Communications

Computers

Drugs

Ecology

Education

Electronics

Energy

Environment

Genetics

Geochemistry

Geology

Geophysics

Medicine
Dentistry
Internal
Surgery

Meteorology

Microbiology

Neurology

Oceanography

Physics
Atomic and Molecular
Elementary Particles
Nuclear
Plasma
Solid State

Psychology

Science Support

Space Exploration

Transportation

Zoology

Agriculture

United States corn production in 1971 and 1972 staged a comeback from the peril of southern corn leaf blight (*Helminthosporium maydis*), which caused huge losses in 1970. America's farmers harvested a corn crop 35 per cent larger than 1970 and 16 per cent larger than the previous high in 1967. They almost completely abandoned the previously grown "T male-sterile" cytoplasm seeds, which are susceptible to the blight, in favor of resistant types. Adequate quantities of the blight-resistant seed corn were produced in 1971 to supply all needs for 1972.

Elsewhere in the world, the dwarf, high-yielding grains continued to perform impressively. India, for example, grew 23 million tons of wheat, an all-time record for that nation. The impoverished and war-torn new nation, Bangladesh, produced a million more tons of rice than it did the year before.

New varieties of plants included an experimental corn hybrid (R802A x R 805) that contains 80 per cent more oil than ordinary corn. It was developed by Denton E. Alexander, University of Illinois corn breeder. The world's first seedless hybrid pickling cucumber (MSU 394G x MSU 4108H) was introduced by horticulturist Larry R. Baker of Michigan State University. The new cucumber represents the ultimate in yields, because every flower is a potential fruit that can develop without any need of pollination.

Tamu Roma, a new sweet sorghum, developed for potential sugar production, was released jointly by the Texas Agricultural Experiment Station and the U.S. Department of Agriculture (USDA) Food Crops Utilization Laboratory at Weslaco, Tex. A completely male-sterile soybean line was announced by Charles A. Brim, a USDA soybean breeder at North Carolina State University. This development may lead to the production of hybrid soybeans.

South Korea had a new cool climate dwarf rice (IR-667) developed by crop breeder Mun Heu Hue of the College of Agriculture at Seoul National University. With this new variety, South Korean farmers anticipated a 30 per cent increase in rice production.

Giant African snail, shown life size, that is causing crop damage in Florida sometimes grows to a length of 12 inches. Three or four of them can devour a head of lettuce overnight. Unchecked, they could cause $11 million a year in crop damage.

Hybrid sunflowers were tested for the first time in Texas as a possible alternate crop to cotton. Sunflower seeds are becoming increasingly important in the production of oil. For example, an estimated 500,000 acres of sunflowers were planted in 1972 in the Red River Valley of North Dakota and Minnesota.

Pest control. Through the pioneering efforts of plant breeder Wendell Roelofs of Cornell University's New York State Experiment Station in Geneva, pheromones became available in 1971 and 1972 for field studies. Pheromones are specific sex attractants that entice insects to a specific place and thus permit early insect-population estimates. This data can be used for more timely, efficient, and effective spraying.

Bacillus thuringiensis, a bacterial insecticide applied in spray solution, is now being tested on more than 20 agricultural crops by entomologist William G. Yendol and his associates of Pennsylvania State University. The bacillus infects and kills such insects as gypsy moths, cabbage worms, and flies.

Researchers reported progress in developing juvenile hormones to control insects that are harmful to man. At the Stauffer Chemical Company, Mountain View, Calif., entomologist Julius K. Menn and chemists Firenc M. Pallos, Peter E. Letchworth, and John B. Miaullis identified new chemicals that resembled and acted like insect juvenile hormones. Such chemical substances prevent the insects from maturing and thus breeding normally.

Plant growth. Verle Q. Hale of the University of California in Los Angeles used the growth regulant triiodobenzoic acid (TIBA) to increase alfalfa seed yields by 60 per cent. Treated plants had more seed-bearing stems.

Ethephon (2-chloroethyl phosphonic acid) was used by growers to regulate the maturity of walnuts so that they would all be ready for harvesting at the same time. The chemical was also used to induce flowering in pineapples, and to regulate the harvesting of sour and sweet cherries.

Plant cleansers. Agricultural Research Service (ARS) scientists in Fort Collins, Colo., declared that the leaves of crop and fruit plants cleanse the atmosphere and can absorb ammonia. Soil scientists Gordon L. Hutchinson,

Lynn K. Porter, and Frank G. Viets of the Soil and Water Conservation Division of the ARS reported in February, 1972, that growing plants may be a natural sink for atmospheric ammonia arising as a pollutant from livestock feed lots. They said that this nitrogen, in turn, contributes significantly to the total amount of nitrogen used by the growing plant community. Corn, soybeans, cotton, and sunflowers all absorb enough ammonia from the atmosphere to promote plant growth.

Plant pathologist Paul R. Waggoner of the Connecticut Agricultural Experiment Station also reported in 1972 that plants purify the atmosphere. He said that all terrestrial plants, including crops, remove ozone from the air.

John H. Hart, plant pathologist at the Michigan Agricultural Experiment Station, and plant pathologist Eugene B. Smalley of the University of Wisconsin independently reported a new chemical that controls Dutch elm disease. This disease, during the past 10 years, has almost wiped out the American elm in the United States. The two scientists said that spraying the chemical on the leaves can eradicate the disease. The spray consists of 8 pounds of Benomyl (methyl 1-[butylcarbamoyl]-2-benzimidazole-carbamate) mixed in 100 gallons of water. Benomyl is a fungicide that has been used to control many other diseases that attack flowers, fruits, vegetables, ornamental plants, and lawn grass.

Feed additives, especially diethylstilbestrol, which are extensively used in animal feeds, came under attack by the Food and Drug Administration (FDA) in 1971. Rations containing diethylstilbestrol, an estrogenic compound, increase the rate of gain in cattle up to 15 per cent and feed efficiency by about 10 per cent. However, a number of limited experiments showed that laboratory animals fed with estrogens subsequently developed tumors.

Legislation has now been introduced in Congress to ban or severely restrict the use of diethylstilbestrol in beef production. This action would undoubtedly result in substantial increases in the price of beef. The use of liquid estrogens in cattle rations was banned in March, 1972. Estrogens can still be used in dry form, which comprises 75 per cent of the total amount used in feed.

Agriculture

Continued

On Feb. 1, 1972, the FDA also proposed the eventual ban of antibiotics from poultry, hog, cattle, and sheep feed. The agency explained that bacteria in the animals might become resistant to the antibiotics and transfer this immunity to other organisms that attack human beings.

Animal disease control was highlighted by a new vaccine developed by veterinarian Charles Mebus of the University of Nebraska for prevention of calf scours, a form of diarrhea. The vaccine is administered to pregnant cows 60 to 90 days before calving.

A new experimental vaccine that may bring leptospirosis under control was developed at the National Animal Disease Laboratory in Ames, Iowa. This disease, often fatal, resembles jaundice and is caused by tiny leptospira organisms. It has long plagued the cattle industry. Leptospira organisms are commonly found in wild animals and are readily transferred to domestic animals and, also, to man.

A widespread tapeworm parasite in cattle, which induces measles, has responded to a new vaccine. The vaccine was developed by Leonard W. Dewhirst, a University of Arizona animal pathologist.

Meanwhile, the eradication of hog cholera in the United States appeared possible. The U.S. government now rates 34 states as free of the disease. The only new outbreak was in the lower Rio Grande Valley in Texas.

The U.S. Department of Agriculture launched an intensified vaccination program to eradicate brucellosis in swine and cattle by the end of 1975. This disease affects the reproductive organs, and leads to abortions and sterility.

The first outbreak of Venezuelan equine encephalomyelitis (VEE) in North America was diagnosed on June 29, 1971. Spread by mosquitoes and biting flies from Mexico, it killed 1,000 horses in Texas. By August 3, 1.1 million horses had been vaccinated in five Southwestern States. Also, in the largest spray program ever attempted in the United States, 8.7 million acres were sprayed with malathion and dibrome for mosquito control.   [Sylvan H. Wittwer]

Anthropology

A 200,000-year-old cranium, face, and massive jawbone found in France are together called Tautavel Man.

Although east Africa continued to be the scene of exciting *hominid* (manlike) fossil finds, Europe and Asia received their share of the headlines in 1971 and 1972. The most interesting European specimens, known collectively as Tautavel Man, were recovered between July, 1969, and July, 1971, from a cave near the village of Tautavel in southern France. The site is being excavated by Henry and Marie-Antoinette de Lumley of the University of Aix-Marseilles.

More than 20 levels, each littered with bone fragments and stone artifacts, have been found. They date back to the beginning of the Riss glaciation, about 200,000 years ago. Hominid remains recovered include many isolated teeth, finger and toe bones, skull fragments, two *mandibles* (jawbones), and the front portion of an adult cranium.

The mandibles are massive and very broad. They seem to combine the features of two other very ancient European specimens, the Mauer (Heidelberg) and Montmaurin jaws, which date back respectively about 500,000 and 200,000 years. The cranium has large brow ridges, a receding forehead, and a flattened brain case. The De Lumleys believe Tautavel Man is more advanced than *Homo erectus*, who lived during the Middle Pleistocene age, about 500,000 years ago, but more primitive than Neanderthal Man, who lived between 200,000 and 40,000 years ago.

A new Neanderthal specimen, the mandible of a 5- to 6-year-old child, was found on Sept. 6, 1970, during road-building operations near Reggio di Calabria on the southwestern tip of the Italian peninsula. It was first described in May, 1972, by Antonio Ascenzi of the University of Rome and Aldo G. Segre of the University of Messina. Dated to the Upper Pleistocene age, which is about 100,000 years ago, the jawbone is described as having a receding chin and other typical Neanderthal features.

The Sangiran area of central Java, Indonesia, famous since 1937 for the recovery of *H. erectus* (Java Man) remains, has produced another fossil brain case, this one known as Pithecanthropus VIII. The specimen was found in 1969, but was not made public outside Indonesia until the spring of 1972. This is only the second specimen of the series to be found outside the geological formation known as the Sangiran Dome in beds that date back to the Middle Pleistocene age. It is still embedded in sandstone, but enough of the thick, low vault of the skull is visible to suggest it is another member of the *H. erectus* species.

The Omo project in southern Ethiopia, directed by F. Clark Howell, University of California, Berkeley, produced several new hominid specimens in 1971. The most interesting is an *ulna* (forearm bone) about 2 million years old. It resembles neither modern man nor any of the living great apes. See Our Earliest Ancestors.

Hominid 24 from Olduvai Gorge, Tanzania, found by P. Nzube in October, 1968, has now been studied by Louis S. B. Leakey of the National Museum of Kenya. The specimen comes from Bed I, just below a level that has been dated to 1.8 million years ago. It was found embedded in a matrix of limestone, badly crushed and distorted, and reconstruction proved difficult. Leakey identifies it as a member of the genus *Homo* but cannot decide if it should be called *habilis* because the occipital bone of the back of the skull is more elongated than that of another specimen he has called *habilis*. Others prefer to call the skull *Australopithecus africanus*, a more advanced version of the hominids found at the Sterkfontein site in South Africa since the 1920s.

One difficulty in interpreting East African hominid remains may have been cleared up in 1971 by Frederick S. Szalay of Hunter College in New York City. The specimen in question is the cranium from Chesowanja, Kenya, dated to between 1.2 and 1.1 million years old. It was originally interpreted as a large-brained australopithecine of the "robust" sort. Szalay reinterpreted the crushing and distortion of the skull, and presents a new reconstruction, a typical, small-brained *A. robustus*.

A *femur* (thighbone) with both ends broken off and a pelvis with the front part missing were taken from Bed IV at Olduvai Gorge in 1970. Called Hominid 28, the specimens were found associated with Acheulean industry tools (hand

"Maybe *you're* descended from a lemur,
but *I'm* not descended from *any* lemur!"

Drawing by W. Miller; © 1972 The New Yorker Magazine, Inc.

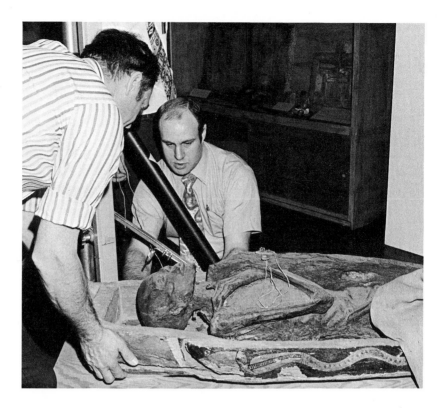

University of Michigan scientists prepare the mummy of an ancient Egyptian noble from their museum of archaeology for study with portable X-ray equipment. These studies can reveal evidence of diseases such as arthritis and arteriosclerosis.

Anthropology

Continued

axes and cleavers) and date to about a half-million years old. They have now been studied by anatomist Michael H. Day of the Middlesex Hospital Medical School in London. The femur, flattened at its upper end and unusually straight, is very similar, according to Day, to the *H. erectus* femurs found at the Choukoutien site, near Peking, China, in the 1930s.

On the basis of this study, Day questions the antiquity of the famous Trinil femur found in central Java in 1891. Thought to be about a half-million years old and attributed to *H. erectus*, it is indistinguishable from modern femurs. The pelvic bone differs in several respects from that of modern man, but no other pelvic remains from *H. erectus* are available for comparison.

Felix, the knuckle-walker. The debate continues as to whether man evolved through a stage of using the knuckles of his hands to support body weight during locomotion, much as the chimpanzee and gorilla do today. The orang-utan, another close relative of man, never developed the specializations of the finger bones for easy knuckle-walking.

However, Felix, an obese old orang-utan at Brookfield Zoo near Chicago, confounds the experts by knuckle-walking around his cage. In March, 1972, Russell Tuttle of the University of Chicago, an opponent of a knuckle-walking stage for man, pointed out that Felix rarely supports most of his weight on his knuckles, and thus cannot be called a true knuckle-walker. See ZOOLOGY.

Eskimo cheekbones. For many years, scientists have puzzled over why the Eskimo's face differs markedly from that of the American Indian despite the fact that the two peoples are closely related. In the Eskimo, the lower margins of the *zygomatics* (cheekbones) are everted and the mandible is lower and angled backward more sharply than that of the Indian. William Hylander of Duke University concluded after studying X rays of 265 adult Eskimo and Indian skulls, that the Eskimo face is an extreme evolutionary adaptation to powerful vertical forces exerted on the teeth. Eskimos are noted for using their teeth to bend kayak ribs, crack bones, and even to open soda pop bottles.   [Charles F. Merbs]

Archaeology

Once again, modern construction projects led to surprising archaeological discoveries in 1971 and 1972. These discoveries and continued excavation of previously discovered sites in Europe, Africa, Asia, and North and South America added greatly to man's knowledge of his early history.

Old World. The grave of an Iron Age chieftain buried with all his possessions was uncovered during quarrying operations in June, 1971, at Garton, in Yorkshire, England. The burial site has since been systematically excavated and studied by archaeologist Tony Brewster of the Inspectorate of Ancient Monuments of Great Britain.

The chieftain, apparently a tall, well-built man of about 35, was buried beneath his wooden chariot, which had 12-spoke wheels with bronze hoops around the hubs. The grave also contained a harness for his chariot horse, a whip, a dagger with a gilded bronze hilt and several other of the chieftain's weapons. In addition, there was a pig's skull that may have been the remains of a food offering.

The burial has been dated to about 200 B.C., and British archaeologists described it as the greatest find of that period made in England during this century. The elaborate nature of the burial indicates that the dead man was an important person. Only eight such graves have been found in the past. Most of these were excavated during the 1800s, before the exacting excavation and dating techniques of today were in use. As a result, archaeologists learned relatively little from them.

Early in 1972, archaeologists, under the direction of Peter Marsden, chief archaeologist of the Guildhall Museum in London, worked feverishly to excavate the remains of Baynard's Castle on the north bank of the River Thames in London. The area was scheduled for the construction of a new City of London Boys School and a new road.

The castle, which had once served as the home for three of the six wives of King Henry VIII, was nearly destroyed in the Great London Fire of 1666. It was originally built by Ralph Baynard, a nobleman who came to England with

Archaeologists seeking the ancient Phoenician city of Sarepta found a small stone seal, *above*, with the name of the city inscribed on it. Site of the find, *right*, is south of Beirut, Lebanon.

Stolen History

In the fall of 1971, a West Coast dealer in antiquities offered an elaborately carved Mayan stele for sale. His price for this stone monument from an archaeological site in Guatemala was $350,000. In January, 1972, Federal Bureau of Investigation agents seized the piece. After court action, it will probably be returned to Guatemala.

This episode highlights one of the most complex and critical problems in archaeology today. Because archaeological objects are selling at such high prices, systematic looting of archaeological sites has become a big business in many parts of the world.

Such looting is not new, however. The Romans robbed Etruscan tombs, and many other generations have had their grave robbers. In fact, archaeologists rarely find a tomb intact. But never has looting reached the proportion it has in the last decade.

Authorities frequently cite among the causes a desire to own rare things, and the theory that art collections are sound investments during a period of inflation. Whatever the reasons, they all contribute to a problem that may soon destroy most of the archaeological sites containing objects of art. The illicit trade in archaeological objects must currently run into at least several hundreds of millions of dollars each year.

The Mexican government placed the problem before the United Nations Educational, Scientific, and Cultural Organization (UNESCO) in 1960. A UNESCO document, prepared in 1970, requests museums not to buy or accept antiquities for which the place of discovery is not known. The document was ratified by the United States and other governments in 1971. But it does not solve the problem completely.

Almost all nations now have strict laws against destroying archaeological sites and exporting antiquities. Yet stopping the looters, in the face of the profits to be made, is an almost impossible task. Complicating the problem is the legitimate trade in art objects that have been in the hands of private collectors and museums for generations. How do you distinguish between an ancient stone sculpture that is being sold from the estate of a 19th-century collector, and one that was recently stolen from a temple at Angkor Wat?

The United States and Mexico ratified a treaty in February, 1970, that ensures the return to Mexico of important sculptures or frescoes stolen and found in the United States after that date. Similar treaties are now being negotiated by the United States with other Latin American countries. However, these treaties cannot hope to affect the trade in small objects that can be smuggled and sold more easily than larger items.

Protecting remote sites, such as those in the Guatemalan jungle, is too much for local police to handle alone. But the seizure of the stele and its return to the country of origin shows that something can be done about the problem. In this case, archaeologist Ian Graham of Harvard University recognized the Mayan stele as one he had found at Machaquila, a remote, little-known site. He surmised that the huge object had been cut up in pieces, smuggled out of Guatemala, and then reassembled for sale in the United States.

The return of the stolen stele serves as a clue to more effective control, if such action can be worked out on an international basis. If every professional archaeologist who knows an area and its important monuments would alert authorities when he finds an illegal sale, dealers would not invest large sums of money in smuggling.

Treaties and conventions are also necessary to establish just what constitutes legal and illegal trade. But to place effective brakes on the illicit antiquities trade, an international system of communication among concerned, knowledgeable persons is also essential. For this reason, archaeologists, museum directors, the heads of national antiquities services, and others from many countries will meet in Cyprus in September, 1972, to form an international communications network. Hopefully, this will help to make illicit trade too hazardous and expensive to continue to proceed at its present critical rate.

For most of us who care about research into man's past, the important issue is not whether a country retains all of the ancient art objects found there. Rather, it is the consideration of whether the present scramble for prized pieces will destroy the irreplaceable records of human history. [Froelich Rainey]

Archaeological assistant Vanessa Knight examines skeleton of an Iron Age chieftain found near Garton, England. The man was buried with his chariot and other possessions, probably in about 200 B.C.

Archaeology

Continued

William the Conqueror in 1066. About 1428, it was destroyed and rebuilt, and historians say that Parliament met there in 1648. Shakespeare refers to the structure in Act III, Scene 5, of *Richard III*.

During the excavation, archaeologists found the remains of a large tower and five smaller ones. There were also some shoes, Roman pottery, and what appears to be a court jester's bell.

Africa. Archaeologist Glynn Isaac of the Department of Anthropology at the University of California, Berkeley, discovered many surface scatters of stone tools, including hand axes, cleavers, and flake tools east of Lake Rudolf in northern Kenya. At two sites excavated, tools were dated between 2 and 2.6 million years ago by a potassium-argon age-determination method. Other rocks, apparently carried to the site from some distance away, and the bones of animals used as food suggest that these were home-base sites. Until this discovery, the earliest dated stone tools, about 1.8 million years old, came from the Olduvai Gorge to the south of this site. See OUR EARLIEST ANCESTORS.

Chinese discoveries. Archaeologists in China reported in March, 1972, that they had discovered two huge tombs of medieval princes near the tomb of Empress Wu Tse-tien in Shensi Province. This empress ruled from A.D. 690 to 705. One of the tombs is that of her son and heir apparent, Chang Huai, and the other that of her grandson, Yi Teh. The largest was more than 300 feet long, 50 feet deep, and 12 feet wide. They looked like underground palaces and included more than 100 well-preserved wall paintings and more than 1,000 valuable relics.

In December, 1971, well-preserved oracle bones made from the shoulder blades of oxen were unearthed near Anyang, in Honan Province. One of them bears an inscription of at least 60 words recording imperial sacrificial ceremonies that were used during the Shang Dynasty in about 1500 B.C.

Japanese murals. Colored murals discovered in a burial mound in central Japan in March, 1972, were hailed as one of the most important archaeological discoveries in that country since

Archaeology

World War II. The richly colored pictures on the tomb walls are believed to have been completed about A.D. 700. They depict white tigers, blue dragons, men and women in Chinese-style dress carrying sticks, the sun, the moon, and constellations showing the North Star.

The burial mound is near the village of Asuka, south of Nara, one of Japan's ancient capitals. It was excavated by members of an archaeological research institute in Nara, under the direction of Masao Suenaga.

Fake objects. In the July, 1971, issue of *Archaeometry*, the journal of the Research Laboratory for Archaeology and the History of Art at Oxford University in England, archaeologists reported that many examples of Neolithic pottery now in museums of Europe and the United States are elaborate modern fakes. Thermoluminescent tests performed by Martin Aitken of the laboratory; Richard Moorey of the Ashmolean Museum, Oxford; and Peter Ucke of University College, London, showed that 48 of a group of 66 objects were counterfeit. [Judith Rodden]

New World. In 1972, the School of American Research of Santa Fe, N. Mex., began systematic excavation at Arroyo Hondo, north of Santa Fe, under the direction of archaeologist Douglas W. Schwartz. Arroyo Hondo is a 1,200-room pueblo that was abandoned many centuries ago. The research team hopes to find artifacts from the initial period of occupation and to determine the ecological effects of the growing civilization on the pueblo's inhabitants.

Raymond S. Baby of the Ohio State University Museum excavated an Indian Hopewell culture house at the Seip Village site near Chillicothe, Ohio, in 1971. He found broken flint knives and many mica chips on the house floor. This suggests that an artisan lived there who made mica figurines similar to those that are found in many of the burial mounds of these early Indians.

Until this excavation, some archaeologists doubted that the Hopewell people lived within, and adjacent to, their great burial mounds and ceremonial earthworks. The posthole pattern

An archaeologist works in the shadow of highway construction vehicles on a prehistoric site in northwestern Florida. Scientists fear that hundreds of such sites have been inadvertently churned into oblivion.

Archaeology

Continued

The remains of a prehistoric Indian chief, with ceremonial flint mace resting on his chest, were found in Missouri's Lilbourn Village in 1971.

indicates that the house was 38 feet long and 35 feet wide. Excavation of other Hopewell houses discovered in the same area in 1971 began in 1972.

Robert P. Graveley of the Archaeological Society of Virginia uncovered part of a prehistoric Sioux village in the Dan River Valley of Virginia in late 1971. The village contained both rectangular and circular houses and a number of graves. In all cases, the dead were buried with the knees drawn up to the chin. This was probably done to make their bodies more compact.

The graves contained small pottery vessels and necklaces made of marine shells. Animal remains as well as the remains of corn and beans gave archaeologists an indication of the diet of the village's inhabitants.

Archaeologists from Arizona State University at Tempe excavated a pueblo at Hunters Point, in 1972, that was occupied from about A.D. 925 to 1325. The record of their findings, including a description of seeds, animal remains, and bits of pottery, were relayed by Sylvia Gains of the university's staff to the uni-

versity's computer, in which it underwent further statistical tests.

R. Gwinn Vivian, an archaeologist from the Arizona State Museum at Tucson made an extensive study of the prehistoric water-control system that was once erected by Indians in the Chaco Canyon, in New Mexico. He found remnants of canals, diversion dams, control gates, and gridded cornfields, indicating a much more developed irrigation system than had previously been thought possible. Aerial photographs revealed traces of a network of paths that once led to the canyon.

Excavations at the Towasahgy State Archaeological Site in Mississippi County, Missouri, in 1972, revealed the settlement plan of this fortified community that was built about 1300. John W. Cottier, an archaeologist for the State Park Board, found the remains of a wall enclosure, bastions of the wall, temples on top of platform mounds, and houses built by the thriving society once there.

A short distance to the south, another fortified prehistoric town, the Lilbourn Village, was being excavated by Carl E.

Archaeology

Chapman, an archaeologist at the University of Missouri. He found human skeletons, as well as the remains of a wide variety of corn, squash, and beans, birds, fish, and other animals. Some of the human bones showed evidence of disease.

On the chest of one adult male, who was buried in a special grave, there was a large flint mace more than a foot long. The mace was probably a prehistoric priest chief's symbol of authority. This is the first find of this type in a Missouri excavation. An earthen mound shaped like the mace, and apparently a ceremonial structure, was also found at the Lilbourn site.

Lyle M. Stone, archaeologist with the Mackinac Island State Park Commission in Michigan, in 1972, found an abandoned campsite of hunters and fishers who lived about A.D. 500. The campsite was found under the ruins of Fort Michilimackinac.

South and Central America. Ramero Matos, an archaeologist at the University of Huancayo in Peru, located about 600 caves, shelters, and open sites with traces of human occupation in 1971.

They were high in the Andes Mountains in Junín. He has since found that 10 of the caves have primitive paintings of hunting and pastoral scenes. Searchers who dug down 10 feet in one cave found bones of two extinct species of animals.

Matos found stone tools at the sites that are similar to the tools used by three of the cultures from the Peruvian highlands. They were first identified in excavations made by Richard S. MacNeish of the Robert S. Peabody Foundation.

Roman Pina Chan, an archaeologist of the Mexican National Museum, began excavating and restoring a pyramid site at Tenango, west of Toluca, Mexico, in 1972. The pyramid and plaza, which were uncovered in 1971, were occupied from A.D. 800 to 1500. It will take several years to restore this large, complex site.

The skull of a man believed to be about 9,000 years old was found in a cave near Tlapacoya in the Valley of Mexico, in 1972. It was discovered by Lorena Mirabell of the Prehistoric Department of the National Institute of Anthropology and History in Mexico City.   [James B. Griffin]

Astronomy

Planetary Astronomy. The epic mission of the Mariner 9 spacecraft dominated the year in planetary astronomy. The craft was inserted into orbit about Mars on Nov. 13, 1971, and returned high-quality scientific data through September, 1972, that has changed many previously held views of Mars. See MARS IS AN ACTIVE PLANET.

Martian dust storm. When Mariner 9 arrived at Mars, it found the surface almost totally obscured by a yellow haze. According to Charles F. Capin and Lowell J. Martin of Lowell Observatory, this dust storm began on Sept. 22, 1971, near Hellas at about 30° south latitude and moved rapidly westward around the planet at a speed of about 19 miles per hour. The storm did not dissipate until early January, 1972. By the time it reached its height in early November, it had completely obscured all surface markings on the planet, much to the dismay of Mariner 9 experimenters.

Several general characteristics of the storm, the first to be spectroscopically observed in detail, are now available.

Truman D. Parkinson and Donald M. Hunten of the Kitt Peak National Observatory in Arizona detected less than half the normal amount of carbon dioxide in the atmosphere in late November. This indicates that the dust must have extended well over 6 miles above the surface. This is also borne out by the early Mariner 9 pictures, which could only discern faint features around Nix Olympica, a 6-mile-high volcanic mountain, and the south polar cap.

Outer planet satellites. Because of a lack of fundamental physical data at very high temperatures and pressures, the internal structure of objects in our solar system remains unresolved. However, in October, 1971, John S. Lewis of the Planetary Astronomy Laboratory of the Massachusetts Institute of Technology (M.I.T.) provided a stimulating discussion on the gross internal properties of the giant satellites of the outer planets. The largest—Callisto and Ganymede, satellites of Jupiter; Titan, a satellite of Saturn; and Triton, a satellite of Neptune—all have planetary dimensions but relatively low mean densities.

Astronomy

Continued

Lewis theorizes that at the time the solar system formed, about 5 billion years ago, conditions in the solar nebula at and beyond the orbit of Jupiter caused a solar-proportion mixture of elements to condense. The resulting material is largely ice of ammonia and methane—a type of cosmic snow. Investigation of internal processes that could reasonably occur in large bodies composed of this ice-and-rock material showed the possibility of large-scale melting and differentiation in the interior. This could occur from even mild sources of internal energy, such as natural radioactivity or dissipated tidal energy. Smaller objects would remain frozen throughout.

Lewis depicts a giant satellite's internal structure as a thin, icy crust enclosing a deep, slushy, liquid mantle surrounding a dense, rocky core. This model clearly could have significance in understanding Saturn's rings.

Occultations. The *occultation* (hiding from view) of a bright star by a planet is a rare and important event. The modulation of the stellar light sweeping through the planet's atmosphere serves as a very fine probe of the structure of that atmosphere. If there is no atmosphere, then measurements of the interrupted starlight provide the planet's dimensions and shape, as well as detailed information about the star.

Jupiter occulted the bright star system of Beta Scorpii on May 31, 1971. As an added bonus, one of the four stars that make up the Beta Scorpii system was occulted by Jupiter's satellite Io.

The University of Texas astronomy department mounted a major expedition to observe these highly improbable events. Eleven scientists watched from observatories in Africa, India, and Australia. Their efforts were amply rewarded with six high-quality occultation light curves caused by Jupiter and two caused by Io.

The occultation data revealed:
- The mean height and a lower temperature limit of the upper atmosphere.
- The base of the Jovian *thermosphere* (the upper region of the atmosphere where temperature increases with increasing altitude).

Hiding a Bright Star

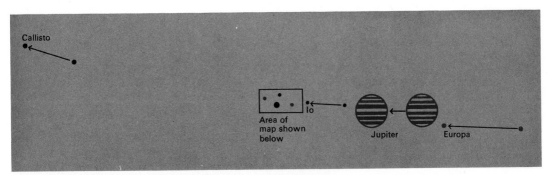

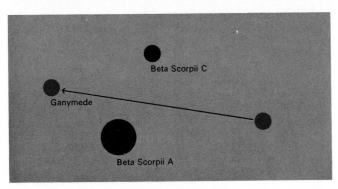

A "near-miss" occultation of the bright star system Beta Scorpii (βSco) was recorded at Kitt Peak National Observatory in May, 1971. Arrows show how Jupiter and four satellites moved in relation to the fixed stars of Beta Scorpii (inset).

The Copernican Celebration

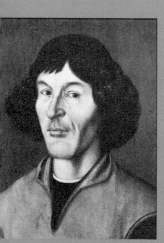

Nicolaus Copernicus

Scientists around the world will commemorate the 500th birthday of Nicolaus Copernicus in 1973. To mark the quinquecentennial, the U.S. National Academy of Sciences will publish a popular book on the overthrow of scientific preconceptions. Scholars will assemble for conferences on the nature of scientific change, on the astronomy of Copernicus, and on science and society. A major international symposium will convene in September, 1973, at Toruń, Poland, the astronomer's birthplace. Many countries, including the United States, will issue commemorative stamps.

The "Copernican Year" is more than the 500th anniversary of an ingenious astronomer. It is nothing less than a celebration of the origin of modern science. It is a festival in honor of man's half-millennium of progress in understanding the physical universe.

The illustrious Polish astronomer was born on Feb. 19, 1473. After his education at the university in Kraków, he spent his adult life in northern Poland serving as a canon for the Cathedral of Frauenburg (now Frombork). There, in 1543, he wrote a book, *Concerning the Revolutions of the Celestial Spheres*, that was destined to overthrow the long-accepted view that the earth was the center of the universe.

Copernicus' idea that the earth revolved around the sun was not new–the Greek astronomer Aristarchus had proposed it nearly 2,000 years earlier. However, Copernicus was the first man to work out the numerical details so that the positions of planets could be predicted in a sun-centered system.

We are so familiar with the idea today that we forget it is not at all obvious. To our common senses, the earth seems very solid and motionless; it is the sun that seems to move across the sky. Copernicus' book, published the year he died, was thus like a delayed time bomb. It was a formidable treatise with little popular appeal. "Mathematics is for mathematicians," Copernicus warned in the preface. Although other scholars soon recognized it was the most competent and profound book on astronomy written in more than a thousand years, few of them were convinced by its novel sun-centered cosmology. "I cannot admire enough those astronomers who accepted Copernicus in spite of the evidence of their senses," wrote the Italian astronomer Galileo in 1632.

Even today, few people can give sound arguments for a sun-centered system. In the 1500s, this was even more difficult. Thus, the second greatest astronomer of that century, the Dane Tycho Brahe, rejected the idea of a sun-centered universe. Although he was one of the greatest observers of all time, he saw no compelling reason to adopt Copernicus' idea.

Curiously enough, a popular legend has grown up about how corrupt and complicated astronomy was in the days before Copernicus, and how this great Polish astronomer simplified and corrected it with his new cosmology. During the reign of King Alfonso X of Spain, the earth-centered system supposedly was so complicated that 40 small circles, or epicycles, were required to explain the motion of each of the other planets. "If I had been around at the time of Creation," Alfonso is reputed to have told his astronomers, "I could have given the Good Lord some hints."

I recently used a large electronic computer to demolish this legend. I carefully recomputed all of Alfonso's astronomical tables to show that his astronomers used a comparatively simple single-epicycle theory that went back to Claudius Ptolemy, an Alexandrian astronomer of the second century A.D. Then, I used Alfonso's tables to find the planetary positions for the century before Copernicus. A comparison with the best almanacs made prior to 1550 proves that only the simple Ptolemaic theory was used.

To be sure, Ptolemy's theory was not very accurate. The positions for Mars, for example, were sometimes wrong by nearly 5°–approximately the distance between the "pointer stars" in the bowl of the Big Dipper. But when I checked the planetary positions predicted by Copernicus, they were nearly as bad as those from the Ptolemaic theory. Furthermore, the Copernican scheme was even more complex.

Why did Copernicus adopt a sun-centered model when it did not agree any better with the observations than the old common-sense earth-centered system did? I believe there are several reasons. In the first place, the earth moving in orbit provided a natural explanation for the puzzling retrograde, or backward motion, periodically observed for the other planets. According to Copernicus, the planets appeared to move backward because they were being overtaken by the faster-moving earth.

Secondly, the dazzling sun was obviously different from the other planets, so a central, fixed position seemed appropriate. And, finally, when the sun was given the central place, the arrangement of the planets became particularly harmonious, with the fastest-moving Mercury nearest the sun and the slower-moving Saturn—the most distant planet then known—at the outer boundary of the planetary system. In other words, Copernicus preferred a sun-centered system because it was more beautiful. As he said, it was "pleasing to the mind."

Although the sun is accepted today as the center of our planetary system, we know it is not standing fixed in space. Rather, it whirls about the center of our Milky Way galaxy at about 125 miles per second. Is our galaxy itself rushing through space? Perhaps, but according to Einstein's theory of relativity, the answer makes no difference. We can consider any reference system as "fixed."

Why, then, does it make any difference that Copernicus said the sun, not the earth, stood still? Before Copernicus, both celestial and terrestrial motions were explained in terms of sympathies and strivings inherent in matter itself. The stars turned in their daily circuits about the earth because a circular motion was the most "natural" one for the eternal heavens. In contrast, the fundamental concept of modern science is the "law of nature" that describes the workings of the physical universe in mathematical terms.

The Copernican theory helped scientists to see the harmonious arrangement of the planetary periods. This gave them a clue that one special physical object, the sun, could be linked to the planetary motions by a mathematical law of gravitation. In this way, modern physical science was born. [Owen Gingerich]

Astronomy

Continued

- Several apparently stratified layers in the atmosphere.
- A very low limit on the total amount of atmosphere on Io.
- The radius of Io and Jupiter far more precisely than any previous estimates.

The star occulted by Io, component C of the Beta Scorpii system, was found to be a double star whose companion is about 10 times fainter.

This data is expected to have enormous impact for understanding processes in the Jovian atmosphere. The more precise radius determined for Io lowers the best mean density estimate enough to suggest it could be more like one of Lewis' "snowball" models than like the earth's moon.

Venus and Mars topography. The thick clouds and atmosphere that surround Venus make it impossible to observe its surface in visible light. However, since the atmosphere is less opaque at radio wave lengths, scientists can use this technique to directly probe the surface. George H. Pettengill of M.I.T. and his associates at the M.I.T. Haystack Observatory and Cornell University's Arecibo Observatory in Puerto Rico reported in February, 1972, that they measured the equatorial topography to an accuracy of ± 150 meters.

The Venus measurements show a peak-to-peak range of about 2.5 miles in altitude, with one mountainous area attaining a height of 1.8 miles above the average surface height. This is considerably less than the 8-mile range on Mars. Nevertheless, the range is surprising since scientists had predicted no substantial topography because of the high surface temperature and the resulting plasticity of the planet's crust.

This radar capability also enabled A. E. E. Rogers and Irwin I. Shapiro of M.I.T. to make exceptionally precise measurements of the topography on Mars, even determining the topographic structure of individual craters. In the region of Iapygia there is evidence for an extremely large crater. It is more than 1,200 miles in diameter and more than 1 mile deep. The crater was scanned at three different latitudes by the radar which gave a detailed map of its structure. [Michael J. S. Belton]

267

Astronomy

Stellar Astronomy. The seventh and largest Orbiting Solar Observatory (OSO) of the National Aeronautics and Space Administration (NASA) was almost ruined by a bad launch on Sept. 29, 1971. A malfunction in the second stage of the Delta rocket placed OSO-7 in an elliptical orbit instead of its planned circular orbit. More seriously, the satellite was spinning rapidly and unable to point at the sun to draw needed solar power.

As the operators at Goddard Space Flight Center attempted to correct OSO-7's control system, they fought a continual drain on the spacecraft's batteries. More than 2,300 commands were sent to OSO-7 during the next $8\frac{1}{2}$ hours until the problem was corrected. When the spacecraft finally locked onto the sun, battery voltage had dropped to within 0.2 volt of failure.

The scientific payload of OSO-7 included instruments developed by astronomers at two government laboratories and at the Massachusetts Institute of Technology, the University of New Hampshire, and the University of California, San Diego.

One of these devices is a white-light coronagraph. It has a boom extending along its optical axis, with a disk at the end of the boom to block out the sun and simulate a solar eclipse. On December 13, a sensitive vidicon camera, in conjunction with the coronagraph, recorded the spectacular eruption of two hot gas clouds into the solar system. The clouds, each about 30 times the earth's diameter, were expelled from the sun's corona, or faint outer atmosphere, at speeds of 600 miles a second.

The events were probably connected with the breakup of a bright streamer in the corona. Nondisruptive streamers have been photographed from the ground during total eclipses of the sun, but have not been studied in detail until now because of the brevity of the natural eclipses.

Solar polar caps. Using another OSO-7 instrument, Werner M. Neupert of Goddard mapped the X rays emitted by highly ionized atoms in the lower corona. Neupert found a systematic pattern in the emissions, which indicated that the corona is much cooler over the poles (about 1 million degrees centigrade) than over the solar equator

(2 million°C.) Thus, even the sun, in a sense, has "polar caps."

Underground solar observatory. Meanwhile, new data from an underground solar observatory led to a major conflict between theory and experiment. Raymond Davis, Jr., of Brookhaven National Laboratory reported in September, 1971, that the number of neutrinos from the sun that were intercepted a mile deep in the Homestake mine in the Black Hills, near Lead, S. Dak., was less than one or two particles per month. The detection rate is roughly 10 times less than expected from the accepted model for the solar energy-generation mechanism.

Davis' equipment, essentially a giant tankful of perchlorethylene (C_2Cl_4) can detect the small percentage of neutrinos that strike atoms of the chlorine 37 isotope, converting them to argon 37. The 105,000-gallon vessel, 20 feet in diameter and 48 feet long, was put in the mine to shield it from the interfering effects of cosmic rays.

Obviously, either the solar interior or the neutrino has properties much different from those calculated by astronomers and physicists.

Binary stars, regarded in recent years as a fairly classical branch of astronomy, assumed new importance during the year. Observers reported strange evolutionary behavior and the discoveries of radio and X-ray radiations from binaries.

In September and October, 1971, Alan H. Batten of Dominion Astrophysical Observatory in Victoria, B.C., and Miroslav Plavec of the University of California, Los Angeles, summarized and deciphered observations taken by many astronomers in the 90 years since the brightness changes of the binary system U Cephei were discovered. Now we know that this object is an eclipsing binary system in which the two member stars seem to be gradually drawing apart and slowing in their orbits. Gas is streaming from the small star, a type-G subgiant, and striking its larger companion. This has accelerated the rotation of the companion, a main sequence B star, until its equator now spins at 310 kilometers per second, five times the expected rate.

Even more remarkable is the conclusion that the relative proportions of the

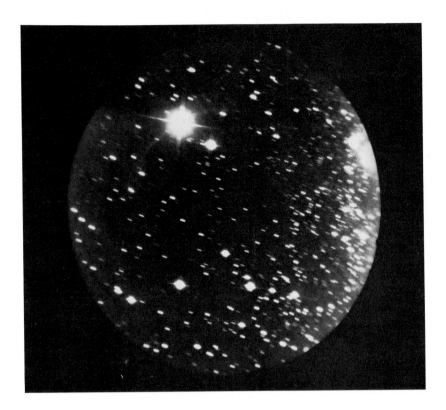

The center of the Milky Way was photographed from the moon's surface in far-ultraviolet light by Apollo 16 astronaut John W. Young.

Astronomy

Continued

two stars have reversed. At one time, the present subgiant was much larger, with about five times the mass of the sun; its companion had about three solar masses. But over a period of some 45 million years, the large star expanded. It lost matter, which accumulated on the companion. As this process continued, the larger star became the smaller one, and vice versa.

In radio astronomy as well, the year's progress was highlighted by studies of binary stars. In June, 1971, radio emission—found earlier to arise in the vicinity of the bright-red star Antares—was shown to originate instead in the faint blue companion of this binary star. Radio signals were also detected from the famous eclipsing binaries Algol and beta Lyrae, and from the binary star HDE 226868, which was also found to coincide with the X-ray source Cygnus X-1. Leading this work were Robert M. Hjellming and Campbell M. Wade of the National Radio Astronomy Observatory.

But perhaps the best news for radio astronomers was the announcement on March 15, 1972, that the National Science Foundation had decided to build the very large array (VLA) project. The VLA will consist of 27 radio reflectors, each one a fully steerable dish 82 feet in diameter. Arranged in a 39-mile-long, Y-shaped pattern, it will be built on the Plains of San Agustin, west of Socorro, N. Mex.

Infrared photo success. Astronomers Karen and Stephen Strom of the State University of New York, Stony Brook, used a two-stage electronic image intensifier to search for groups of stars in the earliest stages of formation. At this stage, they are still surrounded by thick shells of dust and would be radiating most of their energy in the infrared (IR).

The Stroms struck "pay dust" in late 1971 with an IR photograph of the region of the star BD +40°4124. At least four IR stars were recorded. Interpreting the stars' properties in terms of the theory of stellar evolution, the Stroms estimated that this small group of stars formed less than 500,000 years ago.   [Stephen P. Maran]

Astronomy

Continued

During the 200 days the astronomy satellite Uhuru spent scanning the skies, its X-ray detectors observed more than 125 X-ray sources above and below the plane of the galaxy.

High Energy Astronomy. Developments in X-ray astronomy in 1971 and 1972 continued at a rapid pace. A group of scientists at American Science and Engineering (ASE) in Cambridge, Mass., headed by Ricardo Giacconi, compiled the first all-sky catalog of X-ray sources from observations made with the Uhuru satellite, an orbiting X-ray observatory launched in December, 1970.

In joint papers, one on galactic X-ray sources delivered by Harvey D. Tananbaum on May 11, 1972, and one on extragalactic X-ray sources presented in Madrid, Spain, on May 12 by Edwin M. Kellogg, the ASE scientists identified more than 90 new X-ray sources. Approximately 35 sources were found before Uhuru was launched. Of the more than 125 sources known, 15 are associated with unusual galaxies, or clusters of galaxies, and about 40 have been identified with optical and radio objects. These include peculiar stars and double stars; supernova remnants; normal galaxies, including the Andromeda Nebula; radio galaxies; a quasar; and a pulsar.

The catalog off the galactic plane was incomplete, however, covering only 70 of some 200 days Uhuru spent scanning the sky. The scientists suspect that most of the still-unidentified sources above and below the galactic plane will turn out to be extragalactic.

X-ray observations revealed the complex nature of the southern hemisphere X-ray source Centaurus X-3, which has not yet been identified optically. It was found to pulsate intensely every 4.8 seconds. This short period indicates that the X rays come from a small object, possibly a white dwarf or a neutron star. Initially, small variations in the period suggested a pulsarlike source. The periods for pulsars generally increase, however, while both increases and decreases were recorded in Centaurus X-3. Further study has shown that the changes in the period themselves are also periodic.

It is now clear to scientists that these changes are not due to intrinsic changes in the X-ray source itself but are the result of the orbital motion of the X-ray source around another star. The orbit

Some X-Ray Sources Observed by Uhuru

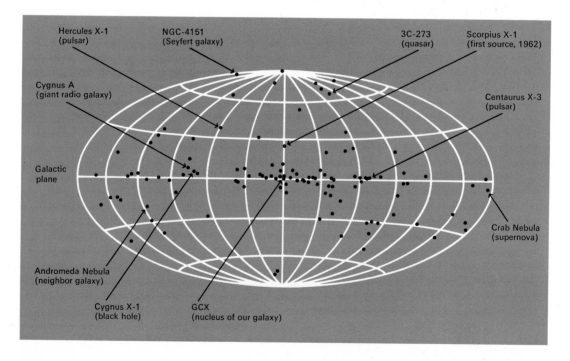

Hercules X-1 (pulsar)
NGC-4151 (Seyfert galaxy)
3C-273 (quasar)
Scorpius X-1 (first source, 1962)
Cygnus A (giant radio galaxy)
Centaurus X-3 (pulsar)
Galactic plane
Crab Nebula (supernova)
Andromeda Nebula (neighbor galaxy)
Cygnus X-1 (black hole)
GCX (nucleus of our galaxy)

appears to be nearly circular with a radius one-tenth of the distance from the sun to the earth. The mass of the other star must be no less than 15 times the mass of the sun.

If we view the system from some point in the plane of the orbit, we would expect that the X-ray source would be occulted (blocked from view) when behind this star and, in fact, this turns out to be the case. During each period, the pulsating X-ray source completely disappears for half a day. Scientists have determined from the duration of the eclipse that the radius of the other star is at least 10 times that of the sun.

A similar pulsating source has been discovered in the constellation Hercules. There, the pulsation period is only 1.2 seconds, while the orbital period again is about two days.

The X-ray variations for Cygnus X-1, a binary system at a distance of about 5,000 light-years, do not show any clear periodic behavior. Thus, we cannot determine the orbital properties from the X-ray data. However, Cygnus X-1 has been optically identified with a B supergiant—a star presumably very much like that associated with Centaurus X-3. The velocity of this star is variable with a period of 5.6 days. This, too, is due to orbital motion, with the big star and the X-ray source apparently both orbiting around their common center of gravity.

From the orbital data, the mass of the X-ray source is greater than 2.5 times the mass of the sun. Because this is larger than a white dwarf or neutron star can be, scientists speculate that they may have finally found one of the long-sought "black holes."

The association between double stars and X-ray sources has given new momentum to accretion models. When one of the components in a double star begins to evolve and expand, it may dump part of its mass onto its companion. If the companion is very compact, the accreting matter will fall on its surface with a large velocity and, as a result, become hot enough to emit X rays.

Crab Nebula emission. Scientists have been unable to explain how X rays are emitted by the Crab Nebula, the remnant of the supernova of the year 1054. Two possibilities have been considered—emission from a hot gas and emission from electrons with relativistic energies in a magnetic field (synchrotron radiation). The latter process, as concluded from its polarization, accounts for most of the optical radiation of the nebula. A rocket experiment conducted in late 1971 by physicists from the Columbia Astrophysics Laboratory under Robert Novick measured the polarization of the X rays. The observations indicated that the X rays result from radiation from relativistic electrons. This confirmed the presence of very energetic electrons (10,000 billion electron volts [GeV] or even more) within the nebula.

Because such electrons lose their energy in about a year in the magnetic fields of the nebula, they must be continuously replenished. The pulsar in the nebula appears to be the only possible source. Thus, one more link has been made between pulsars and energetic particles and cosmic rays.

Double stars also created excitement among radio astronomers. Robert Hjellming and Campbell Wade at the National Radio Astronomy Observatory, in Green Bank, W. Va., discovered that several double stars, among them the companions of Antares and Algol, emit radio waves. The latter star is an eclipsing binary in which the two components periodically occult one another. Their radio emission is strongly variable and some spectacular radio flares have been reported by observers.

Radio emission has also been detected from another binary, Cygnus X-1, as well as from a few other X-ray sources. The origin of these emissions is still obscure, however.

Energetic cosmic rays. Four physicists from Tokyo University reported observing a cosmic-ray event of record energy in late 1971. On the basis of measurements of a shower of electrons and muons that extended over a large area, the Japanese physicists concluded that the event must have been initiated by a cosmic-ray particle with an energy of 4,000 billion GeV. However, the precise value of this energy is still uncertain.

Such energetic cosmic rays are very rare. They are significant because they are not deflected very much by galactic magnetic fields, and, as a result, they may give us clues as to the places of origin of cosmic rays.   [L. Woltjer]

Cosmology. A series of papers published in 1972 by Allan R. Sandage of the Hale Observatories summarized more than 40 years of work on the relationship between red shifts and the distances of remote galaxies. Ever since the first clear statement by Edwin P. Hubble in 1929, this relationship has been the most basic for all cosmology. It shows the most fundamental thing that we know about the universe—that it is expanding—and gives the time scale for its evolution.

The new work of Sandage and others has resulted in new values for the expansion and supports the "big bang" theory of the evolution of the universe.

Except for a few of the very nearest, all galaxies and galaxy clusters show a shift of their color spectra to the red, indicating they are moving away from us. The velocities of galaxies increase in proportion to their distances; thus, velocity = distance x H, the proportionality constant, called the Hubble constant. Except for small local fluctuations, this simple relationship holds in all directions. The motion of galaxies described by these observations is one of general expansion—the distance between any two galaxies increases with time.

Calculated backward in time, the observed motions imply some finite time in the past (whose value is $1/H$) when the distances between galaxies approached zero and the density of matter approached infinity. This was the time of the big bang. Of course, the expansion rate probably has not been constant. Except for very early times, the interaction between the parts of the universe is gravitational attraction, which always tends to slow down the expansion.

The braking effect of gravitation is expressed by the "deceleration parameter," q_o. For the observed part of the universe, it is roughly one-half the ratio of gravitational potential energy to kinetic energy. Astronomers measure it by observing deviations from the Hubble constant for extremely distant objects. Besides giving a correction to the Hubble age, it allows cosmologists to predict the future course of the expansion. If q_o is larger than one-half, the expansion will eventually slow to a halt, then reverse to a collapse. This means a "closed" uni-

verse. If q_0 is smaller than one-half, there is not enough gravitating matter to completely halt the expansion. This would be an "open" universe.

The key to accurately measuring H is to find the distances of galaxies that are up to 200 million light-years away. The general expansion velocity at such a distance is large compared with their random velocities of a few hundred kilometers per second (km/second). Sandage and Gustav Tammann of the University of Basel in Switzerland attacked the problem by placing a sequence of progressively fainter objects along the distance scale from the Cepheid variable stars (within 10 million light-years of earth) to a special class of faint spiral galaxies, which can be measured well beyond 200 million light-years. In late 1971, the two astronomers arrived at a value for H of 17 km/second velocity for every million light-years of distance.

The value of the change in this velocity with distance (q_0) determined from red shifts and the brightness of faint, even more distant galaxy clusters, is less well known. Therefore, we do not yet know whether the universe is open or closed, although the value of $q_0 = -1$, required by the steady-state theory, seems definitely ruled out.

The new value for H gives a Hubble time scale $1/H$ of 18 billion years. When the new value for q_0 is considered, the age of the universe is about 10 billion years, with an uncertainty of perhaps a factor of 2. This agrees with the ages of the oldest known stars (about 10 billion years), the earth (about 5 billion), and the chemical elements (about 10 billion).

The faint cluster work demonstrates that the intensity of the brightest galaxy in a cluster of galaxies is the same, from cluster to cluster, to within about 30 per cent. With improved instruments astronomers promise more rapid measurements of faint galaxy red shifts and brightnesses and, using the brightest cluster galaxy as a standard candle, we can expect a much improved value for q_0 within a few years.

Uhuru measurement. Independent evidence that q_0 may be small (near zero) and the universe open, has been inferred from a measurement, reported in May, by the Uhuru satellite of X-ray emission from a rich cluster of galaxies in the constellation Coma Berenices, near the north galactic pole. J. Richard Gott III of Princeton University and James Gunn of the California Institute of Technology (Caltech) used the measured X-ray intensity to set an upper limit for the mass of hot gas within the cluster. They then used the rate at which the intergalactic medium falls into the cluster to set an upper limit for the density of intergalactic gas. The key argument was that infalling gas would be heated to about 10 million degrees by compression and would emit X rays. They concluded that the intergalactic gas is not dense enough to provide the gravity needed to close the universe. The new result does not by itself establish a small value for q_0, but it argues against one possible form for the required mass if the universe is closed.

Quasars. Margaret and Geoffrey Burbidge, Philip M. Solomon, and Peter Strittmatter, all of the University of California, San Diego, comparing the positions of 47 quasars with the positions of nearby bright galaxies concluded that, in four cases, quasars were near the galaxies. This was too many to be due to chance. They suggested that the quasars are physically associated with the nearby galaxies, even though there are large differences in their red shifts.

This work, however, was followed by a similar study by John and Neta Bahcall of Princeton University and Christopher McKee of the Hale Observatories and Harvard University, in April, 1972, using a larger sample of quasars. This second study concluded that the effect seen by the University of California group did not exist in the larger sample.

The apparent faster-than-light expansion velocity of the radio components of the quasar 3C-279, reported in 1971, was confirmed and a second case was discovered—quasar 3C-273. Both observations were made by a National Radio Astronomy Observatory-Caltech-Cornell University group, using the transcontinental "Goldstack" (the Goldstone antenna in California and the Haystack in Massachusetts) interferometer. A number of possible explanations of these objects have been offered, other than noncosmological red shifts, but until one of these can be tied to the observations in a unique way, the observations will provide perhaps the best ammunition for astronomers who argue that quasars are relatively nearby.   [Jerome Kristian]

Biochemistry

In January, 1972, two biochemists in Canada reported finding some unsettling biochemical properties of nordihydroguaiaretic acid (NDGA). The principal chemical constituent of the leaves and stems of the cresote bush, NDGA is added to human foods at levels as high as 0.2 per cent to prevent spoilage by oxidation.

Chidamdaram Bhuvaneswaran and Krislinamurti Dakshinamurti, working at the University of Manitoba in Winnipeg, tested NDGA on both cells and whole organs in several animals including rats and dogs. The experiments showed in all cases that lower concentrations of NDGA than those added to food inhibit oxygen uptake by *mitochondria* (energy-producing elements in the cell). Thus, the substance interferes with respiration, one of the most important of life's chemical processes. The acid evidently binds to the surface of the mitochondria. The bond must be strong, because washing with sugar solution does not seem to remove the substance.

The scientists warned that there is a potential danger in continuing to use NDGA in human food, particularly when it leads to continuous or gross exposure. However, it appears that NDGA may be of some value in attempts to learn the precise nature of energy-producing reactions in mitochondria.

Several earlier studies had indicated that NDGA is active in cells kept alive in the laboratory. For example, it almost completely stops carbohydrate metabolism in a number of cancer tissues.

Bhuvaneswaran and Dakshinamurti point out that their findings are consistent with these experiments. Any substance that inhibits energy production could well inhibit the rapid cell growth and division of many cancer cells.

Ribosome model. In September, 1971, biologists Yoshiaki Nonomura, Günter Blobel, and David D. Sabatini of Rockefeller University in New York City proposed a model for the three-dimensional shape of a ribosome. Ribosomes are the site of the synthesis of proteins, such as enzymes, in the living cell. All ribosomes appear to be constructed according to the same general plan, with minor variations in size and in the

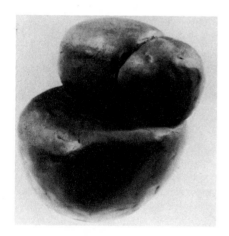

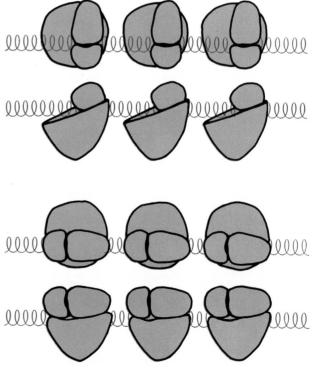

A model that showed a ribosome in three dimensions for the first time, *above*, suggested two ways that these structures might "read" RNA strands. Top and side views, *right*, show how the RNA (color) may be threaded through the ribosomes.

Biochemistry

Continued

quantity of their chemical constituents. They are known to consist of two subunits of unequal size.

The experimenters studied ribosomes from rat liver cells. They treated free *polysomes* (chains of ribosomes commonly found in cells) from these cells with the antibiotic puromycin. This caused the polysomes to separate into their individual ribosomes.

The scientists used negative contrast staining, in which the ribosomes were mixed with uranyl acetate and dried, to prepare the ribosomes for inspection. Uranyl acetate is neither absorbed nor chemically bound to a biological surface. Consequently, it merely surrounded the ribosomes' membranes, outlining their surface. Also electrons do not penetrate it, making it an ideal stain for subjects to be examined through an electron microscope.

Even with this technique, ribosomes are too small to be seen in great detail through the most powerful electron microscopes. However, the Rockefeller University scientists studied many ribosomes trying to relate changes in what little they could see to different ribosome positions. They got additional detail by studying changes in the appearance of a ribosome tilted through 60° angles.

The scientists then constructed a three-dimensional Plasticine model of a ribosome. In this model, the smaller subunit of the ribosome is roughly spherical, elongated, and slightly bent. An indentation, or cleft, divides the subunit across its width into two regions of unequal size.

The larger subunit resembles the larger half of an egg cut across its width. The smaller subunit sits on top and close to the outer edge of the flat surface of the large subunit.

On the basis of their model, the scientists speculated on how a ribosome's structure relates to its function. In protein synthesis, the ribosome somehow holds and "reads" instructions from a molecule of a type of ribonucleic acid (RNA) called messenger RNA. The instructions determine the makeup of the protein being synthesized. The scientists reasoned that the long stringlike RNA molecule is attached to and runs along the ribosome either at the cleft in the smaller subunit or through a channel between the small and the large subunits.

Nerve-muscle junctions. In June, 1971, biologists Alan J. Harris, Steve Heinemann, David Schubert, and Helgi Terakis of The Salk Institute for Biological Studies in La Jolla, Calif., reported a study of the effect of contact between a nerve cell and a muscle cell. Most impressive was their development of a promising system for the detailed study of nerve-muscle junctions.

The normal, intact muscle cell has a restricted area that is highly sensitive to the substance acetylcholine. Acetylcholine transmits the impulse to contract from the motor nerve to the muscle cell it serves. The highly sensitive area of the muscle cell is on its membrane at its junction with the nerve cell. If the nerve is cut, this great sensitivity disappears, and the whole membrane of the muscle cell becomes somewhat sensitive to acetylcholine.

Harris and his colleagues isolated and grew rat muscle cells and nerve cells in separate laboratory cultures. Then, the scientists mixed the cultures, and inserted minute microelectrodes into some of the muscle cells. Next, they applied acetylcholine to various sites along the membrane of the muscle cells, and recorded the resulting changes in muscle membrane electrical potential as registered through the microelectrodes. They found that their cell system closely paralleled what is found in intact animals. The membrane of muscle cells without nerve-cell contact showed a uniform overall sensitivity to acetylcholine. However, the membrane of muscle cells that had become fused with nerve cells was highly sensitive to acetylcholine at the point of contact.

Muscle contraction. Actin and myosin, two proteins, interact with adenosine triphosphate (ATP), an energy-providing molecule, in muscle cells to contract a muscle. In December, 1971, biophysicists Richard W. Lymn and Edwin W. Taylor, both of the University of Chicago, reported experiments that showed how the contraction may be achieved. Within the cell, actin and myosin filaments fold into one another like the fingers of two hands, and ATP is bound to the actin molecules. When a nerve serving the muscle conveys the impulse to contract, calcium ions stored in the muscle cell are released. This triggers projections from

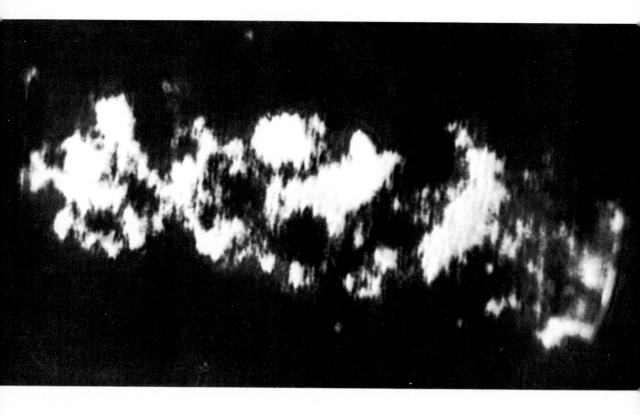

Biochemistry

Continued

This unusual close-up photomicrograph of a virus was taken by Albert V. Crewe of the University of Chicago and deblurred through a new process by physicist George Stroke of the State University of New York at Stony Brook. This may be the first time the true shape of a virus has been observed.

Biology
See Biochemistry, Botany, Ecology, Genetics, Microbiology, Neurology, Zoology

the myosin filaments to alternately attach to and detach from adjacent actin filaments forming transient chemical bridges between the actin and the myosin. These bridges pull the actin filaments into the myosin filaments in such a way that the muscle shortens.

In this position, however, the myosin molecules use energy released from ATP to restore the calcium content of the muscle cells. The muscle fiber then returns to its original, relaxed state.

In November and December of 1971, two scientists reported finding actin and myosin or substances very similar to them in a variety of other cells and tissues, all of which move in some way. Biologists R. E. Fine and D. Bray of the University of Cambridge in England reported that about 20 per cent of the total protein in chick nerve cells is either actin or a substance very similar to it. They also found actin or actinlike proteins in the eye lens, lung, skin, heart, pancreas, kidney, and brain tissue of chick embryos.

Biologist Nadyd Gawadi, also of the University of Cambridge, found actin

associated with threadlike structures that move chromosomes in locust testis, cells that divide by the process known as meiosis. Biologist O. Behnke of the University of Copenhagen and A. Forer and J. Emmersen of the University of Odense in Denmark reported finding actin or an actinlike protein in the same structures in crane fly testis cells and in crane fly sperm tails.

Investigators Robert S. Adelstein, Thomas D. Pollard, and Walter M. Kuehl at the National Institutes of Health in Bethesda, Md., isolated actin and myosinlike proteins from thrombosthenin—a group of contractile proteins in human blood platelets. It is probably involved in coagulation.

Most of the proteins in these cases have not been identified except by their general properties. Further tests will show if they are actin and myosin, or how closely they are related to these contractile proteins. It now seems probable that actin and myosin, or at least actinlike and myosinlike proteins, account for much or all movement in many living organisms. [Earl A. Evans]

Books of Science

Here are 50 outstanding new science books suitable for general readers. The former director of libraries of the American Association for the Advancement of Science selected them from books published in 1971 and 1972.

Agriculture. *To Feed a Nation: The Story of Farming in America* by J. J. McCoy. A history of the advancement, problems, and achievements of American agriculture, with an ecological perspective. The book also provides insights on the control of crop pests and diseases. (Nelson, 1971. 193 pp. illus. $4.95)

Anthropology. *Easter Island: Island of Enigmas* by John Dos Passos. Reviews the observations of people who have visited Easter Island, and it speculates about the great stone heads and about the origin of the island's inhabitants. (Doubleday, 1971. 161 pp. illus. $6.95)

Archaeology. *The First American: A Story of North American Archaeology* by C. W. Ceram. An introduction to archaeology in North America. (Harcourt, 1971. 357 pp. illus. $9.95)

Navaho Material Culture by Clyde Kluckhohn, W. W. Hill, and Lucy Wales Kluckhohn. A new perspective on the life of the Navaho, based on field notes taken by several anthropologists over a period of 25 years. It covers such cultural aspects as their clothing, housing, and forms of recreation. (Harvard, 1971. 488 pp. illus. $25)

Thousands of Years: An Archaeologist's Search for Ancient Egypt by John A. Wilson. A scientific memoir by the former director of the Oriental Institute of the University of Chicago covers a 45-year period, during which general archaeological researchers were replaced by specialized workers skilled in various aspects of archaeology. (Scribner's, 1972. 256 pp. illus. $9.95)

Architectural Engineering. *American Building Two: The Environmental Forces That Shape It* by James M. Fitch. A timely book that shows how modern developments in the social and life sciences—such as anthropology, ecology, psychology, and physiology—are of significance in today's architectural and urban designs. (Houghton Mifflin, 1972. 349 pp. illus. $15)

Astronautics. *Dividends from Space* by Frederick I. Ordway III, Carsbie O. Adams, and Mitchell R. Sharpe, Jr. Three outstanding specialists in the space sciences list the achievements and contributions to knowledge that have resulted from U.S. space programs. (Crowell, 1972. 320 pp. illus. $10)

Astronomy. *The Beauty of the Universe* by Hans Rohr. A superb collection of pictures on the general theme of the universe, covering the moon, the planets, comets, meteors, solar eclipses, and many constellations. (Viking, 1972. 87 pp. illus. $10)

The Discovery of Our Galaxy by Charles A. Whitney. An account of the development of astronomy, beginning with the ancient Epicureans of about 300 B.C. and ending with modern galactic research. It includes the accomplishments of such famous men as Galileo, William Herschel, Johannes Kepler, and Isaac Newton. (Knopf, 1971. 331 pp. illus. $10)

The Life of Benjamin Banneker by Silvio A. Bedini. The story of this American Negro mathematician is set in Maryland during the 1700s. (Scribner's, 1972. 512 pp. illus. $12.50)

Man and His Universe by Zdenek Kopal. The author interprets what is known about stars, their origin, and evolution. He also covers the planets and suggests how and why life seems to have developed only on earth. (Morrow, 1972. 313 pp. illus. $7.95)

What Star Is That? by Peter Lancaster Brown. This unusual guide for stargazers is valuable for observers of all ages and skills. It includes information on instrumentation, seasonal star charts, and 15 color slides. (Viking, 1971. 224 pp. $12.95)

Biochemistry. *Building Blocks of Life: Proteins, Vitamins, and Hormones Seen Through the Microscope* by Roman Vishniac. A most unusual book of photomicrographs of basic substances that show striking similarities to both living cells and abstract paintings. There are also brief explanations of the physiological roles of the organic material shown in the illustrations. (Scribner's, 1971. 62 pp. illus. $6.95)

Botany. *The Nature of Plants* by Lorus and Margery Milne. A survey of the plant kingdom with regard to taxonomy, anatomy, physiology, genetics, ecology, evolution, and economics. (Lippincott, 1971. 206 pp. illus. $5.95)

Drugs. *The Dope Book: All About Drugs* by Mark Lieberman. An informative book on the physiology, pharmacol-

Easter island
island of enigmas
John Dos Passos

The Beauty of the Universe

Books of Science

Continued

ogy, and pathology of drug abuse. It includes the history of some mind-altering drugs and the legal aspects of drug abuse. (Praeger, 1971. 141 pp. $5.95)

Drug Trip Abroad: American Drug-Refugees in Amsterdam and London by Walter Cuskey, Ph.D.; Arnold William Klein, M.D.; and William Krasner. The authors use case histories and interviews to present methods of treating drug addicts in Britain and the Netherlands. (University of Pennsylvania Press, 1972. 205 pp. $6.95)

Marihuana Reconsidered by Lester Grinspoon, M.D. An outstanding treatise on cannabis, which recognizes that all past investigations have been incomplete and their data inconclusive. (Harvard, 1971. 443 pp. $9.95)

Earth Sciences. *America's Natural Treasures: National Nature Monuments and Seashores* by Stewart L. Udall. The former U.S. secretary of the interior takes his readers on a guided tour of national nature monuments, national wildlife refuges, seashores, and wild and scenic rivers. A map is included. (Rand McNally, 1971. 224 pp. illus. $14.95)

The Earth We Live On (2nd ed.) by Ruth Moore. A complete revision of the 1957 edition presents man's efforts to explain the origin and changes of the earth, from ancient mythology to contemporary geophysics. (Knopf, 1971. 459 pp. illus. $8.95)

The Moving Continents: The Story of the New Geology by Frederic Golden. A review of the various theories of the origin of the continents, beginning with the first book of the Bible. A rich bibliography encourages further study. (Scribner's, 1972. 128 pp. illus. $5.95)

Water: The Wonder of Life by Rutherford Platt. This is a fundamental account of water and its essential role in maintaining the earth's resources. (Prentice-Hall, 1971. 274 pp. illus. $8.95)

Ecology. *The Arena of Life: The Dynamics of Ecology* by Lorus and Margery Milne. Many outstanding photographs help to explain 12 scientific disciplines that are related to ecology. (Doubleday, 1972. 325 pp. illus. $15)

The Closing Circle: Nature, Man, and Technology by Barry Commoner. The author explains ecology, gives case histories of certain polluted areas in the United States, and discusses the social and technological factors that led to the pollution of the environment. (Random House, 1971. 326 pp. $6.95)

One Earth, Many People: The Challenge of Human Population Growth by Lawrence Pringle. A discussion of the basic facts of human population dynamics that presents both the optimistic and pessimistic sides fairly. (Macmillan, 1971. 83 pp. illus. $4.95)

Genetics. *The Mysteries of Heredity* by John J. Fried. This story of the development of our knowledge of heredity and of inheritance mechanisms includes an up-to-date account of gene action and a summary of current research projects. (John Day, 1971. 190 pp. illus. $6.95)

Medical Sciences. *Cortisone* by Edward C. Kendall. The late Nobel prize laureate's memoir is a readable narrative of his voyage of scientific discovery, detailing failures as well as successes. (Scribner's, 1971. 175 pp. illus. $7.95)

My Life and Medicine: An Autobiographical Memoir by Paul Dudley White. A detailed account of the famous cardiologist's achievements and experiences. (Gambit, 1971. 272 pp. illus. $6.95)

Mineralogy. *The Gem Kingdom* by Paul E. Desautels. The author describes the history and qualities of natural and synthetic gemstones. (Random House, 1971. 252 pp. illus. $17.95)

Oceanography. *Two Hundred Million Years Beneath the Sea* by Peter Briggs. A story of the research vessel *Glomar Challenger*, its personnel, and some of their scientific findings relating to theories about the origin of the oceans and the continents. (Holt, Rinehart and Winston, 1971. 238 pp. illus. $7.95)

The Sun Beneath the Sea by Jacques Piccard. A history of the submersibles *Auguste Piccard* and *Ben Franklin*, from concept to design and reality. It gives details about their construction, testing, and operation, and includes a documentary of operations in the Gulf Stream during 1969. (Scribner's, 1971. 444 pp. illus. $12.50)

Physics. *Coulomb and the Evolution of Physics and Engineering in Eighteenth-Century France* by C. Stewart Gillmor. Describes the work of physicist Charles Augustin de Coulomb within the framework of engineering in France. (Princeton, 1971. 328 pp. $12.50)

Discovering the Natural Laws: The Experimental Basis of Physics by Milton A. Rothman. A summary of the experi-

Books of Science

Continued

mental basis of many of the laws of physics and how they developed. (Doubleday, 1972. 227 pp. illus. $5.95)

Einstein: The Life and Times by Ronald W. Clark. This intellectual journey along Albert Einstein's life trail, ending in a world threatened by nuclear war, shows the interaction of science and politics. (World, 1971. 718 pp. $15)

A History of Electricity and Magnetism by Herbert W. Meyer. An interesting, nontechnical history of electricity and magnetism, with enough biographical detail to bring out the human aspect. (M.I.T. Press, 1971. 342 pp. illus. $10)

Man and the Atom: The Uses of Nuclear Energy by Frank Barnaby. The author traces the development of nuclear energy, its use in warfare, in research, medicine, agriculture, and industry, and for electric power. (Funk & Wagnalls, 1972. 216 pp. illus. $6.95)

Physics and Beyond: Encounters and Conversations by Werner Heisenberg. The author, a Nobel laureate, recalls discussions he had with Einstein, Max Planck, and other famous scientists. (Harper & Row, 1971. 247 pp. $7.95)

Psychology. *Anger* by Leo Madow, M.D. A specialist tells how to recognize anger and understand its causes and how it shows itself in normal and abnormal situations. (Scribner's, 1972. 132 pp. $6.95)

Human Sexual Behavior: A Book of Readings by Bernhardt Lieberman. Particularly suitable for mature high school students and college undergraduates. (Wiley, 1971. 451 pp. illus. $11.50)

Science in General. *Asimov's Guide to Science* by Isaac Asimov. A book that deals with all of the sciences. (Basic Books, 1972. 945 pp. illus. $15)

Zoology. *American Entomologists* by Arnold Mallis. Biographical sketches of 203 entomologists who lived between 1749 and 1966 emphasize their professional accomplishments. (Rutgers, 1971. 566 pp. illus. $15)

Animals Nobody Loves by Ronald Rood. Brief life histories of some animals, such as snakes and mosquitoes, that many persons consider to be pests or predators, but which have useful roles in the ecology of fields, plains, forests, and bodies of water. (Stephen Greene, 1971. 215 pp. illus. $6.95)

The Ape People by Geoffrey H. Bourne. A factual report on work with non-human primates, much of which is fundamental to an understanding of human problems. (Putnam's, 1971. 364 pp. illus. $7.95)

The Cougar Doesn't Live Here Any More by Lorus and Margery Milne. The authors tell how man's greed and ignorance of ecological processes have placed the wildlife of the world in danger of extinction. (Prentice-Hall, 1971. 258 pp. illus. $9.95)

The Desert Locust by Stanley Baron. Describes scientific research in progress and programs being developed to control plagues of desert locusts through international cooperation. (Scribner's, 1972. 224 pp. illus. $6.95)

The Last of the Ruling Reptiles: Alligators, Crocodiles and Their Kin by Wilfred T. Neill. The author predicts that these species will be extinct by about the year 2000 and gives an excellent review of their biology. (Columbia, 1971. 503 pp. illus. $15.95)

The Life of Sharks by Paul Budker. A balanced account of sharks and their activities, including details about classification, myths and legends, and their commercial uses. (Columbia, 1971. 239 pp. illus. $12.50)

Lost Leviathan by Francis D. Ommanney. A solid reference work on the biology, life history, migration, and economic exploitation of whales, which are near extinction in many waters. (Dodd, Mead, 1971. 280 pp. illus. $5.95)

Mosquito Safari: A Naturalist in Southern Africa by C. Brooke Worth. The author relates his experiences and observations while he was in Africa investigating mosquitoes as carriers of viral diseases (Simon & Schuster, 1971. 316 pp. illus. $8.95)

The Oxford Book of Invertebrates: Protozoa, Sponges, Coelenterates, Worms, Molluscs, Echinoderms and Arthropods (other than insects) by David Nicols and John A. L. Cooke. An outstanding illustrated book on animals without backbones, describing British specimens. Similar species inhabit North America. (Oxford, 1971. 225 pp. illus. $11)

The World Wildlife Guide edited by Malcolm Ross-Macdonald. Gives general descriptions of the animal life in major geographical areas and describes parks, reserves, and sanctuaries throughout the world. (Viking, 1972. 416 pp. illus. $8.95)   [Hilary J. Deason]

Botany

Several discoveries of natural insecticides headed the list of advances in botanical research in 1971 and 1972. Botanists Cyro P. Da Costa of the University of São Paulo in Brazil and Charles M. Jones of Purdue University demonstrated that one species of insect may be repelled, another attracted, by the same chemical substance in a plant. Through evolution, some insects apparently adapt themselves to respond positively to chemical substances that were once undoubtedly active as natural insect repellents.

Da Costa and Jones studied cucurbitacins, natural chemicals found in cucumber plants. Cucumber beetles have evolved so that they are now attracted to plants containing cucurbitacins instead of being repelled by them. However, cucurbitacins still repel the two-spotted mite and probably many other pests. On the other hand, types of cucumbers that have no cucurbitacins are very susceptible to mites, yet do not attract cucumber beetles.

S.V. Amonkar and A. Banerji of the Bhabha Atomic Research Centre in Bombay, India, announced in December, 1971, that they had isolated chemical substances in garlic that kills insect larvae. The biochemists identified the larvae-killing chemicals as diallyl disulfide and diallyl trisulfide. These substances destroy aphids, cabbage butterfly larvae, and Colorado beetle larvae. At a concentration of only 5 parts per million, they kill the larvae of several species of mosquitoes, including the common garden mosquito and the species that carries yellow fever. They also act as insecticides against the potato tuber moth, red cotton bug, red palm weevil, and houseflies.

Garlic is not toxic to higher animals and has been a part of man's diet for centuries. As a chemical pesticide, therefore, garlic has great advantages over substances such as DDT.

The wood of the western red cedar (*Thuja plicata*) has long been known to repel various species of insects. In October, 1971, V. Hach and Elizabeth C. McDonald of MacMillan Bloedel Research, Ltd., in Vancouver, Canada, prepared compounds from cedarwood

The sweet-gum tree on the right was shaken for 30 seconds each day. The one on the left was not disturbed. Botanists think that motion triggers some mechanism, perhaps a hormone, that hinders growth in plants.

Botany

extracts that repelled the disease-carrying mosquito, *Aedes aegypti*. The compounds were effective for up to 50 days, much longer than standard repellents. The new substances, amides of thujic acid, also very effectively repel two very common species of cockroaches.

Terrance R. Green and Clarence A. Ryan of Washington State University reported in February, 1972, that they had discovered that "Inhibitor I," a protein from potato tubers, also occurs in potato and tomato plant leaves. The protein apparently affects the digestibility and palatability of the leaves for insects. The plants are evidently able to regulate the quantity of inhibitor that they produce. If Colorado potato beetles are allowed to feed on the leaves of young tomato plants, the concentration of Inhibitor I in all the leaves rises dramatically. The concentration also increases when a leaf is wounded by a paper punch.

Lawrence E. Gilbert of Stanford University demonstrated in 1971 that a common tropical American passionflower, *Passiflora adenopoda*, produces hooked trichomes, or hairs, on its leaves that immobilize and even kill the larvae of common tropical butterflies.

Tree size. P. L. Neel and Richard W. Harris of the University of California, Davis, reported in 1971 that moderate shaking greatly affected the growth of young sweet-gum trees, *Liquidambar styraciflua*. Trees that were shaken gently by hand for 30 seconds each day for a month grew only one-fifth as high as trees that were not touched. They also developed terminal buds, and had significantly fewer nodes and lateral branches.

The botanists concluded that wind, in addition to its well-known effects of increasing *transpiration* (the passing of moisture through leaves) and carbon dioxide intake, triggers a built-in mechanism, probably a hormone, for regulating tree growth.

Stomata. In 1971, it was learned that the opening in stomata, the tiny pores in plant leaves, is controlled by potassium in their guard cells. Guard cells are kidney-shaped cells that surround the stomata and determine their shape. Stomata permit the movement of carbon dioxide, oxygen, and water vapor in photosynthesis, respiration, and transpiration of plants. Artificial control of these pores could make it possible to control the rate of water loss and carbon dioxide intake, and thus also the rate of photosynthesis.

Brij L. Sawhney and Israel Zelitch, biochemists at the Connecticut Agricultural Experiment Station at New Haven found that the guard cells of a fully opened stoma contained more than $2\frac{1}{2}$ times as much potassium as those of a closed stoma. They suggested that the movement of potassium into guard cells is caused by a natural potassium-ion pump, which gets its energy from cell processes that are triggered by light.

Primitive plants. The evolutionary ancestry of angiosperms, or flowering plants, is still one of the great mysteries of botany. Consequently, the structure of primitive families of angiosperms is of great interest. Botanist James Walker of the University of Massachusetts reported in 1971 that plants in the primitive family *Annonaceae* have a pollen grain that is unique among flowering plants because of its large size and distinctive structure. The pollen grains in the genera *Annona* and *Cymbopetalum* are the largest known in flowering plants. In addition, its highly distinctive clustered pollen units, polyads, are found in separate stamens.

Jane Gray of the Museum of Natural History at the University of Oregon and geologist Arthur J. Boucot of Oregon State University reported in September, 1971, that they had found fossilized spores of a land plant in Silurian rock deposits along the Niagara River in New York state. The Silurian Period lasted from 435 million to 405 million years ago. Although no other plant parts were found in these deposits, the spores indicate that there were primitive plants in North America at a very early date.

Plant life on Mars. Whether or not there is life on Mars has long been a matter of debate. In April, 1972, Norman H. Horowitz of the California Institute of Technology studied the dry valleys of South Victoria Land, Antarctica, the world's coldest and driest desert. He found that soils there contain no living microorganisms. Conditions on Mars are much more rigorous than those in Antarctica. From this, Horowitz concluded that fears that terrestrial microorganisms carried to Mars could multiply and contaminate the planet are unfounded.   [William C. Steere]

Chemical Technology

A wide range of new developments in chemical technology found their way into use during 1971 and 1972. The innovations came in such varied fields as plastics, coatings, power generation, nutrition, and medicine.

Better plastics. Engineers have long known that reinforcing ordinary plastic materials with such additives as tiny glass fibers gives them greater strength. Now, advances in both materials and processing technologies are producing even stronger reinforced plastics (RP) that can be used in new ways.

Graphite fiber, a form of carbon, is the workhorse in the new stable of RP products. Replacing some or all of the glass fibers with graphite fibers makes the RP much stronger. For instance, RP beams and columns are practical replacements for steel or aluminum in chemical processing operations troubled with corrosion or electrical insulation problems. Adding only 5 per cent graphite fiber to the conventional glass-and-polyester RP formula can more than double the load a structural member can bear. Similarly, epoxy resin RP panels containing graph-

ite compete successfully with honeycombed metal units in aircraft.

Even more exotic materials are on the way. Researchers at Pfizer, Incorporated, of New York City reportedly have developed an ultrathin graphite ribbon that is strong and stiff in all directions. The more this material is stressed, the stiffer it appears to get. In mid-1972, however, Pfizer said the material was not yet ready for commercial use.

Du Pont Company of Wilmington, Del., began marketing an organic fiber as strong as stainless steel wire but only one-sixth as heavy. The new fiber can be used alone or in combination with other fibers for RP components. Du Pont announced details of the organic fiber in February, 1972.

Painting with powder. Restrictions on the use of solvents that cause air pollution spurred the use of polymer-based powder coatings to finish industrial goods. Powder coatings do not use the volatile solvents that are common in conventional liquid paint. And they are easy to apply. One coat of the powder is usually enough; excess coating material

A new water-soluble, plastic container, that can hold oil-based or solid materials, is also edible. As one use, a package of frozen vegetables can be boiled, package and all. The container disintegrates in 15 minutes in water.

Chemical Technology

Continued

Mosquito pupa breaks the surface of water to get oxygen, *top*. But it cannot break through a surface layer of lipid one molecule thick, *bottom*, and eventually suffocates. Such films may provide useful insect control.

can be reused; and appropriate polymers can be selected for the specific coating job. Popular coatings include such materials as epoxies, vinyls, nylons, and polyethylene.

In one application method, charged particles of the coating are shot at a grounded object with an electrostatic spray gun. The coating is then melted or otherwise cured in place. In another method, the object is heated and immersed in a bed of powder fluidized by gas flow. Some polymer systems, however, must still be preceded by primer coats, most of which contain solvents.

A new, specialized system, introduced in 1971 by Grow Chemical Corporation, combines powder coatings and conventional painting techniques. Water containing particles of insoluble powder, either an acrylic or urethane compound, is painted on the part. Then the water is allowed to evaporate. The powder forms a continuous film when the part is cured in an oven.

Coal to the rescue. To avoid the energy shortage predicted for the United States (see ENERGY), planners and researchers increased their efforts to make pipeline gas from coal that can be used interchangeably with natural gas.

Almost all of the new gasification processes are aimed at boosting the gases' heat content to the level of natural gas by first treating them to produce carbon monoxide and hydrogen, then converting these substances to methane, the principal component of natural gas. Such projects were stimulated when President Richard M. Nixon chose to step up federal funding for coal gasification in his proposed 1973 budget.

In the 1950s, natural gas replaced low-heat-value fuel gases then made from coal in the energy market. Economical and clean-burning, it seemed the ideal fuel for energy-hungry generating stations. But natural gas reserves are now dwindling, and relatively abundant coal reserves have regained importance as an energy source.

Because burning coal directly would worsen air pollution, coal gasification is appealing. Older commercial gasification methods, however, produced a gas with only about half the heat content of the natural pipeline gas to which the U.S. economy has been geared. Plans now include the construction of three

big pilot plants, two-thirds of the cost financed by the government and one-third by private industry.

At least five private ventures are also planned. Furthest along is one the El Paso Natural Gas Company plans to have operating in northwestern New Mexico. It will produce 250 million cubic feet of gas per day by mid-1976.

Coal is not the only possible source for replacement gas. At least four plants will begin in 1973 to supply pipeline gas. They will use light naphtha, a petroleum liquid, as a starting material.

Fuel from wastes. A pilot plant near San Diego, Calif., was producing both gas and fuel oil from solid municipal wastes. The Garret Research and Development Company said the fuels were produced at a savings of 16 per cent. The waste is shredded and dried and metal, glass, and other inorganic materials are removed. After being shredded again, the waste is heated in an inert gas atmosphere to produce 2 barrels of oil and 6,000 cubic feet of gas from a ton of dry waste material.

The gas is now being used to fuel the pilot plant, but a small laboratory version is designed to generate pipeline-quality gas instead of oil. The process converts about 80 per cent of the organic waste to gas.

Making protein. Progress continues in efforts to make inexpensive protein. A fish protein concentrate (FPC) announced in March, 1972, at Texas A&M University is made as a dry powder, but forms a gel when combined with water. The powder is more than 90 per cent protein. Except when made from fish caught in oil-polluted water, it is nearly odorless and tasteless and accepts a variety of artificial flavorings.

Rehydrated FPC was substituted for beef in sausages normally half beef and half pork. None of 34 judges in a taste panel detected any fish taste and one FPC batch "was considered close to acceptability." Judges scored color, flavor, texture, and overall satisfaction. The scores were lowest for texture.

Cattle manure is being converted to a high-protein dietary supplement for animals at a General Electric Company pilot plant at Casa Grande, Ariz. The manure is first mixed with water and dormant bacteria. When the mixture is heated to 140°F., the bacteria begin

A new plastic bag, *top,* breaks down when exposed to weather in outdoor tests conducted by a Swedish laboratory. In 17 days, the bag had cracked, *middle.* After 34 days it was partly pulverized, *bottom.*

digesting the manure and producing protein. The process may also work with sawdust and other organic wastes.

Activated carbon is playing an increasing role in cleaning up the nation's wastes. In either granular or powdered form, it has a very large surface area that makes it an effective adsorber. It is often used to remove organic contaminants from small waste streams. The granular type has recently been used to clean up effluents from processes such as those in petroleum refineries, food processors, and textile mills.

Activated carbon is considerably cheaper to make as a powder but, until recently, only the granular form could be regenerated and reused. One of the most successful of several newly discovered powder-regenerating methods is used at Westvaco Corporation's Covington, Va., plant. In mid-1971, it began regenerating 8 to 10 tons of carbon per day from corn syrup refineries.

In the process, the sticky, spent carbon is suspended in a high-velocity, low-pressure stream of air and steam and carried into a reactor heated to about 1800°F. Oxygen is added and adsorbed organic foreign matter is burned. After the reactivated carbon is suddenly cooled, about 85 to 90 per cent of it is recovered.

Helpful hulls. Oil company chemists looked forward in mid-1972 to an analysis that would confirm success in using rice hulls to control spills of oil on water. Standard Oil Company (Ohio) demonstrated in November, 1971, that the hulls quickly soaked up a slick when they were blown over the surface of a small pond where oil had been spilled. The hulls combined with the oil to form a mix that was sucked up with a pump. At another demonstration with oil in a small tub of water, observers followed it by drinking some of the water. They said there was no oil taste.

Bone repair. Scientists at the Columbus, Ohio, laboratories of Batelle Memorial Institute have developed a promising new method to repair bone using ordinary calcium phosphate formed into a porous ceramic. Almost immediately after the ceramic is implanted, blood and other body fluids penetrate it and bone starts to regenerate. Within two or three months, said Thomas D. Driskell of Batelle-Columbus, the old bone is completely remodeled. "One finds *corti-*

cal (hard, outer) bone where there is supposed to be cortical bone, *cancellous* (spongy) bone where there is supposed to be cancellous bone, and marrow where there is supposed to be marrow."

The real significance of the development is that the ceramic material is only a temporary bridge. It gradually disappears as the new bone forms. "In six weeks, microscopically, you can hardly see any ceramic particles," Driskell said.

The work is sponsored by the U.S. Army Medical Research and Development Command's Dental Research Division. It has focused on repairing animals' jawbone defects. Adding length to long bones seems clearly possible, but no joint replacements are anticipated. In addition to carving ceramic "bridges" from blocks of the material, scientists have packed ceramic into voids in bone. New bone later filled the void.

Steelmaking progress in Europe contributed to American industry. The United States Steel Corporation imported a new German method that improved on the "basic oxygen process" (BOP), itself a step ahead of older open-hearth steelmaking. In the new process, called Q-BOP, oxygen and powdered lime or other additives are blown through nozzles in the bottom of the furnace, instead of being fed in at the top.

Q-BOP improves yield, allows more scrap to be used, and reduces air pollution. Three BOP furnaces being built in Gary, Ind., will be converted to the Q-BOP process. When these begin operating in late 1972, they will be the first in North America.

Sulfur paving. Sulfur is added to asphalt paving material, Thermopave, being developed by Shell Canada, Ltd. As much as 13 per cent sulfur, mixed with petroleum fractions and sand, causes the pavement to set rapidly. It closely envelops individual grains of sand, adding strength and stiffness, and allows inexpensive sand to replace crushed rock in the mix. Shell researchers found that a 4-inch-thick Thermopave layer over a weak subgrade successfully supported heavy road-construction equipment in a delta area. Paving by older methods had been impractical on that type of soil.   [Frederick C. Price]

Chemistry

Chemical Dynamics. The "organic age" of molecular beam chemistry – the study of individual reaction-producing collisions between organic molecules – opened in 1971 and 1972. The new experiments have two important, bellwether aspects. One is energetic: They reveal the energy flow within the molecules to bonds near the reaction site. The other is stereochemical: The results virtually realize the chemist's dream of observing the configuration of the interlocked reacting molecules that exists for only a few trillionths of a second.

The first organic reactions studied using molecular beams add an atom or ion to a hydrocarbon molecule that contains a carbon-carbon double bond. The double bond is thereby converted to a single bond and a new bond is formed to the attacking atom or ion. The bond conversion frees a large quantity of energy, which appears as vibrational excitation of the addition complex. In an ordinary chemical environment, these vibrations would be quickly damped by collisions with other molecules. However, subsequent collisions do not occur in a highly rarefied molecular beam. The vibrational energy remains trapped in the addition complex, shuffling among various bonds until one bond ruptures and the complex dissociates to form the reaction products, which fly apart.

Energy flow. Chemists have often assumed the energy flows randomly among all the bonds. The fraction that appears in the kinetic energy of the products can then be calculated from simple statistical theory. This kinetic energy is directly measured in the beam experiments. Comparison with the statistical calculations shows the number of bonds that participate in the energy flow.

These experiments give direct evidence that energy flow is random in some reactions and very specific in others. Chemists John M. Parsons and Yuan T. Lee at the University of Chicago studied a markedly specific case, the reaction of fluorine atoms with ethylene, $F + CH_2 = CH_2$. This reaction forms a C_2H_4F complex which dissociates to $CH_2 = CHF + H$. The energy released to propel the products apart is

285

"I, Professor T. Harrington Filmore, MS, PhD, ScD, world-renowned member of the scientific fraternity, Nobel Prize winner for brilliant work in the field of advanced chemo-nuclear research, author of a dozen masterful tracts and papers, have my finger stuck in a flask."

Chemistry

Continued

high, indicating that there is little energy flow beyond the C-F and C-H bonds at the reaction site.

Another convincing example of limited energy flow occurs in the reaction of ethylene ions with ethylene, $C_2H_4^+ + C_2H_4 \rightarrow$ [complex] $\rightarrow CH_3 + C_3H_5^+$. Richard L. Wolfgang and his co-workers at Yale University studied this reaction. In this work, the energy of the incident ethylene ion was increased over a wide range to enhance the vibrations of the reaction complex, and hence decrease its lifetime before dissociation. The spectrum of kinetic energy of the products indicates that less than two-thirds of the bonds are active in the energy flow. As the vibrations increase, this active fraction decreases appreciably because of the shortened lifetime of the intermediate complex.

Reaction geometry. In addition to the energies of the product molecules, the angles at which they emerge from the colliding beams of reactants are also measured. These angles often provide information about the shape of the complex in its "transition state," the

critical configuration that leads to the breakup into product molecules. The product angular distribution is determined primarily by the rotational properties of the complex.

For example, consider first a complex in its transition state having all its atoms lined up in a row. The rotation of this complex is like a drum major's spinning baton, whirling about an axis perpendicular to the line of atoms. The complex rotates a few times before dissociating. The products fly out almost uniformly over all angles about the rotation axis, like water from a lawn sprinkler.

The rotation axes of all the complexes formed are perpendicular to the direction of approach of the reactants, and distributed at all angles about it. The net result is an angular distribution of product molecules that corresponds to the lines of longitude on a globe. The product intensity becomes very high in the "polar" regions, parallel or nearly parallel to the initial reactant approach, and low in the "equatorial" regions.

But if the complex does not have its atoms in a line, the products can

Chemistry

Continued

emerge, for example, sideways as the spinning complex breaks apart. The product intensity then becomes higher in the equatorial regions, lower in the polar regions.

Thus, whatever the shape of the transition-state complex, it leaves an imprint in the angular distribution of its products, even though it lasts only a few trillionths of a second. Analysis of such data has given interesting results. For instance, Jefferey T. Cheung and J. Douglas McDonald at Harvard University studied the reaction of chlorine atoms with bromoethylene, $Cl + CH_2 = CHBr$. They obtained a large yield of $CH_2 = CHCl + Br$, with angular distribution peaked in the polar regions. This indicates Cl attacks the C atom to which Br is attached and the transition state has the Cl-C-Br configuration. Reactions of Cl with $CHBr = CHCH_3$ and $CH_2 = CBrCH_3$ give the same result. A different result is obtained, however, when Br is not attached to a doubly bonded C atom. For example, in the reaction $Cl + CH_2 = CH-CH_2Br$, the angular distribution indicates Cl and Br are at opposite ends of the complex.

Laser excitation. Another important new method first reported in 1972 enables chemists to study how exciting molecules vibrationally changes their reactivity. Truman J. Odiorne, Philip R. Brooks, and Jerome V. V. Kasper conducted the prototype experiment at Rice University on the reaction of potassium atoms with hydrogen chloride, $K + HCl \rightarrow KCl + H$. Like many chemical reactions, this one needs energy to occur, and unexcited HCl molecules give a very small yield. The researchers constructed an HCl chemical laser. When a beam of unexcited HCl molecules is irradiated by a pulsed beam of light from this laser, the molecules increase their vibrations.

By synchronizing detection of the reaction product, KCl, with the laser pulses, the experimenters directly observed the difference in reactivity of excited HCl (produced during the "on" cycle of the pulses) and unexcited HCl ("off" cycle). The excited molecules were a hundred times more reactive. The sensitivity is dramatic. Typically, only 5 molecules of KCl were detected for each laser light pulse. [Dudley Herschbach]

Structural Chemistry. The structure for deoxyribonucleic acid (DNA) proposed in 1953 by James D. Watson and Francis H. C. Crick, and for which they shared the 1962 Nobel prize, was based on the meagerest of X-ray diffraction data. Since then, no alternative models of this genetically important molecule have been throughly tested.

Crystallographers have, however, continued to study simple fragments of DNA, from which ample data are obtainable, and these have produced some surprises. The fragments are purine or pyrimidine bases, sugars, phosphates, and various combinations of these. Several dozen such structures were reported in 1971 and 1972, including that of the dinucleoside phosphate, uridylyl-3', 5'-adenosine phosphate (UpA).

The structure of UpA was solved by a team of chemists consisting of Joel Sussman, Nadrian Seeman, Helen Berman, and S. H. Kim, working at Columbia University in New York City, the Institute for Cancer Research in Philadelphia, and the Massachusetts Institute of Technology in Cambridge, Mass. It is the first determination of the structure of a natural dinucleotide phosphate.

All genetic material, whether it be DNA or one of the varieties of ribonucleic acid (RNA) has alternating phosphate groups and ringlike sugar groups as its main strand, or backbone. The overall shape of the molecule must depend on the most stable angles of rotation about the single bonds in the phosphate group as well as on the shape assumed by the sugar ring. The researchers found that UpA exists in two quite different molecular shapes. Interestingly, neither of these shapes corresponds to the helical structures usually assumed for DNA and RNA.

Furthermore, all four nucleosides, or genetic code units, have several structural features in common. If future work shows these features to be general, then nucleotides in nucleic acids are much more limited in their structural configurations than scientists have assumed. New models of DNA and RNA, or modifications of existing models, must then take these limitations into account.

Protein structure. As in the case of DNA and RNA, much work continued to be reported on small fragments of other large molecules. In the area of

Chemistry

protein structure, these fragments are amino acids in the simpler studies and polypeptides in the more complex ones. More than a dozen such structures were determined in 1971 and 1972. Knowledge of the bond angles, bond distances, torsion angles, and nonbonded interactions of these fragments is essential in order to interpret the data obtained from the much more complicated protein molecules themselves.

The Laboratory of Molecular Biology, Cambridge, England, continued to examine various hemoglobins, extending the pioneering work of Max Perutz, who first worked out the crystal structure of these oxygen-carrying blood proteins. Recent work has concentrated on determining the structural abnormalities in mutant hemoglobins. Scientists must measure thousands of X-ray diffraction intensities for each compound and calculate thousands of points of electron density in order to determine their structural differences.

J. Greer, working at Cambridge, reported on no less than six hemoglobin mutants in 1971 and 1972. The mutants, unfortunately, are named after the geographical locality in which they were first discovered, so the names contribute nothing to identifying the nature of the mutant, but serve only for general bookkeeping purposes.

Greer found that in hemoglobin Chesapeake, which has drastically reduced oxygen-carrying ability, one of the amino acids in the protein chain, an arginine, had been replaced by leucine. But when this same arginine had been replaced by glutamine, in hemoglobin J Capetown, the mutant is almost normal. The researchers do not know the structural basis for the milder abnormality of J Capetown.

Greer found not too dissimilar results for hemoglobins Kansas and Richmond. Threonine in Kansas and lysine in Richmond replace a single asparagine. Kansas displays markedly abnormal physiological properties, while Richmond is virtually normal. These studies vividly demonstrate that very complex structural changes occur when only 1 out of 287 amino acid links that make up this molecule are modified, and that these distortions defy prediction. Researchers in this area of structural chemistry know at present what some of the trees look like, but they have yet to discern the nature of the forest.

Enzymes. The results of studies of the structures of two digestive enzymes were announced in 1971. The first of these, α-chymotrypsin, was worked on by Alexander Tulinsky and his co-workers at Michigan State University. The researchers discovered that this protein, which has a molecular weight of 25,500, crystallizes with two different subunit molecules. These have been observed with X rays down to a resolution of 3.5 angstroms (A). Electron density maps show that certain regions of the two independent molecules are strikingly similar, but that significant differences exist between the subunits.

The second digestive enzyme to be studied is a 1:1 complex of trypsin-trypsin inhibitor, studied by Frank Cole and Rengachary Parthasarathy at the Roswell Park Memorial Institute in Buffalo, N.Y. These two researchers also found two independent molecules in the repeating unit.

Although the diffraction pattern is detectable to a resolution of 2.5 A, suitable heavy atom derivatives of the subunits have not yet been prepared. As a result, the electron density functions, which would enable comparison with α-chymotrypsin, are not available.

Metallic compounds. Chemist Sten Samson at the California Institute of Technology investigated alloy structures exhibiting nearly the same degree of complexity as biologically important molecules, but of vastly simpler atomic constitution. Some of these alloys have deceptively simple formulas, such as Mg_2Al_3, Cu_4Cd_3, and Mg_6Pd. Actually, their structures are enormously complex. The repeating unit of Mg_2Al_3 contains 1,168 atoms, and it consists of 672 icosahedra (20-faced geometrical figures) and 252 other complex polyhedrons. Cu_4Cd_3 has a 1,124-atom repeating unit with 568 icosahedra and 124 complex polyhedra. The repeating unit of Mg_6Pd is somewhat smaller, with 396 atoms. Icosahedra are also important in this structure. Studies of the geometrical features of the prominent polyhedrons in these compounds have provided valuable hints as to the geometrical factors that determine their stability, but the complete nature of alloys is still uncertain. [Jerry Donohue]

Chemistry

Continued

The remarkable change in chemical architecture from starting compound, left, to gibberellin A₁₅ emphasizes this 1972 synthesis achievement by Japanese chemists.

Chemical Synthesis. Japanese chemists reported the synthesis of an important plant hormone, *dl*-gibberellin A_{15}, in 1972. The synthesis scheme is lengthy, and the remarkable change in chemical architecture in transforming a relatively simple starting molecule to the gibberellin A_{15} molecule emphasizes the magnitude of this accomplishment.

In 1938, two Japanese chemists, T. Yabuta and Y. Sumiki, first discovered the cause of the frequent, abnormally tall growth of young rice plants. The unusual plant growth was caused by the metabolites of a fungus, *Gibberella fujikuroi*, from which a series of plant hormones known as gibberellins were isolated. Later work proved that many higher plants contain gibberellins along with other plant hormones, such as the auxins and kinins. Apparently, these hormonal systems help to regulate such developments as root growth, budding, and fruit formation in plants. The gibberellins, typified by gibberellic acid and gibberellin A_{15}, cause cells to grow longer, and thus increase the height of the plants.

The molecular structure of gibberellic acid was determined in 1962 using X-ray diffraction methods with gibberellic acid crystals. Later chemical work in 1967 provided the structure of the related A_{15} gibberellin molecule.

Now an account of the complete chemical synthesis of *dl*-gibberellin A_{15} has been published in the *Journal of the American Chemical Society* by Wataru Nagata and his co-workers at the Shionogi and Company, Ltd., research laboratory in Tokyo. The synthesis procedure is stereospecific – it produces the correct structural configuration at each of seven critical points, the asymmetric carbon atoms present in the molecule. Tests of biological activity with rice shoots proved that synthetic gibberellin A_{15} was as effective as the natural material that was isolated from fungal metabolites.

New synthesis tool. Much effort has been devoted over the last decade to the use of transition metal atoms, such as iron, nickel, and cobalt, to aid organic syntheses. Unfortunately, few such reactions were being discovered that

A Gibberellin Synthesized

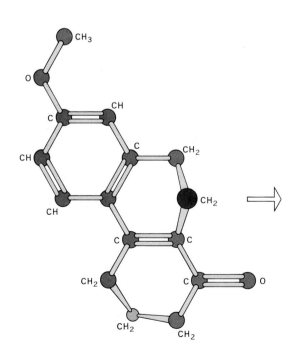
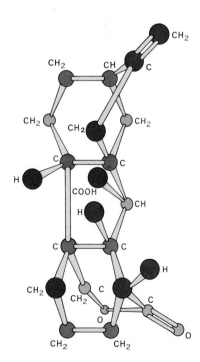

New Ketone Synthesis

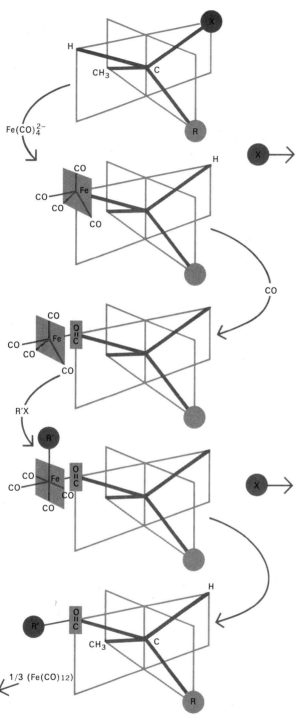

Fe(CO)$_4^{2-}$

R'X

1/3 (Fe(CO)$_{12}$)

In a new application of transition metal chemistry to organic synthesis, Stanford University chemists used an iron (Fe) compound to build a specific arrangement of atoms and chemical groups (R and R') around the asymmetric carbon atom (C).

lend themselves to useful organic syntheses. A new, successful reaction may have reversed this trend. It was reported in 1972 by James P. Collman and his co-workers at Stanford University.

Iron pentacarbonyl, Fe (CO)$_5$, which is readily available by reacting iron metal (Fe) and carbon monoxide (CO), is easily reduced to the iron tetracarbonylate ion, Fe(CO)$_4^{2-}$. The Stanford group used this ion in a variety of useful organic syntheses.

For example, it was used in the synthesis of a molecule, a ketone, having a specific structure. The scientists first reacted Fe (CO$_4$)$^{2-}$ with an octyl tosylate ester (RX) that is optically active—it rotates the plane of polarized light passing through it to the right—and then with methyl iodide (R'X). The stereospecific reaction induced by the iron tetracarbonylate ion produced the ketone (RCOR'), all the molecules of which have a completely reversed configuration at the asymmetric carbon atom producing the optical activity. Thus, the product rotates polarized light to the left.

A reaction mechanism which easily explains this unusual synthesis involves the reaction of Fe(CO)$_4^{2-}$ with RX to form an RFe(CO)$_4^-$ intermediate and X$^-$. This causes a complete inversion of configuration at the asymmetric carbon atom. This intermediate then reacts with CO, which is inserted directly between the iron atom and the asymmetric carbon atom.

This species then is reacted with a second alkyl halide (R'X) to form RR'Fe(CO)$_4$. The RR'Fe(CO)$_4$ collapses to form RCOR', still with 100 per cent total stereochemical inversion of configuration at the asymmetric carbon atom that was bonded to the iron atom. The collapse step must completely retain the stereochemical configuration. This general synthesis method should find wide applicability in many organic reactions having rigorous stereochemical requirements.

Xenon-metal bond. The first compounds of the noble gases, the relatively stable xenon fluorides XeF$_2$ and XeF$_4$, were reported in 1962. This was a momentous event in chemistry because chemists had thought it impossible to react noble gas atoms (He, Ne, Xe, Kr) with any reagent, since their valence

Chemistry

shells are filled. In fact, this total chemical inactivity had been given the status of a chemical law.

More recently, chemists have synthesized compounds such as the xenon oxides, represented by XeO_2, an unstable white solid that decomposes explosively to Xe and O_2. Now, chemists have synthesized a new type of noble gas compound in which a noble gas is bonded to a metal atom. In 1972, Charles T. Goetschel and Karl R. Loos of the Shell Development Company reported their successful synthesis of what may be the first example of a xenon-to-metal bond.

The reason that the nonmetals F and O bond to Xe lies in the great electronegativity, or ability to attract electrons, of neutral fluorine and oxygen atoms. Since fluorine is the most electron-attracting element known with an electronegativity of 4.0, while oxygen has an electronegativity of 3.5, it is easy to understand the greater stability of xenon fluorides over that of XeO_2.

The synthesis of xenon compounds other than the fluorides and oxides had seemed unlikely, however. There are no other atoms having electronegativities of the order of 3.5 to 4.0.

Goetschel and Loos allowed an unusual oxygen salt, $O_2^+BF_4^-$, to react with xenon gas at $-100°C$. A white solid product formed that decomposed above $-30°C$. to give Xe and BF_3. The method of formation of the white solid and the products observed upon decomposition, when coupled with the spectroscopic properties shown by the compound, strongly suggest that the white solid is $F-Xe-BF_2$.

This material would be expected to be unstable, since the $-BF_2$ group has a calculated electronegativity of about 3.0. Boron has an electronegativity of only 2.0 and the bonding of two strongly electron-attracting fluorine atoms to boron accounts for the calculated electronegativity value of 3.0 for the $-BF_2$ group. As a result of this work, other new noble gas compounds may soon be synthesized that contain a chemical bond between the noble gas atom and a metal atom that has been partially fluorinated.   [M. Frederick Hawthorne]

Communications

The Federal Communications Commission (FCC), in July, 1972, granted permission to the American Telephone and Telegraph Company (A.T.&T.) to place a sixth transatlantic cable in service. The cable will have solid state amplifiers every 5 miles and will be able to transmit 4,000 two-way voice conversations over a single coaxial structure.

A.T.&T. will be responsible for overall system design and will furnish deep-sea amplifiers and necessary shore-based power systems. Great Britain will furnish the cable, a new design that has less signal loss. France will provide the multiplex terminals. Present plans call for installation of the new cable and the beginning of service in 1976.

Long negotiations within the International Telecommunications Satellite Consortium (Intelsat) seemed near conclusion in mid-1972. Intelsat was established in 1964 with a tentative organization in which the Communications Satellite Corporation (Comsat) was both the U.S. member and manager for the space components of the system. The 1964 organization was to be reviewed beginning in February, 1969. In 1972, after more than two years of debate, a final structure proposed by Japan and Australia was slowly winning ratification by Intelsat's member governments.

Under the proposed system, an "Assembly of Parties," with each of the 83 member governments having one vote, would decide broad policy matters and long-term objectives. An international Board of Governors would control day-by-day operations. After six years, the Board of Governors would appoint a director-general as chief executive. Until then, Comsat will continue as manager. Later, the director-general will hire contractors for the technical and operational functions. Presumably, the eligible contractors would include Comsat.

Local satellite systems also appeared to be growing. Russia continued to expand its domestic Molniya system, which now serves over 35 Soviet ground stations in both Europe and Asia. It appeared this system, which has used satellites in elliptical inclined

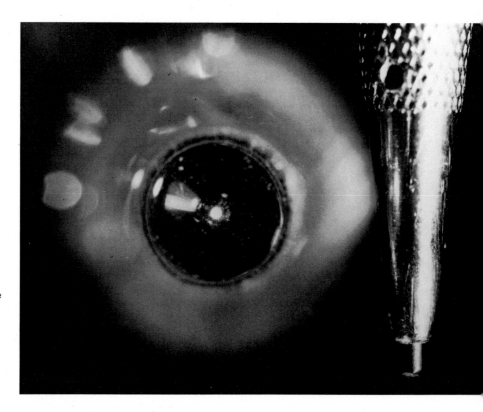

Miniaturized waveguide gas laser seen end-on, may someday be used to amplify signals that travel on light waves.

Communications

Continued

orbits, may soon shift to equatorial synchronous satellites. These orbit in such a way that they appear to stay in one position over the equator.

Canada was about to enter the field with a system by Telesat Canada. It would furnish television to remote northern territories and coast-to-coast telephone service to the more populated south. Hughes Aircraft Company will build the satellites, which will be similar to Intelsat 4s.

Other regional systems are being considered by Australia, Brazil, and a western European consortium. Brazil proposes to lease capacity on Intelsat's satellites to transmit and receive signals.

In the United States, the FCC ended the long wait with a ruling setting the conditions under which communications by satellite could be provided. Such a system has been under discussion since the mid-1960s. In June, 1972, the commission said it would open the field to all qualified applicants, but it set conditions on the domestic satellite activity of A.T.&T. and Comsat. Further discussions seemed likely before all

the questions are settled and a system can be started.

Deep space links. The Apollo 15 and 16 lunar missions spectacularly demonstrated U.S. mastery of deep space communications, both voice and video. Millions watched and listened to the astronauts' lunar surface journeys a quarter of a million miles away.

In what may well be the ultimate stretching of space communications, the Pioneer 10 spacecraft was launched toward Jupiter on March 2, 1972. Its planned trajectory will bring it within 87,000 miles of the huge planet by early December, 1973, when it will be 500-million miles from Earth.

A miniature nuclear reactor will furnish power to operate the instruments that will take pictures, gather data, and send the results back to the Earth.

In whipping past Jupiter, Pioneer will gain more velocity than needed to carry it beyond the solar system's farthest planets. Scientists estimate that communications can be maintained and data gathered for about seven years. By then, the craft will be about a

Communications

billion miles from the Earth. At this point, radio signals will take 1½ hours to reach the Earth.

Pioneer also carries a primitive means of communication—a plaque giving pictorial information about the appearance of humans. It also gives elementary information about the location of our solar system in the galaxy. The plaque was included on the possibility that the spacecraft might someday be recovered by intelligent beings of another world. See SPACE EXPLORATION, Close-Up.

On the road. Telephones can reach only a few of the millions of automobiles in the United States. Such calls are made by radio transmission. The main reason for the limited service is that only 33 radio channels with a total band of less than 2 megahertz have been allocated in the United States for this purpose. By reusing these channels in widely separated areas, about 35,000 mobile subscribers are being served, but the system is overloaded and service is not up to wire telephone standards.

A new system proposed by Bell Telephone Laboratories will have up to 800 frequency assignments for transmitters and an equal number for receivers. A wide, high-frequency band would be assigned to this purpose by the FCC. All the channels could be reused in each cell of a geographic pattern, with cell size adjustable according to need.

Simultaneous use of a channel in cells near each other would normally cause high interference. However, the system provides for the stronger signal to "capture" the receiver and thus become the only signal heard.

Small cells would permit both fixed and mobile transmitters to be low-powered. A central control office will locate the mobile unit by measuring its signal strength. All mobile units can operate at any of the 800 frequencies within the band controlled by this office.

During a call, the control office keeps track of the mobile unit's location and assigns it appropriate frequencies. When the unit crosses the cell boundary, it reassigns frequencies that are affected and passes the call over to the next cell's control office without interrupting conversation.   [Eugene F. O'Neill]

Computers

Computer technology took a big step forward in 1971 and 1972. Large-scale integration (LSI)—large and complex electronic networks on a single silicon chip—moved into the market place to make better-performing computers possible at lower cost. The popular dream of a "computer on a chip" remains unfulfilled for full-sized, general-purpose computers. But small, specialized systems consisting of little more than a few LSI logic and memory modules are now available to the public.

Each of these modules performs a complete set of arithmetic, logic, control, or memory functions on one or two silicon chips a fraction of an inch square. The computer system has relatively few connections between modules. Mass-production economies in basic functional modules let system designers by-pass the laborious, detailed engineering of logic and memory circuits. Instead, they use mass-produced modules and concentrate on overall system architecture and organization to suit the buyers' needs.

Portable calculators. LSI's contribution to the computer field is well illustrated by the HP35 portable calculator, mass-produced in 1972 by Hewlett-Packard Advanced Products of Cupertino, Calif. Weighing only 9 ounces and measuring only 1 x 3 x 6 inches, it can, with single key strokes, perform all the calculations commonly used by scientists and engineers. The HP35 can add, subtract, multiply, divide, extract square roots, and find a variety of trigonometric and logarithmic functions.

Answers are displayed instantly by light-emitting diodes on the top of the calculator. Any number between 10^{-2} to 10^{10} is displayed in 10 digits with the decimal point properly placed. A number outside of this range, but within the broad range of 10^{-99} to 10^{99}, is displayed in scientific notation.

A set of LSI electronic networks on several silicon chips make up the heart of the HP35. LSI and metal-oxide semiconductor (MOS) technology give the chips the equivalent of 30,000 transistors to perform all the various logic and memory functions.

A set of fixed programs for the various mathematical functions are built into its

Pocket-sized electronic calculator can do all that a slide rule does, and with far greater speed and accuracy.

Computers

Continued

memory, which makes the HP35 in effect a preprogrammed computer. A single key stroke executes a specific program to compute the needed function. The portable calculator can perform all the calculations that scientists and engineers traditionally have worked out on slide rules, and do them faster and easier.

Answers are accurate to 10 significant digits, far beyond the capability of any precision slide rule. The MOS circuits operate more slowly than the more familiar bipolar transistor circuits, but they are still much faster than man. The HP35, for example, can add or subtract two 10-digit numbers in 60 thousandths of a second.

The MOS/LSI technology made it practical to produce complex electronic networks containing thousands of transistors on a single chip at a cost of less than a penny per transistor. The chip's tiny size, however, sharply limits the number of terminal leads available for interconnection; thus, it limits the device's operating speed.

Modular design was greatly improved by LSI technology. In late 1971,

for example, Intel Corporation of Santa Clara, Calif., introduced its MCS-4 microcomputer set for general computing and control applications. The set of four functional modules, each fabricated on an LSI chip, is priced at less than $100. The set—the heart of a computer—contains a central processor module, a data storage module, a control memory module, and an input/output module.

Equipped with its adder circuit, a number of data and address registers, and the associated control logic, the central processor can execute up to 45 machine instructions. This is enough to carry out all the operations commonly associated with modern, full-sized computers. The control memory, preprogrammed to the customer's specifications, stores the microprograms and data tables needed for routine operation of the computer and its peripheral equipment, such as teletypewriters, printers, and card readers. Multiple modules can extend the capacity of an MCS-4 system.

The system of functional modules can be updated by inserting new microprograms, or it can be expanded by adding

Computers

more functional modules to meet changing needs. The specialized, modular design computer is expected to be used widely in laboratory instrumentation and industrial process control.

Advanced computers. One of the largest and fastest new computers is being built by Texas Instruments, Incorporated, in Austin, Tex. It will be used to study global weather patterns and establish techniques for long-range, worldwide weather prediction. The new device will carry out computation of changing conditions at each of the thousands of points in all parts of the world through a mathematical model of weather dynamics. Because the conditions at each point are affected by changes at surrounding points, large volumes of information from many sources must be processed simultaneously at very high speed. The new machine is expected to simulate global weather patterns for a full year in only 60 hours. Such a task is beyond the capability of any of today's computer systems.

Computer network. A nationwide computer network, ARPANET, linking facilities from coast to coast, virtually doubled its size from mid-1971 to mid-1972. Developed by the Advanced Research Project Agency (ARPA), the network had linked computers in nearly 30 universities and other installations by mid-1972. Special equipment developed for ARPANET allows computers of different design to communicate with each other with great flexibility and speed. They are tied together by a network of high-fidelity telephone lines.

A computer message is routed automatically to its destination in tenths of a second. When facilities are not available at one location, they can be borrowed from another network computer.

Through ARPANET, research groups will thus have access to a wide range of computing facilities. A researcher will also be able to make quick use of computer programs and reference data that are available at other institutions. As it became a functioning entity rather than an advanced research project, consideration was being given to the proposal that the network be made independent of ARPA.   [Arthur W. Lo]

Drugs

The Food and Drug Administration (FDA) removed the artificial sweetener saccharin from the "Generally Recognized As Safe" (GRAS) list of food additives on Jan. 28, 1972, and restricted its use while additional safety reviews were being completed. Previously, in 1969, the FDA banned the general use of cyclamates, another popular sugar substitute after reports that they produced cancer in mice.

The new FDA order recommends that the average adult use no more than one gram of saccharin a day. This amount is found in seven 12-ounce bottles of the standard diet drink or 60 saccharin tablets, each the equivalent of one teaspoonful of sugar.

Saccharin has been widely used in foods for over 80 years without evidence of harm to human beings, but some test rats given about 100 times the maximum dose of the new regulation developed tumors of the bladder.

In January and February, 1972, the FDA also proposed severe restrictions on hexachlorophene, an antibacterial agent used in many soaps, shampoos, lotions, ointments, powders, and deodorants. Hexachlorophene can be absorbed into the blood stream through the skin, especially if it is scraped or burned.

In an experimental study by the Sterling Drug Company, reported to the FDA in November, 1971, baby monkeys who were bathed daily for 90 days with 3 per cent hexachlorophene developed *edema* (swelling with water) of the brain. In human beings, use of large amounts of hexachlorophene to cleanse extensive burn areas of the skin has caused damage to the central nervous system, and death in some cases.

On Feb. 2, 1972, the FDA advised hospitals against the standard procedure of bathing infants in a 3 per cent hexachlorophene solution. They recommended the baths only when the hospitals are faced with outbreaks of staphylococci infection, against which the substance is particularly effective. The FDA approved the general hospital practice of hexachlorophene hand washing by hospital personnel.

Over-the-counter drugs. Although many drugs are available in the United

Drugs

Continued

Drug-dependent veterans returning to the United States will be treated at one of 60 drug-abuse centers. The Veterans Administration estimates that the new centers can handle 10,000 patients.

States only on a doctor's prescription, thousands of remedies for minor illnesses and symptoms can be purchased at will. In 1972, the FDA announced a major scientific and regulatory program designed to assure that all over-the-counter drugs are safe, effective, and accurately labeled. A panel of physicians will evaluate these agents under the supervision of a National Drug Advisory Board, chaired by the FDA commissioner. The study, which will require two to three years to complete, is aimed at establishing a scientific basis for the regulation of over-the-counter drugs in ways comparable to those used for prescription drugs.

Drug bioavailability—the amount of a drug available at its point of action in the body—has received considerable attention. Investigators at New York City's Columbia University College of Physicians and Surgeons and Harlem Hospital Center measured the blood serum levels of digoxin in human volunteers after the volunteers had received pills produced by a number of different manufacturers.

Digoxin is a popular cardiac drug. Its chief therapeutic effect is to increase the force of the heart's contraction when it fails to pump adequately due to damage caused by various forms of heart disease. An estimated 3.5 million Americans spend about $40 million each year on this drug, which is manufactured by at least 30 companies.

The physicians found that the level of digoxin in the blood varied by from four to seven times with the same dosage of the drug from different manufacturers. A number of factors, including drug particle size, disintegration and dissolution rates, and the effects of various inert ingredients, probably account for the wide differences.

Similar discrepancies in bioavailability have been reported for other medicines. This is significant, because many drugs, such as digoxin, produce toxicity at blood levels that are not much greater than those necessary for the beneficial effect. Thus, if a patient's prescription for one of these compounds is filled using pills from a different manufacturer, or even a different lot produced by

Veterans Administration Drug Addiction Centers

- ■ **Specialized Addiction Centers (32)**   □ **Neuropsychiatric Services (28)**

Drugs
Continued

the same manufacturer, a relatively greater or lesser amount may be "available" in his blood.

The FDA published a list of drugs in 1971 for which biologic availability as well as chemical potency must be demonstrated as a requirement for use in patients. Digoxin may now have to be added to that list.

Vitamin E. Many recent articles in medical and lay publications have recommended substantial doses of vitamin E for a variety of disorders including sterility, habitual abortion, and various forms of vascular disease. It has also been said to retard the aging process.

Unfortunately, there is insufficient evidence to substantiate these claims in human beings. In fact, there is so little known about vitamin E that nobody has been able to figure out what constitutes vitamin E deficiency. Thus, although it is considered an essential nutrient in man, the actual vitamin E requirement for humans is unknown.

Some premature infants who developed hemolytic anemia, a breakdown of red cells, while being fed on certain for-mulas improved when given vitamin E. Some patients who absorb an abnormal amount of dietary fats also may benefit from supplemental vitamin E. However, most nutritionists feel that there seems to be little to be gained from its use except in special circumstances.

New drugs introduced during 1972 included:

• Trobicin (spectinomycin hydrochloride), a new antibiotic for use in the treatment of gonorrhea, especially when the maximum dose of penicillin has proved ineffective or the patient is allergic to penicillin. The new drug is not effective against syphilis, however. This is a potential drawback, because patients frequently suffer from these two venereal infections simultaneously.

▪ Narcan (naloxone hydrochloride), a narcotic antagonist that reverses many of the effects of narcotic agents, especially respiratory depression. Unlike other agents in this group, Narcan does not produce narcotic effects when given alone and will reverse the effects of Talwin (pentazocine), a relatively new narcotic. [Arthur Hull Hayes, Jr.]

Earth Sciences
See Geochemistry,
Geology,
Geophysics,
Meteorology,
Oceanography

Ecology

Three ecologists reported in March, 1972, that the rainfall over the northeastern United States has become acidic. Gene E. Likens of Cornell University in Ithaca, N.Y.; F. Herbert Bormann of Yale University in New Haven, Conn.; and Noye M. Johnson of Dartmouth College in Hanover, N.H., conducted tests that revealed pH levels of 3.5 to 4.5, about the same as the acidity of orange juice. Normally, rain water is neutral, with a $pH7$.

The scientists blame the large quantities of sulfur dioxide and other pollutants being released into the air by electric generating stations and industrial plants that burn fossil fuels—coal, natural gas, and oil. Several pollutants from these fuels form acids in water.

Acidic rainfall had been detected previously by scientists in Sweden and Norway. It causes much concern because if it continues, the acidic rainfall may damage forests and crops either directly, by burning and otherwise harming or killing young, tender leaves, or indirectly, by increasing nutrient drain off and otherwise altering soil chemistry.

Peat bogs. Marshy areas covered by highly organic turf called peat—are interesting to ecologists because they preserve materials that fall into them while remaining virtually unchanged themselves. Living things, such as pollen grains, that fall into such bogs are preserved because of certain natural conditions—acidic soil virtually devoid of oxygen. Thus, by examining peat at varying depths, ecologists can trace some of the history of a bog area.

Ecologist LaVerne H. Durkee used this method to work out the history of the postglacial vegetation of a bog area in Hancock County in north-central Iowa. The Grinnell (Iowa) College ecologist reported his work in the late summer of 1971. He analyzed cores of material from as deep as 30 feet.

Durkee determined the abundance of different plant species at particular times by checking the pollen at various depths in the core. The age of the soil at various levels was found through the radiocarbon dating technique.

The earliest pollen he found came from spruce fir trees. These trees evi-

Analyzing Ecosystems

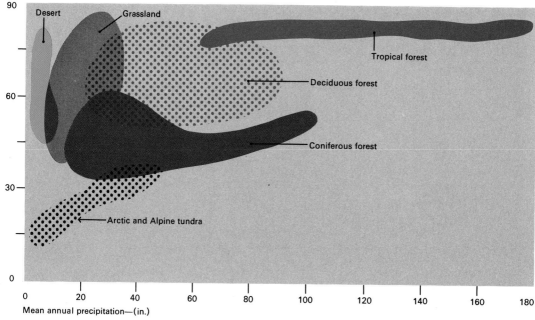

Mean annual temperature — (°F.)

Source: National Science Foundation

Ecology

Continued

A chart created by ecologists relates mean annual temperatures to mean annual precipitation for various ecosystems throughout the world.

dently began to grow in the area about 12,000 years ago. A forest of deciduous trees, such as oak, hickory, and maple, replaced the spruce firs about 10,500 years ago. From 9,000 to 7,500 years ago, there was a progressive increase in low, leafy herbaceous plants such as dandelions and other wild flowers. By the end of this time, the prairie flora, such as the daisies and wild carrots, or Queen Anne's lace, that dominate the bog today was well established.

Ecologists Eville Gorham and Ronald H. Hofstetter of the University of Minnesota reported in the late summer of 1971 that the peat bogs in northern Minnesota did not retain radioactive tritium, a fallout by-product of nuclear weapons tests.

The scientists dug out long cores of peat from the bogs and tested sections of them for tritium. They found no evidence of tritium in sections from about 13 feet or more below the surface of the bog. Much of the tritium, which gets into the bog in precipitation, is picked up by plants from the soil along with water. The plants then return it to the

air, along with water, by releasing it in a natural process called evapotranspiration. Most of the rest of the tritium is carried away in run-off into low-lying areas around the bog.

Postglacial vegetation. In early 1972, Alan J. Craig of the Limnological Research Center at the University of Minnesota reported research that had traced the history and development of plants in a portion of the Boundary Waters Canoe area of northeastern Minnesota. Craig analyzed sediment cores from the area's Lake of the Clouds.

Old lake sediments contain laminations, or layers, composed of alternating dark organic and light inorganic matter that provide a time scale for the material they contain. Craig's cores contained 9,349 laminations representing the same number of years. Through analyses—mostly of pollen—of the laminations, Craig could describe the vegetational history of the region around the lake.

After the glacial ice retreated from the region, plant species from the surrounding areas colonized the deglaciated terrain, but the vegetation was predomi-

Ecology

Continued

Ecologist Samuel M. McGinnis of California State College, Hayward, checks a tortoise's temperature by listening to the number of clicks per minute on a radio. The radio signal comes from a tiny pill-sized, temperature-sensitive transmitter that the ecologist fed to the tortoise. The signal booster on the animal's back allows monitoring for up to a quarter-mile.

nantly mosses and other species that do not produce pollen. These early colonists were succeeded by low, grassy vegetation like that found on tundras. Willows also grew there. Next, as the climate warmed up, a boreal, or cool-weather, spruce forest entered the region from the south. This forest lasted until about 9,200 years ago, and as the climate continued to warm, it was replaced mostly by jack pine and red pine trees. Alder began to grow in the area about 8,300 years ago, and white pine began about 7,000 years ago.

At about the same time as the arrival of white pine, the climate turned drier. This caused prairies to develop closer to the lake than before. About 3,000 years ago, wetter, cooler climate returned and pine trees gave way to boreal spruce and cedar–similar to today's conditions.

Feeding ecology. In winter 1972, ecologists Christopher C. Smith and David Follmer of Kansas State University of Agriculture and Applied Science, Manhattan, reported on their study of the feeding ecology of two common species of squirrels, the gray squirrel

(*Sciurus carolinensis*) and the fox squirrel (*Sciurus niger*). The scientists had set out to answer several questions including:
- Do the two squirrel species differ in the efficiency with which they use different foods?
- Is there a logical reason for the food preferences of squirrels?
- How do the squirrels' food preferences affect the plant species they feed from?

The scientists tested squirrels of both species in the laboratory for nut preferences and rates of feeding and correlated the data they obtained with data on efficiency of digestion of the different nuts. The results indicated that the two species of squirrels have the same food preferences and digest their foods equally well.

Both eat the hard-shelled nuts of hickory and walnut, except in the winter. Then, during brief interruptions of their hibernation, they prefer acorns. This fits in with their seasonal activity patterns and resultant nutritional needs. The hard-shelled nuts provide more energy and protein than do acorns, but they are harder to find and to crack. Acorns are more plentiful and ideal in winter, when

Ecology

the animals need little nutrition because they are hibernating.

The squirrels' habit of burying many nuts of the types they like to eat has an interesting result: They "plant" and, thereby, promote mixed stands of the species of trees that bear their food.

Fox squirrels live in open forests and along the edges of forests, whereas gray squirrels inhabit dense forests. Since their food preferences are alike, the squirrels' habitat differences are probably based upon differences in how they forage for food and escape from predators. For example, the fox squirrel is larger than the gray squirrel and, therefore, is better able to defend itself. It is also more likely than the smaller squirrel to try to avoid danger by hiding. The gray squirrel is swift and agile. Its first line of defense is to scurry into the trees and leap from branch to branch to make its escape.

A more complex problem in feeding ecology is tracing various substances through food webs, or food chains. One such web exists among the insects and other tiny creatures that live in the organic layers of the upper soil, especially in forests. These creatures are important factors in some decomposition, the means whereby all dead organic matter and the nutrients it contains are recycled for the ultimate production of new living tissue. The creatures eat the material and their digestive systems put it through an initial breakdown, readying it for final decomposition by fungi and bacteria.

In the late spring of 1971, ecologists Norman E. Kowal and DeReyee A. Crossley, Jr., of the University of Georgia in Athens reported new data on ingestion rates of pine litter by such creatures, and on the consumption of the creatures by predators. The scientists used pine litter that had been made radioactive by injecting radioactive calcium into the trees upon which the living material had grown. Then, they raised natural populations of the creatures and their predators in microcosms—standard test-tube-sized vials to which some of the litter had been added.

By regularly measuring the amount of radioactive calcium in the living organisms and the litter, the scientists determined rates of consumption and elimination of the litter. The tests showed that the types of creatures tested consume only about 1 per cent of the pine litter that drops on the ground in the forest. Their predators consume about 17 per cent of that 1 per cent. The rest is eaten by other creatures or broken down directly by bacteria and fungi.

Allelopathy. Ecologists Roger del Moral and Rex G. Cates of the University of Washington, Seattle, reported in late 1971 that allelopathy played a role in the distribution of vegetation in many areas they studied throughout the state of Washington. Allelopathy is a process in which one plant releases a chemical, or chemicals, that retard the germination, growth, or metabolism of another.

The process is distinct from competition, in which one plant may eliminate another by depleting a vital resource such as nitrogen or water. Studies of allelopathy have now been carried out in several parts of the world in a variety of climates. These studies have revealed that the process is widespread and often considerably alters the makeup of plant communities.

Del Moral and Cates tested 40 plant species in the laboratory for production of inhibitors. They also compared the effectiveness of the plants' airborne inhibitors to that of their waterborne inhibitors in soil. Then, the scientists checked their results against the distribution of plants growing under natural conditions in the same area as the species tested. In several cases, especially with maple trees, arbutus, and rhododendrons, high inhibition values in laboratory experiments were associated with distinct changes in ground-cover plants around these species.

The scientists also found that in the dry zones in eastern Washington, allelopathic inhibition due to airborne inhibitors was more common than that due to waterborne inhibitors. They found that the opposite was true in wetter western Washington.

The experimenters also confirmed something that ecologists have suspected but not proved about allelopathy. The inhibitory substances involved are principally natural waste products of the plants that produce them. Thus, allelopathy, though it plays a prominent role in eliminating some species in favor of others, does not seem to have been developed as a weapon to eliminate competition. [Stanley I. Auerbach]

Education

Two novel attempts in science education were tested during the 1971-1972 school year. Science educators were searching for new ways of teaching that will encourage students to explore their own interests and learn by building upon them. This effort was an outgrowth of a statement by James B. Conant, the president emeritus of Harvard University. Conant estimated that only 15 per cent of the students in the United States learn well in the traditional classroom, where they attempt to reach rigid goals through traditional book learning and study in an unimaginative academic setting. Under these conditions, the 15 per cent who learn does not include many bright students.

The minischool. A new experimental out-of-school science program was operated in New York City. It was financed by a National Science Foundation (NSF) grant to Frances Behnke of the Department of Science Education at Teachers College, Columbia University, to direct the program. During the year she and general science teacher Melanie Barron co-directed a unique coopera-tive effort with Manhattan's Joan of Arc Junior High School, Teachers College, and the New York University Medical Center. Eight Joan of Arc teachers and four graduate-student interns from Teachers College formed an instructional team for 120 ninth-grade students selected at random. In workshops and seminars, the teachers and interns developed investigative, interdisciplinary curriculum materials that were specially designed for the program.

Just being in a conventional schoolroom is oppressive and uninspiring to many inner-city students, so the program included establishing a minischool in a dilapidated Manhattan storefront. Under the gentle direction of the interns, the students decided how to fix it up. The minischool, therefore, became their school.

Four days a week, each student pursued his own interests in science at his own pace. The teacher was neither an authoritarian nor a decision maker, but was available to facilitate learning by guiding, suggesting, and providing insights into difficult problems.

By colliding carts that roll on casters, students experience the forces that they study in physics classes. This innovative method of teaching was devised by physicist Robert Johns of the Academy of the New Church in Bryn Athyn, Pa.

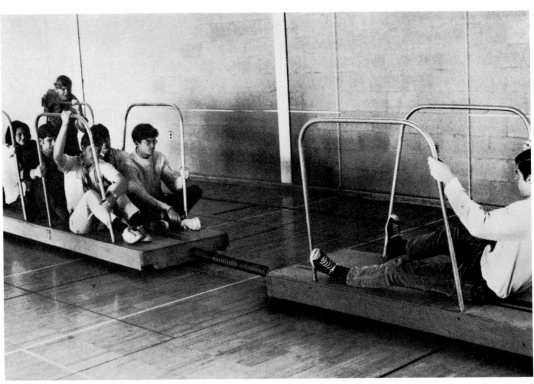

Sexism
In Science

Affirmative action by the American Association for the Advancement of Science (AAAS), the nation's largest general scientific organization, highlighted a year of intense efforts to improve opportunities for women in science. At the AAAS's 138th annual meeting, held in Philadelphia in December, 1971, the following resolution was introduced by Hazel M. Fox of the University of Nebraska and immediately adopted:

"Whereas the talents and contributions of women in science are not fully recognized, and whereas there is no central listing of women in science, I move that the council request the board of the American Association for the Advancement of Science to consider establishing an Office for Women's Equality to work toward full representation and opportunity for women in scientific training and employment affairs of the association, and in the direction of national science policy."

That women are underrepresented in all scientific disciplines had been shown clearly by the 1970 Manpower Survey of the National Science Foundation. Although women make up about 38 per cent of all undergraduate students in American colleges and universities, only 4 per cent of all physicists, for example, are women. In 1970, women earned these percentages of doctorate degrees: engineering, 0.4; physical science, 5.7; life sciences, 12.9; psychology, 23.5; and other social sciences, 11.9.

This may reflect the desire of some women to enter more traditionally "feminine" fields. However, it may also result from discriminatory practices. In December, 1971, Commissioner Ethel Bent Walsh of the Equal Employment Opportunities Commission said such discrimination "is accomplished by admission quotas in undergraduate and graduate schools; higher admission standards for women than for men; and discrimination in financial assistance for graduate study."

Although a presidential executive order issued in October, 1968, prohibits federal contractors from discriminating on the grounds of sex, the Office for Civil Rights of the U.S. Department of Health, Education, and Welfare (HEW)

Median Salaries of Scientists

Scientific and Technical Field	Doctorate		Masters		Bachelors	
	Male	Female	Male	Female	Male	Female
Agricultural Sciences	16,300	—	12,500	—	12,000	—
Anthropology	14,800	12,300	—	—	—	—
Atmospheric & Space Sciences	17,600	—	15,800	—	15,000	—
Biological Sciences	16,000	13,000	11,300	10,000	10,800	9,000
Chemistry	17,400	12,800	15,000	10,600	13,800	9,900
Computer Sciences	19,700	—	16,500	13,100	16,000	13,300
Earth & Marine Sciences	15,600	13,500	14,000	10,000	15,000	9,600
Economics	17,400	14,800	14,200	12,000	16,300	13,400
Linguistics	13,500	12,000	10,300	9,300	10,200	—
Mathematics	14,700	12,300	12,900	9,600	17,400	9,200
Physics	17,300	13,400	14,100	11,100	14,600	10,800
Political Science	14,500	12,000	10,500	10,100	10,700	—
Psychology	16,000	14,000	13,100	12,000	14,600	13,500
Sociology	15,000	13,000	10,100	9,500	10,000	9,500
Statistics	17,400	13,200	16,200	14,300	17,000	14,000

Source: 1970 Manpower Survey of the National Science Foundation

did not begin to enforce it until January, 1970. Then feminist groups, such as the Women's Equity Action League and the National Organization for Women, began to file charges. By July, 1972, about 360 colleges and universities had been charged and more than 80 had been investigated. The slow pace of change may be because HEW approves hiring goals that often fall short of achieving a representation of women in universities proportional to the number of women with Ph.D. degrees in current and future labor pools.

The lack of a national roster of women scientists further hampered the appointment of women to academic positions. The usefulness of a roster was demonstrated when the National Institutes of Health was challenged to achieve equal representation on its study panels. Several proposals to create national and regional rosters are under consideration.

In the long run, equal opportunities depend on removing the admissions barriers in colleges and universities. Eliminating quota systems and discrimination in awarding fellowships and scholarships could be achieved by extending existing legislation to cover students in federally assisted programs. The education bill signed by President Richard M. Nixon on June 23, 1972, includes amendments that forbid such discrimination, but only at the graduate level.

More difficult problems for women scientists involve interruption of their careers at potentially the most productive stage to meet family obligations. In recognition of this, the Radcliffe Institute was established at Radcliffe College, Cambridge, Mass., in 1960. In 1971 and 1972, more than 200 women benefited from its fellowship programs, informal career counseling, and research on the educated woman's motivations and role in society. Mid-career programs of this type help women return to scientific pursuits or maintain them on a part-time basis.   [Arie Y. Lewin]

Education

Continued

The four classes of ninth-graders went to the medical center one day a week for an individualized life science program. Their day was divided into three parts. In the morning, they studied science in the medical school laboratory. A film and discussion hour followed. In the afternoon, each student had a work-study session with a member of the medical center professional staff.

The students worked on projects in microbiology, anatomy, physiology, care of laboratory animals, and many other areas of study. Thus, at the medical center, science was not just classroom learning but real experience.

The entire program was aimed at creating a high degree of motivation in the students, and preliminary evaluation by a team of researchers under the direction of Professor O. Roger Anderson at Teachers College shows that it was quite successful.

Environmental Studies (ES), a new and still incomplete set of multidisciplinary curriculum materials, tries to build self-esteem in all students. The program is funded by the NSF and sponsored by the American Geological Institute in Boulder, Colo. ES is environmental science in its broadest sense. It stresses the student's inner environment (relationships within himself), his immediate environment (classroom, schoolyard, community), and the global environment.

A complete set of ES materials for each classroom costs $20 and consists of 100 assignment cards and a booklet called "ESSENCE–ES sense,"–a teacher's "nonguide." The teacher may use the cards as a curriculum or as supplementary materials. ES is being used from kindergarten through graduate school. Only the most commonly available school equipment is necessary (old magazines, masking tape, clay, water-based paints, etc.). However, Polaroid cameras and portable tape recorders are highly recommended.

"ESSENCE–ES sense" is an idea booklet that attempts to acquaint the teacher with the philosophy of the program and encourage in him an attitude that will make use of the ES materials successful. The guide stresses human qualities and open, interpersonal rela-

Education

Continued

tionships based on mutual trust between students and teachers. Teachers are reminded that cameras and tape recorders are tools to be used, rather than museum pieces. Students must be trusted to use the equipment on assignment outside the classroom without supervision, and to invent criteria for evaluating their own learning.

The assignment cards are deceptively simple. They are problem-oriented, and therefore cross-disciplinary, as opposed to traditional, rigidly separated disciplines such as geology, reading, or chemistry. Each card has a visual metaphor of the assignment on the front. For instance, a picture of a lawn with the caption "Show Me a Million" refers to a million blades of grass. Teacher cues on the back of the card are headed by "The Action," which gives the primary assignment ("Go outside and find a million of something and prove it.").

Because the student makes his own decisions based upon his own personal interpretation of the assignment, the answers he brings back are likely to be correct. This results in success, which in turn increases his self-esteem. Instead of being rewarded for conforming, students are rewarded for their inventiveness and diversity.

Most of the test population of more than 50 schools has consisted of inner-city students from New York City to Los Angeles, but ES has also been used effectively in rural and suburban schools. In addition, the program provides excellent training for college students in teacher-preparation courses because they are forced to closely examine their relationship with students.

As a general rule, teaching materials must withstand a rigorous evaluation before they are widely accepted. But, because the primary objectives of ES lie mainly in the area of feelings, rather than in learning facts, it will be difficult to get hard data to judge its success.

ES was first tested during the 1970-1971 school year. A study of comments by teachers and students involved in the program was underway in 1972, and an evaluation of what factual information was learned in the traditional areas of study is planned. [Linda Gail Lockwood]

Electronics

Developing practical applications for the charge-coupled device (CCD) was one of the leading advances in electronics during 1971 and 1972. When the device was announced by Bell Laboratories scientists in 1970, its potential in computer memories, image sensors, and other functions was promptly recognized. After two years of laboratory development, it was ready for use.

The CCDs' major feature is that they are simple to make, compared to the transistors and diodes that do the same job. Basically, a CCD consists of a silicon layer treated to have either positive or negative characteristics and covered with a thin layer of silicon oxide. A series of metal electrodes is laid down over this. Voltage applied to the electrodes forms "potential wells" under them. Information stored as a sequence of charges within the wells is retrieved by electronically shifting the charges' position from well to well. In this way, the information moves along the line of electrodes.

The CCD device has a number of useful applications. In computers, for example, a shift register is a commonly used circuit in which binary numbers are stored temporarily and shifted out as they are needed for further processing. CCD shift registers, because they are simpler and smaller, can process 10 times more information than conventional transistor types.

The CCD can also function as an image sensor. A silicon CCD is an excellent optical detector throughout the visible spectrum—potential wells can be created by shining a light on it. Thus a completely solid state version of a television camera pickup tube, or vidicon, can be made of a CCD array.

With the proper arrangement, electrons are freed in the silicon and captured in the potential wells under the electrodes when the silicon layer is exposed to light. In effect, the image is stored electronically. Then the stored image can be shifted out, line by line, much as in the conventional vidicon.

Bell Laboratories is particularly interested in developing a practical CCD vidicon because of its stake in Picturephone, the Bell System service that shows users a television image of each

Disk-shaped metal oxide varistor, center, saves diode from simulated lightning. The device protects any electronic or electrical equipment against power surges.

Electronics

Continued

other during a call. Widespread commercial use of Picturephone has been delayed by problems with vidicon reliability and manufacturing. Bell Labs scientists have made a demonstration CCD vidicon. It does not have quite enough resolution for Picturephone, but the scientists see no great problem in achieving the necessary level. Other versions of the CCD vidicon have been developed by RCA Corporation, Fairchild Semiconductor, and General Electric Company.

Electronic medicine. Electronic heart pacemakers must be worn for long periods of time by patients with severe cases of cardiac irregularity. The pacemaker stimulates the heart regularly to keep it beating at a steady pace.

But the pacemaker's batteries, which must be implanted within the patient's chest along with the pacemaker, run down and must be replaced about every two years. This calls for costly and risky surgery. One solution to this problem is a nuclear-powered pacemaker in which a small amount of plutonium 238 provides enough power to run the circuit

for many years. Medtronics, Incorporated, a Minneapolis firm, is developing the nuclear-powered devices using a power supply manufactured in France. In July, 1972, the federal government approved tests in which up to 10 of the devices will be implanted in humans.

Robert I. Bernstein, professor of electrical engineering at Columbia University, and a team of Israeli researchers propose another way to avoid the surgery. In their design, a radio frequency transmitter would be held against the patient's chest for about eight hours every four to six months to recharge the pacemaker's nickel-cadmium batteries.

Laser protection for teeth. Man's age-old enemy—tooth decay—may yield to the laser, according to a dentist at the school of dentistry of the University of California at Los Angeles. Ralph H. Stern reported that he had successfully prevented decay with a laser.

Stern aimed a carbon dioxide laser emitting 25 watts of pulsed infrared power at the surface of teeth that had been extracted. Both treated and untreated teeth were then inserted into bridges worn by other patients. Thin gold plates were placed over the bridges to trap food particles and debris. After four weeks, the untreated teeth indicated cavities, but the laser-treated teeth did not. Stern explained that the laser actually glazed the outer surface of the tooth, sealing the enamel against decay.

More hi-fi channels. The high-fidelity field saw the beginning of a debate over which new recording technique should win general acceptance. Such catchwords as "Quadrasonics," "Quadraphonics," and "Quadraline" proliferated as names for systems designed to expand conventional two-channel stereo broadcasting and recording to four channels. The campaigns were reminiscent of battles over phonograph record speeds and color television techniques.

The issue hinges on whether discrete or matrixed channels will become standard. In the discrete system, four separate tracks are broadcast or recorded. In the matrixed system, two additional channels are superimposed on a conventional two-channel medium, such as tapes or disks, using an electronic encoding technique. An appropriate decoder is needed in receiving or playback equipment to separate the channels.

RCA Records, teamed with the Japan Victor Company, is producing discrete four-channel disks for use on equipment to be marketed by Panasonic and Japan Victor. Columbia Records Division of CBS, Incorporated, is teamed with Sony Corporation to produce matrix-type disk recordings and playback equipment.

Pictures from space. A new electronic technique dramatically improved television reception from space during the Apollo 16 lunar mission. It involved a method of electronically compensating for the poor low-frequency response of the transmitting equipment used during the mission. Simultaneously, the new method corrected for noise or "snow" that marred earlier pictures.

Although details were not available, a computer is apparently used to predict the extent of the transmission noise and correct for it. The method was developed by John Lowry, a vice-president of Image Transformers, Incorporated, a small California firm. [Samuel Weber]

Energy

Americans were warned repeatedly during 1971 and 1972 of a fast approaching energy supply crisis. Government, industry, and university sources predicted that energy resources will be in painfully short supply in the years just ahead, that the problem will grow more severe in the late 1970s, and that long-term prospects are, at best, uncertain.

Federal Power Commissioner John A. Carver, Jr., summed up the situation in March, 1972, saying, "... our shortage is not only endemic, it is incurable. We're going to have to live with it the rest of our lives."

Other experts were less pessimistic, but all agreed that the failure to anticipate the crisis had been a major contributor to the problem. Most observers recognized, however, that no imaginable combination of technical innovations could now eliminate the problem completely or forestall the need for drastic adjustments in industry and in individuals' living habits.

Natural gas supply shortages typified the dilemma and drove the energy industry to reach for new sources of both natural and artificial gas. Natural gas, essentially pure methane, is an almost ideal fuel for many uses. It has almost no sulfur or other environmental pollutants. It now supplies about a third of the nation's fuel needs, and might be used in many other ways if cheap, abundant reserves were assured. It could serve, for example, as a clean, efficient substitute for gasoline.

The supply is already so low, however, that several eastern utilities began to turn down requests to provide it for new home furnaces in 1971. The Department of the Interior estimated in January, 1972, that the total amount available, including imports, could fall 17 trillion cubic feet short of the annual demand, which it put at 38 trillion cubic feet, by 1985. Gas produced from naphtha by well-established processes cannot begin to fill this gap.

Finding substitutes. Coal is still relatively abundant, but industry has not yet produced high-quality gas from it. In addition, burning coal adds to air pollution. Flue gases from virtually all coal- or oil-fired generating plants con-

Energy

Continued

A lithium-chalcogen storage cell is being assembled in the inert atmosphere of a sealed glove box. Larger versions of this electrochemical power cell may help to solve peak energy demands.

tain sulfur dioxide. In an effort to reduce this problem, industry has tried scrubbing the offending sulfur from the flue gas and making clean-burning gas from the coal. As a short-range solution, some utilities have been burning desulfurized oil and coal or oil that contained little natural sulfur. But such fuels are scarce, costly, and often unable to meet the more rigid anti-pollution standards that are expected to take effect soon.

Industry tested many sulfur dioxide scrubbing techniques in 1971 and 1972, but they promised no quick, universal solutions. A lime slurry scrubber, placed in service in a Commonwealth Edison coal-burning station near Chicago in early 1972, was described as the world's largest, but the plant's small size indicated clearly that the process was not yet far advanced.

The Tennessee Valley Authority (TVA) planned to install a full-scale scrubber at Widows Creek, Ala. Experience at a Muscle Shoals, Ala., pilot plant, however, convinced TVA officials that scrubbing with water suspensions of lime or limestone, the most popular mediums, involved problems having "no ready solution," despite nearly 40 years of development.

Gas from coal. Sulfur scrubbing may continue too costly to be of much help in reducing pollution. Several industry groups planned instead to use low-heat-value gas made from high-sulfur coal. One such group, led by Westinghouse Electric Corporation, proposes to make the gas at plants near the mines. The gas will be cleaned before it is burned in a combined-cycle generating plant at the site. Such plants use heat discharged by a primary gas turbine section in a secondary steam-powered section. They are about 10 per cent more efficient than conventional plants. Utilities began buying intermediate sized combined-cycle plants, which are well suited to burning the lower-heat-content gas, in 1971.

Economical ways to make a high-heat gas from coal to rival natural gas is a goal of several projects. One such project, sponsored by the U.S. Bureau of Mines, uses a nickel-based catalyst. Even a pilot demonstration of this "Synthane" process is at least three or four years away, however, and no competing process seems nearer.

Nuclear explosives detonated deep in submarginal gas-bearing formations offer another way to supplement gas supplies. This method's technical validity was demonstrated by 1967's Project Gasbuggy in New Mexico and 1969's Project Rulison in Colorado, both conducted by industry and the Atomic Energy Commission (AEC).

Submarginal gas-bearing formations occur widely in the western United States. The process seemed economically promising, and the earlier experiments suggested the gas's low radioactivity poses no hazard to users.

Widespread concern about nuclear explosions' effects, however, raised doubts that this technology can relieve shortages soon. Opposition to the blasts threatened the next scheduled AEC-industry experiment, Project Rio Blanco. That test is planned for late 1972 or early 1973 about 50 miles north of Grand Junction, Colo., in the Mesa Verde geological formation. It will focus on the advantages of using three almost-simultaneous detonations. If the test goes well, the industrial sponsors plan to use this technique to tap important gas deposits 5,000 to 9,000 feet below the surface in that region. The Rio Blanco project's general manager estimated in April, 1972, that 300-trillion cubic feet of natural gas were "potentially recoverable" by such means.

Fast breeders. Construction of a medium-sized fast breeder nuclear reactor gained impetus with the recognition of the energy supply problem. The project was identified by President Richard M. Nixon in June, 1971, as one of the highest priority energy programs. Preliminary arrangements to build a demonstration plant in eastern Tennessee were announced Jan. 14, 1972, but construction is not expected to begin before 1974. Such sources can hardly become significant until the late 1980s.

The great advantage the fast breeder holds over current big power reactors is its extremely efficient use of nuclear fuel. As it uses plutonium 239, it converts natural uranium 238 into more fissionable plutonium. But in April, 1972, 31 scientists and other professionals unsuccessfully urged Congress to refuse funds to start building the breeder. They questioned the safety of the plants, the plutonium handling methods,

Energy

Continued

and the disposal of radioactive wastes.

Breeder reactors were nearing completion in France, Great Britain, and Russia, however. British and Russian stations were expected to start generating electricity before 1973.

A permanent place for nuclear wastes continued to be sought by the AEC. The agency planned to bury radioactive wastes in old salt mines and had spent from $7 million to $8 million preparing an old mine near Lyons, Kans. Strong political and environmental pressures opposed the project, and the Kansas Geological Survey was asked to study possible locations. In January, 1972, it called the Lyons area the worst of eight it had studied. The AEC dropped its underground storage plan.

Other possibilities for waste disposal were also proposed. Scientists from the AEC's Lawrence-Livermore Laboratory in California discussed the use of atomic explosives to create chimneys of broken rock about a mile under nuclear fuel reprocessing plants. Radioactive waste would be stored there and its heat eventually should melt the rock and seal the chamber. (See GEOLOGY) The AEC also said rocketing waste into the sun was being studied.

Nuclear power from atomic fusion had more appeal than that from fission because fuel would be abundant, the process could be safer, and pollution danger would be low. For those reasons, the decision by the Nixon Administration to concentrate on fast-breeder development instead of on fusion raised objections from some environmentalists.

Fusion's feasibility as a controlled energy source had not been scientifically proved, however. Breeders thus held more immediate promise.

New batteries. Other energy programs aim at developing highly efficient, compact electric storage batteries. One such program, at the Argonne (Ill.) National Laboratory, seeks to develop a lithium-anode, sulfur-cathode battery for use in electric vehicles. The laboratory believes that large batteries of this kind could act as local electricity reservoirs to help meet peak demands, blamed for power "brownouts" in many areas. [Ed Nelson]

Environment

Extravagant publicity for environmental problems subsided in 1971 and 1972 and the difficult task of finding solutions began in earnest. Scientists in many disciplines struggled to define the nature and causes of environmental problems, while the public tried to relate each new crisis to the many older problems clamoring for attention. Scientists warned that environmental cleanup will be costly and some argued that fundamental changes in industrial societies are needed.

Vietnam environment. The war in Southeast Asia ranked as one of history's most massive experiments in environmental change. The United States stopped spraying herbicides over much of the countryside early in 1971 following widespread protests by scientists. Spraying denied cover to the enemy, but also destroyed food crops. But, other forms of destruction continued.

In August, 1971, biologists Egbert W. Pfeiffer of Montana State University and Arthur H. Westing of Purdue University visited Vietnam for the Scientists' Institute for Public Information. Later, they reported these findings: About 12 million tons of explosives had been used in Vietnam up to the time of their visit. Pfeiffer estimated that Indochina's landscape is "more or less permanently rearranged by more than 20 million craters . . . the holes alone would cover a combined area of about 325,000 acres."

Herbicide spraying ended, but forest cover in Vietnam is now being removed by bulldozer teams. Westing estimated that bulldozers were stripping all vegetation from about 1,000 acres per day at the time of his visit. About 750,000 acres had already been denuded.

Tentatively, the biologists concluded that cultivated areas hit heavily with conventional high explosives would be very difficult or impossible to recultivate. Bombing, defoliation, and land-clearing operations had eliminated 3 million acres of forest cover, 12 per cent of the nation's total. Increased erosion and flooding appeared likely to result. In addition, craters may provide breeding areas for disease-carrying insects.

UN conference. Efforts to resolve environmental problems on an interna-

Environment

Continued

tional level ran afoul of competing national interests. Political clashes plagued preparations for the United Nations (UN) Conference on the Human Environment, scheduled for Stockholm, Sweden, from June 5 to 16, 1972.

A major planning meeting, held in Prague, Czechoslovakia, in May, 1971, was downgraded in status at the last moment, and almost canceled, because of a disagreement concerning the status of East Germany. The United States, Great Britain, and France objected to full participation by East Germany, which is not a UN member. The Soviet Union announced it would boycott the conference.

Political considerations so heavily dominated planning that the conference seemed likely to produce little progress. It did, however, approve Earthwatch, a global program to monitor the earth's habitability. Ironically, the conference's importance was undercut just before it began as the United States and Russia agreed on a joint environmental research program. The pact was signed May 23 in Moscow.

Nuclear power. Hearings conducted by the Atomic Energy Commission (AEC) produced masses of new information regarding the safety of nuclear power plants. The hearings were expected to continue at least through the summer of 1972.

The uranium fuel "core" of a nuclear electric power plant can be heated to the melting point by its stored radioactivity, even when the plant is shut down. If this happened, catastrophic quantities of radioactive materials could be released. An emergency core cooling system (ECCS) is built into each nuclear plant to guard against this, but the system's adequacy has been under fire for some time.

In June, 1971, the AEC issued temporary ECCS standards. Some plants had to modernize their ECCS and others had to observe peak generating temperatures of no more than 2300° F. The AEC also convened public hearings to consider permanent standards. The hearings produced extensive criticism of the AEC safety program, from within the commission itself and from outsiders. Some highly placed staff mem-

Environment

Continued

Young volunteers load scrap paper at one of three collection depots of Ecology Action of Florida, Incorporated, of Miami. Such groups, which also collect glass and metal for recycling, have sprung up in many other communities in the United States.

bers urged suspending further plant construction or operations until safety questions could be satisfactorily resolved.

Computer-simulation techniques to predict the systems' effectiveness were called inadequate. Most had never been tested under realistic conditions. The safety dispute, along with criticism of the AEC stance on environmental policy, slowed plant construction.

The AEC stiffened its requirements after a federal court called its implementation of the National Environmental Policy Act a "mockery" in July, 1971. It required far more detailed statements from electric power companies regarding nuclear plants under construction. Most of the nuclear plants under construction face considerable delays. By summer, 1972, the AEC had not granted a new construction permit for a nuclear plant in over a year, although at least 40 applications are before it. About 40 other plants are also being planned by utilities (see ENERGY).

The DDT battle. A special examiner appointed by the Environmental Protection Agency (EPA) held long hear-

ings on DDT. Lasting until April 25, 1972, the hearings included scientists' speculation that DDT may damage human health as well as the environment. Witnesses said that animal experiments show the insecticide may cause cancer.

The hearings examiner recommended no further restrictions on the use of DDT. But, in June, 1972, William D. Ruckelshaus, EPA administrator, banned its use in the United States almost totally, effective on Dec. 31, 1972. The industry promptly appealed the ban.

Detergents dispute. Scientists became more divided over phosphates in water pollution. The U.S. government had warned against high-phosphate detergents. Research had shown, however, that in many areas, especially on seacoasts, phosphates were much less important than nitrates in causing eutrophication, or overnourishment, and the resulting pollution of bodies of water. Other research indicated that carbon dioxide could sometimes trigger this destructive process.

But, particularly in fresh-water lakes, phosphates caused at least part of the

Environment

Continued

problem. High-phosphate detergents, the main source of phosphate pollution, were banned in several communities whose populations totaled over 30-million. At least one such community reported that water quality then improved.

The lack of an accepted substitute for phosphate made life difficult for detergent manufacturers. Alkaline compounds were restricted as too caustic. Nitrilotriacetic acid, or NTA, was banned as a potential health hazard. There were reports that it seemed to cause birth defects in laboratory animals, that it might cause cancer, and that it helped heavy metals get into the food chain. All this led the heads of four federal agencies to recommend the continued use of phosphates on Sept. 15, 1971.

Automobiles and air. The National Academy of Sciences released a study on Jan. 1, 1972, concluding that automobile manufacturers would be unable to comply with air pollution regulations that are scheduled to go into effect in 1975 and 1976. It supported industry requests that the deadlines be extended. The academy felt that the industry had

not yet developed the technical means to reduce auto tail pipe emissions by the required 90 per cent. The EPA denied the request.

Hydrocarbon and carbon monoxide emissions are to come under strict control in 1975, and nitrogen oxide emissions in 1976. Observers predict that a catalyst, apparently not yet developed, will be required to remove the nitrogen oxides. The industry has laboratory techniques to remove the nitrogen oxides, but does not have an economically feasible system that will work for the 50,000 miles or five years required.

Environmentalists tended to agree with manufacturers that adequate pollution controls for the internal-combustion engine will be extremely costly and difficult to perfect. Many of them said that the best solution is to replace the internal-combustion engine with a cleaner power source such as a turbine or steam engine. While Japanese manufacturers have begun to market a steam car, developed in the United States, American manufacturers have so far declined to consider this option. [Sheldon Novick]

Genetics

A new technique for staining human chromosomes was reported in September, 1971, by scientists in laboratories in Denver, Colo.; Houston, Tex.; Edinburgh, Scotland; Vienna, Austria; and Paris, France. The new technique, described at the Fourth International Congress of Human Genetics, in Paris, makes it possible to identify each of the 46 human chromosomes, even though many of them are essentially the same size and the same shape when viewed through the microscope.

The various investigators developed somewhat different approaches. However, all of the methods depend on partial denaturation, or breakdown, of the chromosomes with heat or acid, followed by a period of renaturation, or reconstruction, and then, treatment with a standard stain called the Giemsa stain. Each chromosome then appears with a characteristic series of dark-staining bands of varying width and intensity. The methods are simple and require no special equipment. Other new techniques, reported only months before the Giemsa technique, involve fluorescent

dyes. However, they require special equipment and considerable experience.

Why the chromosomes stain in patterns in the new methods is not known. It may be that all of the methods selectively remove some sections of protein that are associated with the chromosomes' deoxyribonucleic acid (DNA). This would expose more DNA to the stain in some regions than in others.

In November, 1971, Francis H. C. Crick of the MRC Laboratory of Molecular Biology in Cambridge, England, proposed a structure for chromosomes that fits in with this possibility. Chromosomes are composed of DNA and proteins. DNA is a double-stranded, coiled molecule. Crick suggests that the double strand is separated by protein at various points along a chromosome. If this is so, denaturing might remove some of these proteins from the regions of DNA they separate, permitting that DNA to stain more intensely.

Structural gene decoded. Structural genes are those that specify the production of an enzyme or other protein. Proteins are composed of chains of inter-

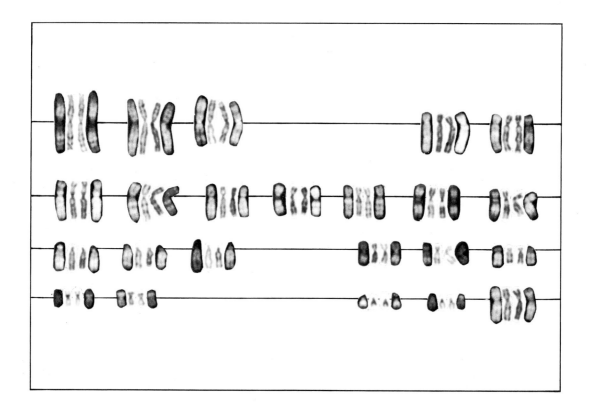

Genetics

Continued

Pairs of chromosomes from two cells of the same person were treated by two techniques that reveal bands by which each of the chromosomes can be identified. The pairs in the center of the group were stained by the new Giemsa technique, while the pairs that flank them were stained with a fluorescent substance. The banding is almost identical, but the Giemsa technique is both easier and less expensive to perform.

connected substances called amino acids. The sequence of the amino acids is determined by the sequence of portions of the gene known as bases. In May, 1972, molecular biologists W. Min Jou, G. Haegeman, M. Ysebaert, and W. Fiers at the State University of Ghent in Belgium established the complete sequence of the bases in the gene that specifies the production of the protein forming the outer coat of a virus known as MS2.

Because a complete gene is a very long structure containing at least several hundred bases, it is hard to determine the sequence in which the bases appear. The scientists overcame this problem by using some enzymes that broke the viral gene into smaller fragments that can be more readily sequenced.

The genes of MS2 are all in a single strand of ribonucleic acid (RNA), a portion of which is the gene that codes for the coat protein. Such strands of RNA are often thought of as simply a single straight strand. However, many of them, including the viral RNA strands, have a substantial amount of chemical pairing

between their bases. This pairing occurs when the RNA folds back on itself, producing a helical, or coillike, structure. Such pairing and the resulting shape is responsible for some of the physical and chemical properties of RNA, such as its mechanical strength and its resistance to mutation.

On the basis that approximately 75 per cent of MS2 RNA is in the helical form, the Belgian scientists proposed a structure for the RNA included in the coat-protein gene. They nicknamed this structure the "flower model" because of a fancied resemblance to a flower when it is drawn in two dimensions. Within the virus, of course, the gene is folded in three dimensions and, therefore, is far more complex.

Knowing the complete base sequence and structure of such a gene opens the possibility of learning a great deal about gene function. It also may help solve such problems as why some parts of genes are more subject to mutation than are other parts.

Structural genes synthesized. The genes of all living things are portions of

Genetics

Continued

long strands of DNA (or RNA in the case of some viruses), at least one of which is present in each cell of every individual. A DNA gene functions through RNA intermediaries which have a base sequence that complements the sequence of the gene.

Although nobody knows the complete structure of any structural DNA gene, three teams of U.S. scientists, in 1972, independently synthesized the DNA gene that specifies hemoglobin production. Using essentially the same technique, each team implemented reverse transcriptase, an enzyme that triggers RNA to produce DNA strands with bases complementary to its own.

The scientists added the enzyme to the RNA intermediary of the hemoglobin-producing gene along with previously available DNA raw materials. The resulting DNA strands were complementary to and the same length as the RNA. Thus, they were replicas of the gene that produced the RNA.

All three research teams got their RNA from rabbit cells, so they synthesized the rabbit hemoglobin gene. However, a group led by molecular biologist Sol Spiegelman of Columbia University in New York City also used RNA from human cells and mouse cells. As a result, they also produced the genes for human and mouse hemoglobin.

The prospect of synthesizing or otherwise obtaining individual human genes is exciting because it would open up new avenues of genetic research, and ultimately such genes might be used to cure genetic deficiencies. However, the technique used in the hemoglobin-gene syntheses cannot readily be used with other genes. In most cells, hundreds or thousands of genes are producing RNA, and isolation of the RNA from any specific gene is very difficult.

The red blood cells of mammals and other vertebrates are exceptions to this, however. They produce mostly RNA from the hemoglobin gene, and all three teams of scientists obtained their RNA from these specialized cells. If other genes are to be synthesized in this manner, laboratory techniques that control the rate at which cells produce RNA will have to be developed. [H. Eldon Sutton]

Geochemistry

Geochemists continued to make important advances in lunar research in 1971 and 1972. The research included new discoveries about the distribution and amount of radioactive elements on the moon, better knowledge of the lunar highland rocks, and new data on the ages of lunar rocks and soils.

Scientists at the Lamont-Doherty Geological Observatory of Columbia University were the first to measure the amount of heat escaping from within the moon. The heat flow is an indirect measure of the amount of the radioactive elements uranium, thorium, and potassium inside the moon, because their radioactivity is the principal heat source.

The lunar heat flow at the Apollo 15 site was half as large as that of the earth. Scientists expected a much smaller heat flow because the moon's mass is only about $\frac{1}{80}$ that of the earth and the earth's radioactive elements are generally thought to be concentrated in the outer crust. If the heat flow at the Apollo 15 site proves to be typical of other sites on the moon, then geochemists will have to drastically revise their estimates of the amounts of radioactive materials in the moon.

Studies of lunar soils and breccias (rock fragments that are cemented together) suggest that the lunar highlands contain much anorthosite, a granular igneous rock consisting mainly of plagioclase feldspar. This was confirmed by a novel experiment conducted from the Apollo 15 mission's orbiting command module. The chemical composition of large areas of lunar surface was determined by an X-ray fluorescence experiment in which the X rays from the sun were utilized. These cause secondary radiations to emit from lunar rocks. The command module measured the intensity of the secondary radiations.

A team from the Goddard Space Flight Center in Beltsville, Md., studied the fluorescence measurements and found that the aluminum-silicon ratios were distinctly higher in the highlands than in the lunar maria. Because this ratio is also much higher in plagioclase feldspar than in basalt, the scientists concluded that the lunar highland rocks are probably richer in anorthosite.

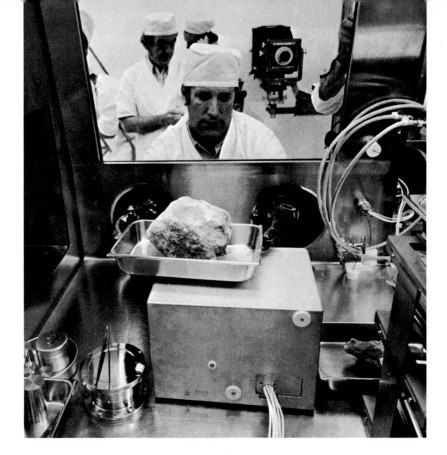

Weighing a large lunar rock brought back by Apollo 16 was the first step in a series of geochemical tests that have revealed its age and its composition.

Geochemistry

Continued

Lunar rock dating. New age determinations were made in 1972 for basaltic crystalline rocks recovered from three different locations on the moon. The rubidium-strontium and potassium-argon dating methods both produced the same results for the rocks.

The rocks from the Apollo 14 site are 3.9 billion years old. Scientists at the California Institute of Technology (Caltech) dated a small fragment of basalt recovered by Luna 16 at 3.4 billion years. Basalts recovered from all sites, including those of Apollo 11 and 12, range from 3.2 to 3.9 billion years old.

The age of a piece of anorthosite recovered by Apollo 14 was of particular interest because it was believed to be one of the earliest rocks formed on the moon—a so-called "genesis rock." According to some theories, the crust of the moon was initially all composed of anorthosite. If this is true, then some anorthosites must be nearly as old as the moon—about 4.6 billion years. However, according to potassium-argon dating, the Apollo 14 anorthosite fragment is only 4.1 billion years old.

Dust from all four Apollo and Luna sites has been found to be about 4.6-billion years old by the rubidium-strontium method. Caltech geophysicist Gerald J. Wasserburg and his colleagues believe the agreement indicates that rocks on the moon's surface probably have an average age of 4.6 billion years. The dust generally contains much more rubidium than the neighboring rocks, so that the age of the bulk sample is strongly influenced by the age of rubidium-rich components.

Many lunar observations, including the age data, can be explained in terms of a model proposed by geophysicists Paul Gast of the Manned Spacecraft Center near Houston and Robert McConnell of Earth Science Research, Incorporated. They suggest that the moon's outer layer melted early in its history, and when the rock solidified, it formed a plagioclase-rich anorthosite crust. The lunar highlands represent remnants of this layer. Later, some of the material under the crust melted partially and solidified as basalts rich in potassium, rare earth elements, and phosphorus.

Geochemistry

Continued

These basalts were produced about 4.1 billion years ago, or perhaps earlier. Some of them were brought to the surface by volcanic action. From 3.9 to 3.2-billion years ago, melting took place still deeper in the moon. The basalts that were produced during this period also came to the surface as lava and are now found in the maria. Since that time the surface has been altered mainly by meteorite impacts.

Mars. Mariner 9 was launched from earth on May 30, 1971, and entered orbit around Mars on November 13. Although artificial satellites had previously flown past the planet, this was the first one to be placed in orbit around Mars. A planetwide dust storm, in progress when the satellite arrived, handicapped a number of experiments, but was turned to advantage in others.

Very significant in this regard were infrared spectroscopy experiments. These used the absorption spectrum of infrared radiation by the dust particles (the fraction of the radiation absorbed plotted against its wave length) to draw inferences about the chemical composition of the particles.

Scientists compared the absorption spectrum of the Martian dust with the spectrum obtained from dust particles of various earth rocks. They found that the Martian dust most closely matches that of earth andesites. Andesite is formed as certain lavas harden and is a major constituent of volcanic mountains such as Mount Rainier in Washington and Mount Hood in Oregon.

Scientists at the Goddard Space Flight Center interpreted the Martian dust data as indicating that there is a substantial amount of andesitic material on the surface of the planet. By analogy with the earth, they believe this implies that considerable geochemical differentiation has occurred on Mars. This would mean that the surface of Mars does not simply preserve an unaltered record that goes back to the time the planet was formed. Lunar research has shown that the same is evidently true of much, if not all, of the moon's surface.

Television pictures from Mariner 9 revealed several large craters that are surrounded by what appear to be lava flows. These craters are probably volcanic, and could be sources of some of the geochemically differentiated rocks

identified by the dust studies. Three of the craters are situated along a ridge, much like many volcanoes that are found on the earth.

The infrared and ultraviolet spectroscopy experiments confirmed that the Martian atmosphere consists predominantly of carbon dioxide. Small amounts of water vapor were also noted, but by no means is there enough water to fill the fabled "Martian canals." Analysis of Mariner 9 data is continuing. Meanwhile, Russia has placed two satellites in orbit around Mars and landed a third one on the planet. See MARS IS AN ACTIVE PLANET.

Oldest terrestrial rocks. Geologists from Oxford University in England and the Geological Survey of Greenland have discovered rocks in the Godthåb district of southwestern Greenland that are 3.6 to 4.0 billion years old. Rocks older than 2.7 billion years are rare on earth, although a few rocks with ages of 3.2 to 3.5 billion years have been found in Swaziland on the African continent and in the Minnesota River Valley in North America.

L. P. Black, Noel H. Gale, Stephen Moorbath, and R. J. Pankhurst of Oxford and V. R. McGregor of Greenland used two methods to establish the age of the rocks, one based on the radioactive decay of rubidium to strontium and the other on the radioactive decay of uranium to lead. The measurements were made on the total rock samples rather than on separated minerals. This was possible because the rocks were very old and because their composition is that of granite, which generally contains high proportions of rubidium and uranium with respect to strontium and lead. The rubidium-strontium system gave an age of 3.98 ± 0.17 billion years, while the uranium-lead system gave an age of 3.62 ± 0.10 billion years. Work is continuing, using separated minerals and additional rock samples, to try to resolve the differences between the two ages.

In any case, these rocks are the oldest yet found on earth—as old, in fact, as many of the lunar crystalline rocks. It is interesting that the Greenland rocks have a granitic composition, which indicates that the geochemical differentiation processes must have produced granite at a very early stage in the earth's history.

[George R. Tilton]

Crystals that grew in a small cavity in a lunar rock found at Apollo 14 landing site were photographed through a scanning electron microscope. The largest is apatite, a calcium phosphate. They grew from a hot vapor as the rock cooled.

Geology

Kilauea, an active volcano on the island of Hawaii, has long been an excellent laboratory for the study of igneous, or heat produced, processes occurring at the earth's surface. Observations there by Wendell A. Duffield of the U.S. Geological Survey in 1971 may have provided new insights into the formation, movements, and destruction of the large plates that make up the earth's crust. See EARTH'S HEAT ENGINES.

Kilauea has been in continuous eruption since May 24, 1969, principally around a vent at Mauna Ulu, a broad spatter and lava shield that has grown to a height of more than 300 feet. At its top, a steep-walled oval crater about 300 by 500 feet had developed by the end of 1970. A column of highly fluid lava filled the crater, its level fluctuating between 50 and 160 feet below the rim.

At times, however, the level remained fairly constant while lava welled up at one end of the vent and drained at the opposite end. A black, glassy crust formed as the upwelling lava came into contact with the much cooler atmosphere. This thin crust separated into thin, relatively rigid plates that moved toward the drainage area. The plates eventually melted again when they rejoined the lava column below the drain.

For brief times, these plates were "rafts" moving on a nearly flat surface of circulating lava. They were only fractions of an inch thick, and most of them were from 50 to 100 feet wide. The plates moved relative to one another, striking and sliding along their edges, or, more commonly, moving apart with new crust forming along their trailing edges.

The new material grew at an average rate of about 4 inches per second at the margin of a typical spreading plate. The plates moved away from their neighbors in directions ranging from perpendicular to acutely oblique, and this was accurately reflected by surface striations on the new crust.

Duffield carefully photographed all these motions. They may well be similar to the much slower processes in large parts of the earth's crust. Zones of plate enlargement could be closely likened to the zones of spreading in the earth's ocean floors. The Mid-Atlantic Ridge

Ocean Floor Chalk Deposit

Chalk deposits taken from the Pacific Ocean floor show the floor has moved north over the past 100 million years. The chalk came from tiny shells that sank to the ocean floor at equatorial regions and were buried before they could dissolve.

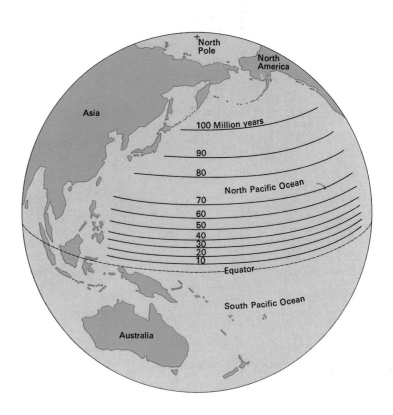

Geology

Continued

A colorful natural museum was created in 1972 on the terraced sides of Interstate 70 in the Rocky Mountains west of Denver. The terracing bares a cross section of rock strata telling the geological history of the last 130 million years.

and the East Pacific Rise, for example, are thousands of miles rather than tens of yards long, and the crust is miles rather than fractions of an inch thick. They spread inches per year rather than inches per second, and they have been active for tens of millions of years rather than tens of seconds. Yet, their resemblance to the Mauna Ulu lava features is remarkable.

Among the similarities are numerous transform faults–discontinuities that reflect differences in rates of crustal growth. Duffield found zones of strike-slip offset on the crater floor only between adjacent zones of crustal formation, where parts of two adjoining plates were moving in opposite directions.

He observed numerous triple junctions where three equally spaced zones of plate formation radiated from a single point to form sectorlike domains of expanding crust. Such junctions are strikingly similar, except for their scale, to major features beneath the central Indian Ocean and the northeast Pacific Ocean. Also comparable is the great triple junction in the Pacific basin

southwest of Central America, where the Pacific, Cocos, and Nazca plates meet.

Whenever lava entered Kilauea's Mauna Ulu crater from a new direction, the expanding plates of new crust pressed against older plates, creating zones of convergence and destruction. The colliding plates either both plunged downward, presumably to be converted back into lava, or one plate was thrust beneath the other at a relatively shallow angle. The rate of plate destruction was about equal to that of crustal formation elsewhere in the crater. The zones of convergence could well be similar to zones along the edges of the earth's oceanic basins.

The assemblage of crustal plates, spreading zones, transform faults, and zones of plate destruction observed in Kilauea volcano could be viewed as global plate tectonics in microcosm. Whether the thin crust at the Mauna Ulu vent constitutes a true scale model of the earth's crust is not clear, but their similarities are astonishing.

"There are many obvious direct parallels between the two situations," Duffield said. "Possibly an account of the behavior of the lava column can generate new ideas and working hypotheses for the study of global plate tectonics, whose processes and features are less accessible for examination."

Isolated mammals. The spreading of plates in sea-floor areas provides the mechanism for the drifting apart of continents. The pattern of their drift indicates that, as recently as 200 million years ago, during Mesozoic times, a single supercontinent existed. This gigantic land mass, commonly referred to as Pangaea, supported a variety of primitive mammals that ultimately evolved into the forms known today.

A puzzling feature, however, is that relatively primitive aboriginal land mammals tend to be concentrated today in three widely separated regions—Australia and New Guinea, South America, and Madagascar. This distribution generally has been explained as a result of three independent episodes of the dispersal of living things across water gaps. The theory assumes that continents and islands were fixed in position during the time of dispersal.

Another explanation, much more in accord with current knowledge of continental drift, was advanced in February, 1972, by Jack Fooden of the Field Museum of Natural History and Chicago State University. He proposed that the primitive animals in these regions represent successively isolated samples of the world's evolving mammal fauna as it existed when the land masses separated during the breakup of Pangaea.

According to this thesis, aboriginal *monotremes* (egg-laying mammals) and marsupials in Australia and New Guinea were isolated during the separation of east Gondwonaland (including present India and Antarctica) from the remainder of Pangaea. These groups have survived in isolation to the present time. However, monotremes in the main part of Pangaea became extinct as placental mammals began to develop.

At this stage, South America, with its sample of marsupials and primitive placentals from the world fauna, separated from Africa and the rest of Pangaea as the southern Atlantic Ocean began to form. Marsupials, in turn, became extinct in the main part of Pangaea, but they survived in the isolation of South

America. Finally, Madagascar detached from Africa, isolating a sample of the second stage of primitive placental evolution in the world's fauna.

Fooden's hypothesis should help to explain the mammalian history of North America after its separation from Europe and during its subsequent union with Asia via the Bering bridge and with South America via the Isthmus of Panama. It also should apply to India, which separated from east Gondwonaland at an early stage, was isolated for millions of years, then ultimately became joined to the Asiatic continent.

Radioactive waste disposal. Safe disposal of the highly radioactive waste materials from nuclear reactors requires special efforts. An imaginative new concept, currently under study and development by Jerry J. Cohen, Arthur E. Lewis, and Robert L. Braun of the University of California's Lawrence-Livermore Radiation Laboratory offers considerable promise for effective waste disposal on a large scale.

The operation envisioned by the scientists begins with creation of a rubble chimney deep in the earth by a controlled nuclear blast in granite or other relatively massive rock. Such a chimney of broken material, about 65 feet in diameter and 6,500 feet deep, would hold about 14 million gallons of fluid.

Highly radioactive liquid wastes and water would be pumped directly into the lower parts of this chimney, and the heat released by the radioactivity would quickly cause boiling. Steam and accompanying rare gases would be processed at the surface and then vented. When a predetermined total amount of waste materials is added over a period of years, the chimney would be allowed to boil dry, and then would be sealed. The rubble lying at its base would melt promptly thereafter.

Melting of the surrounding rock should proceed at markedly decreasing rates until thermal equilibrium is attained. Then, slightly less than a century after original activation of the chimney as a disposal unit, a globular mass of magma about 650 feet in diameter will have formed. With continued escape of heat, it would gradually solidify to form new rock and permanently entrap the waste radioactive material.   [Richard H. Jahns]

Geophysics

Efforts continued in 1971 and 1972 to refine the plate tectonics model and study its implications. The model pictures the earth's outer shell as made up of about 20 large, moving plates. The plates converge along deep earthquake zones, diverge along mid-ocean ridges, and slide along each other in places such as the San Andreas Fault in southern California.

Studies over the last decade of magnetic irregularities in crustal rocks beneath the oceans suggested that the rocks' ages increase with distance from the ridges where they are produced. This was confirmed by the Joint Oceanographic Institutions for Deep Earth Sampling (JOIDES) deep sea drilling project, which began in 1968.

Geophysicist John G. Sclater of the Scripps Institution of Oceanography and others noted in 1971 that there was also a systematic relationship between elevation of these ocean rocks and their age. Preliminary work by Walter Pitman of the Lamont-Doherty Geological Observatory at Columbia University suggests that this may have implications for continental geology. If the rate of sea floor spreading had increased in the past, more of the ocean floor would have been at a higher elevation. As a result, the sea level should have risen and encroached upon the land areas. This may explain past spreading of the sea over the land.

Plate motion. Direct measurements of plate movement are difficult to make because the movements average only a few inches a year. It is not yet clear whether the motions are steady or spasmodic. Measurements made in Iceland and Ethiopia from 1967 to 1972 have revealed motions just large enough to be measured, and these appear to agree with the plate tectonics model.

Scientists working with the National Aeronautics and Space Administration (NASA) began using two laser satellite tracking stations in 1972 to measure the distance between two points along the San Andreas Fault. They will measure the motion of the two plates at the fault over a period of years.

The tracking stations will make simultaneous range measurements on a

The San Andreas Fault trace near Hollister, Calif., is visible as a faint diagonal line in the dark field. During the earthquakes of July 17, 1971, monitoring devices showed a record creep there of more than a third of an inch.

Geophysics

Continued

Chairman James R. Schlesinger, *right,* of the Atomic Energy Commission was one of the first to check the effects of Cannikin, the nuclear explosion set off under Amchitka in the Aleutian Islands in November, 1971. Geophysicists reported the blast produced some of the most precise seismic data recorded for any earth tremor.

number of geodetic satellites equipped with laser reflectors. Continued measurements over the years should enable observers to determine the ratio of movement along the fault. A third tracking station could enable them to measure the rate of opening of the Gulf of California.

The current plate tectonics model may not explain crustal deformation prior to 180 million years ago. Evidence from the oceans disappears at about that time, and geophysicists must rely on data from the continents. However, in 1972, John Dewey and John Bird of the State University of New York, Albany, analyzed the development of the Appalachian-Caledonian Mountain system prior to 300 million years ago. They demonstrated that the geological features and the history of the mountain development were compatible with the plate tectonics model.

Polar wandering. In a related study, Edwin Larson of the University of Colorado analyzed the magnetic record in rocks from North America, Africa, and western Europe for the last 3 billion years to determine the manner in which the continents wandered around the North Pole over that time. The data suggests that North America rotated in a clockwise direction for about 1.5 billion years, then counterclockwise the next 1.5 billion years.

The simplest explanation for this continental movement is that the continents were grouped together through Precambrian and Paleozoic times and that continental drift, related to differential plate movement, first occurred about 200 million years ago. If this proves to be the case, then a different mechanism must be sought for the tectonic events prior to that time.

Driving mechanism. The driving mechanism for plate tectonics is still very poorly understood. Suggestions about the mechanism include thermal forces and gravitational instability. See EARTH'S HEAT ENGINES.

Leon Knopoff of the University of California, Los Angeles, created a model in 1972 that was based on nonhomogeneous heat production in a single plate. He assumed that the heat that flows from the surface of the continent can be accounted for by the radioactive heat in crustal rocks while the heat that flows

from the ocean bottom originates deeper in the earth.

In his tabletop laboratory model, a lithospheric "raft" driven by such heating under the continents moved toward the cooler oceans, as occurs in the zones of plate convergence. In accord with Sclater's model relating ocean floor age and elevation, the motion was away from the hotter areas. This would correspond to the actual movement of plates away from the ridges.

Earthquakes. Man-made earthquakes continued to concern seismologists. For more than five years, earthquakes have been produced at Denver and Rangely, Colo., by pumping liquid wastes into deep holes in the earth. People have objected to plans to dispose of liquids in this way in other areas.

Lynn Sykes of Columbia University established a network of portable seismograph stations in the Buffalo, N.Y., area to assess earthquake risk related to deep disposal of fluid wastes. Several hundred small earthquakes per day were recorded at a station less than a mile from a high-pressure injection well at Dale, N.Y. After the well was shut down in November, earthquake activity dropped markedly, and within three days was back at the preinjection rate of about one quake per week.

Lunar geophysics. Preliminary heat flow measurements on the moon, in 1971, yielded results indicating higher temperatures than were expected (see GEOCHEMISTRY). At the same time, the presence of paleomagnetism suggested a molten core once existed there, while lunar quake dynamics indicated that such a core is still present.

Sufficient seismological data now exist to chart the outer 60 miles of the moon with some accuracy. Near the surface, extremely low seismic velocities indicate that there are only lunar soils and broken rocks. Below this, to about 6 miles, pressure increases the seismic velocity. This layer may consist of a series of lava flows or basalt.

From 15 to 40 miles below the surface, seismic velocities indicate a composition of anorthositic gabbro rock. There is a major discontinuity at 40 miles, and below this the apparent velocity is higher than the mantle velocities found just beneath the earth's crust. [Charles L. Drake]

Medicine

Dentistry. Research on implants accelerated during 1971 and 1972. Implants are appliances that are inserted into the jaw to anchor false teeth.

Many materials have been used for implants in an attempt to find a substance that will be compatible with the surrounding living tissues and that will give prolonged use. To date, chromium-cobalt alloys have worked best.

Subperiosteal implants were the most widely used during the past year. These are placed in the hard, dense cortical bone of the lower jaw. However, the procedure requires complex surgery to take bone impressions of the cortical bone and cast the metal implant accurately. In addition, there is danger of infection, inflammation, or other negative reactions in tissues around the basic metal framework and around posts that jut into the mouth and secure the false teeth. Another problem is that connective or fibrous tissue forms around the implant. It would be more desirable for the bone to regenerate.

The use of endosseous implants is a newer concept. These are placed direct-ly in the much less dense medullary bone of the upper jaw. Many other types of implants have also been tried—from screws, tubes, and pins, to vented blades. Again, the typical response has been regeneration of dense fibrous tissue around the implant,

Many oral implantology problems remain to be solved—among them varying tissue response from one individual to another, and technical laboratory and surgical difficulties. In addition, further research is needed on implant metals. Most important, dentists must be trained to perform the complicated procedures.

Dental plaque retardation. At a meeting of the American Chemical Society in Boston on April 12, 1972, researchers reported laboratory tests in which two new chemicals applied to extracted sterilized human teeth prevented the formation of dental plaque, which causes decay. The new chemicals are modifications of the chemical chlorhexidrine, which previously had been reported to offer a lesser amount of plaque inhibition. [Robert J. Hilton]

The posts of two metal implants jut through the gums of a patient's lower jaw, *below*. False teeth will be anchored to them. An X-ray picture, *below right*, shows similar anchor-shaped implants in the upper and lower jaws of another patient.

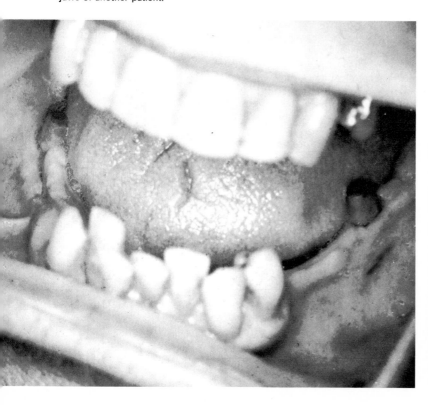

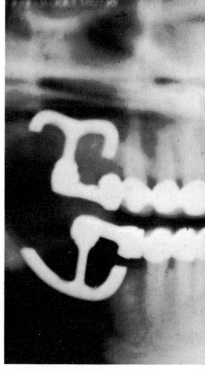

Medicine

Continued

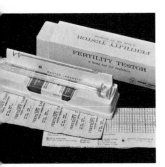

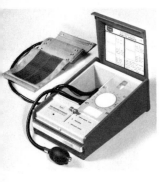

The Fertility Testor,
top, which shows the
period each month that
a woman is fertile, and
the sphygmometrograph,
above, which measures
blood pressure, were
among the various
medical home testing
kits marketed in 1972.

Internal Medicine. Although American medical science has made impressive advances since World War II, decreases in the overall U.S. death rate and in the infant and maternal mortality rates have leveled off. Nevertheless, American life expectancy has increased markedly since the early 1900s. A person born at the turn of the century lived an average of 49.24 years, while the average life expectancy of a person born in 1970 was 70.2 years.

The U.S. death rate declined steadily from 1900 to about 1950. About then, however, the improvement slowed to a virtual halt. For reasons that are still obscure, this is true in all of the world's economically advanced countries. This unfavorable new trend, however, started five years earlier in the United States than it did in northern Europe.

The problem is complex. In the first place, people living in the United States want good health more than they want health care. Moreover, an American's chances of getting good medical care often depend on his ability to pay. Throughout American life, poverty and ill health have been closely linked.

The improvements in medical care and public health have also brought into sharper focus some of the effects on personal health that are more related to a person's life style and his socioeconomic status than to the medical care he receives. For example, narcotics and drug addiction, alcoholism, overeating and insufficient exercise, and cigarette smoking are all endemic in the American life style.

Malnutrition of the affluent. Although man now has the technological capacity to wipe out hunger, affluence has brought with it its own form of malnutrition. Dr. Aaron M. Altschul, professor of community medicine and international health at the Georgetown University School of Medicine in Washington, D.C., discussed this new problem at the American Chemical Society meeting that was held in Philadelphia on Feb. 15, 1972.

"In general," he said, "as societies become more affluent and they change their eating habits, the proportion of obesity increases, as does the incidence of diabetes, coronary artery disease, hypertension, and other diseases. However, evidence of a cause-and-effect re-

lationship between diet and changes in the disease pattern is not available and, clearly, these diseases are from multiple causes."

The problem of affluent malnutrition stems from eating too much food, the wrong foods, and excessive amounts of sugar and other carbohydrates.

Obesity. New research is revealing a fact that bewilders medical men and that seems certain to dishearten obese people—diets and willpower are useless prescriptions for those who have been obese since infancy. There is no known treatment, other than a life of near starvation, that will enable this vast group of fat people to keep themselves at reasonable weights.

A research group at the Medical College of Wisconsin called obesity a "disease" of "epidemic proportions" in 1971. They say it is probably more critical to national health than malnutrition. From 20 to 25 per cent of Americans who are considered obese run the risk of premature death from a myriad of diseases, including heart disease and diabetes. See A MATTER OF FAT.

Diet and heart attacks. In November, 1971, Dr. Robert Wissler of the University of Chicago offered impressive experimental evidence to the annual meeting of the American Heart Association in Anaheim, Calif., that the typical American diet fosters the development of arteriosclerosis, or severe hardening of the arteries. Arteriosclerosis accounts for more than a third of the deaths among American men between the ages of 40 and 45. It is the main cause of heart attacks in the country.

Wissler's study was done with rhesus monkeys because their body metabolism handles food very much like the human system. When middle-aged male rhesus monkeys were fed the average U.S. table diet for two years, they suffered three times as much aortic arteriosclerotic disease as monkeys that were fed a more "sensible" diet. In addition, arteriosclerotic deposits were four times greater among monkeys eating the average American diet than in monkeys who were fed more sensibly. Wissler said that his findings supported what studies in human populations "have already strongly suggested—that diet is extremely important to the development of arteriosclerosis."

Solution To the Sickle Cell?

Normal, doughnutlike red blood cell, *below*, is transformed into the elongated, deformed sickle shape, *right*, in sickle cell anemia. These cells clog blood vessels, causing pain and tissue destruction.

Sickle cell anemia is not yet curable, but the success of new medical treatments and diagnostic tests in 1972 quickened the cadence of the march against this inherited disorder. The disease afflicts blacks and, to a lesser extent, Puerto Ricans and Cubans. One type occurs when both members of a pair of chromosomes carry the gene. More than 50,000 Americans have this type. Many of them may never reach 20.

A milder form of sickle cell anemia occurs when only one chromosome carries the sickle cell gene. Most persons with this trait live normal lives, but they can transmit the defective gene to their children. More than 2 million American blacks (about 1 in 10) carry the gene in this way. If both parents have the sickle cell trait, there is a 25 per cent risk that their children will be born with the severe form of the disease; a 50 per cent chance that they will be born with the mild form.

Sickle-cell symptoms include severe and sometimes excruciating bone, joint, or abdominal pain—often accompanied by fever—and can last for hours or days. The discomfort occurs when the red blood cells become misshapen and take the form of a crescent, or sickle. Makio Murayama, of the National Institutes of Health, in Bethesda, Md., has suggested that this is triggered by hemoglobin molecules that bond together into chains within the cells. If the condition goes untreated, blood vessels become clogged with the elongated cells, and tissue dies.

Pathologist Robert M. Nalbandian of Blodgett Memorial Hospital in Grand Rapids, Mich., found in 1971 that urea, a substance that scientists believe is formed in the liver, effectively blocks and reverses sickling. However, urea therapy is complex and expensive.

Researchers believe the key to conquering sickle cell anemia is early testing and genetic counseling. The first step is getting a sample of blood for a computerized procedure that can screen thousands of samples simply and painlessly. Persons with the disease are treated and counseled about the chances of passing their genetic disorder to their children.   [Richard P. Davies]

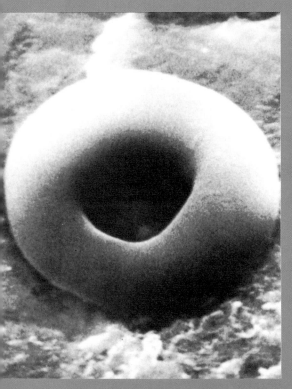

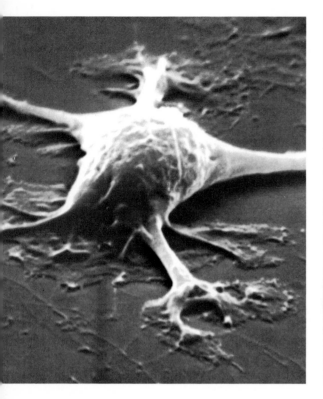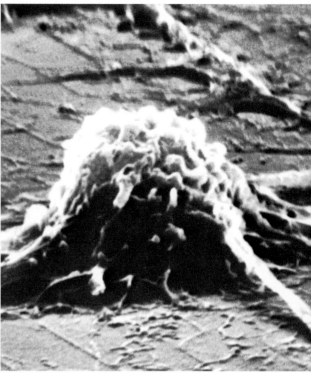

Medicine

Continued

A macrophage, one of the army of scavenger cells that roams the body to engulf foreign invaders was photographed for the first time while at rest, *above*, and while in action, *above right*.

The study also indicated that a "prudent" modification of the American diet—with a reduction in saturated fats, cholesterol, and refined sugar—could avoid arteriosclerosis.

Vaccination. Because more and more Americans are ignoring the vaccines available to them, total victory against some all-but-conquered diseases appears to be becoming more elusive in the United States. Public health experts are alarmed because these diseases are preventable and could be virtually eliminated by the use of vaccines that are available. The best examples include measles and the relatively rare poliomyelitis and diphtheria, which still cause some deaths among children.

Poverty and poor motivation to take advantage of preventive medical care appear to be the major causes of the low vaccination rates. The problem has been compounded in some communities by a cutback in funds for a federal vaccination program that ceased in 1969.

Black lung disease. The first International Conference on Coal Workers' Pneumoconiosis, or black lung disease,

in September, 1971, ended on an optimistic note. Dr. Irving Selikoff, chairman of the meeting, sponsored by the New York Academy of Sciences, was optimistic about several new findings, especially the success of a 25-year-old Australian program to control coal dust in the mines. Australia has almost no black lung disease today. Dr. Selikoff also noted that coal dust has been reduced in many mines in the United States to below 3 milligrams of respirable dust per cubic meter of air, the maximum set by the Federal Coal Mine Health and Safety Act of 1969.

The conference stressed that more research is needed to focus on the diagnosis, treatment and, especially, the prevention of pneumoconiosis. Selikoff pointed out that $350 million is currently being spent each year in the United States to compensate miners afflicted with the disease. "That's several hundred times what we are spending for research," he said.

Occupation and cancer. An extensive study of workers who are exposed to suspected hazardous substances

Medicine

Continued

Photos taken through a scanning electron microscope reveal the dramatic difference between the orderly parallel bundles of fibers in a normal cartilage, *below*, and the twisted whorls of partly calcified fiber in osteoarthritic cartilage, *below right*.

was jointly announced Nov. 15, 1971, by the American Cancer Society, Mount Sinai Medical Center in New York City, and leading trade unions. The study will analyze the death records of thousands of workers in various industries—including asbestos workers, typographers, printing tradesmen, and chemical workers. The project will cost an estimated $1 million a year.

Pollution. When the pollution in New York City air increases to levels commonly reached many times a year, healthy people are much more likely to feel sick and even to get sick, according to a government report presented in Minneapolis in October, 1971.

Following two three-day "pollution episodes" last year, residents in the Westchester section of the Bronx and the Howard Beach section of Queens reported far more symptoms—including eye and throat irritation, chest pain, and shortness of breath—than after three-day periods of "acceptable" air pollution levels. By comparison, symptom levels were unchanged among residents of Riverhead, L.I., where

pollution for all of those three-day periods remained at the same relatively low level of concentration.

This is one of the first studies documenting the ill effects that result from increased, but not unusually high, pollution levels. The report was presented in October, 1971, by Dr. Arlan Cohen, who was formerly with the air pollution control office of the Environmental Protection Agency, which financed the pollution study.

"Pharmacogenetics has important practical implications," geneticist Arno G. Motulsky of the University of Washington Medical School in Seattle told the Fourth International Congress of Human Genetics, in Paris in September, 1971. "To do the necessary studies and to learn more about pharmacogenetics," he said, "we need more pharmacologists who are also trained in medical genetics."

Pharmacogenetics is the interaction between drugs and each person's body chemistry—an important new discipline in medical practice. Doctors have discovered that one man's miracle drug

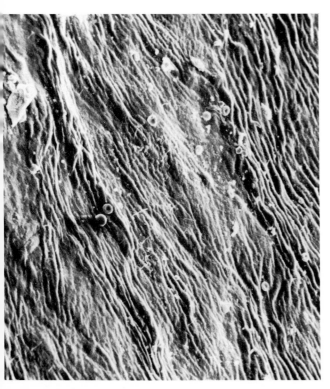

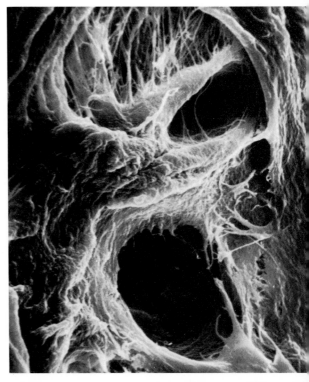

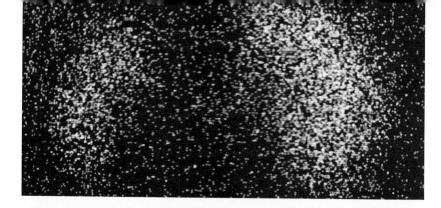

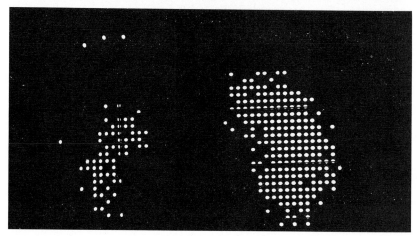

Scintillation camera photo of patient's kidneys, *above right*, reveals that kidney on left is smaller and generally poorer in function. Multichannel analyzer picture of the kidneys, *right*, shows the difference in more easily measured form.

Medicine

Continued

can be another man's poison. The reaction to a drug can be severe enough in some cases to cause death, and yet the same drug can produce symptoms so mild in another person that it is completely overlooked by both patient and doctor. Doctors are finding with increasing frequency that the reason appears to be pharmacogenetic.

Of course, many drug reactions are not pharmacogenetic. They are caused by allergies, excessive dosages, or as yet unknown mechanisms. Pharmacogenetic reactions result from inherited variations in enzymes that cause them to produce abnormal reactions when a given drug enters the body. These biochemical reactions generally occur within the liver, kidney, or blood.

Trauma, a medical category of casualties resulting from a wide range of injuries, is the leading killer of Americans through the age of 38 and has become a public health problem of growing importance. About 114,000 deaths were attributed to trauma in 1970. In order to reduce this toll, a panel of doctors agreed at the American Medical Association meeting in New Orleans in November, 1971, that hospital emergency-room facilities, professional staffs, and public services must be better organized.

Trauma includes bleeding, shock, fractures, bruises, and the conditions, such as loss of consciousness, that accompany them. It results from burns, injuries received in automobile accidents, muggings, gunshots, stabbings, electrocutions, and other injuries.

A spokesman for the National Institutes of Health (NIH) in Bethesda, Md., said such injuries cost a total of $27-billion in the United States in 1971. The NIH pointed out that trauma patients account for 22 million hospital bed days, or 12 per cent of the national total, and occupy four times as many hospital beds as do cancer patients.

The emergency room is "the first line of defense" in treating traumatic injuries, said Dr. Isadore Yager of East Jefferson General Hospital in New Orleans. It must provide adequate space and equipment for treating trauma patients. [Theodore F. Treuting]

Medicine

Continued

Surgery. Important progress in surgery was made in 1971 and 1972 through some new, simpler approaches to complicated surgical problems. Chief among them was a novel new approach to gallstones surgery.

An estimated 15 million Americans have gallstones. To get rid of them, surgeons remove more than 300,000 gall bladders annually. The traditional belief that a diseased gall bladder produced these gallstones has changed, however, in the last few years. Doctors now believe that the secretion of an abnormal bile in the liver of some individuals may be responsible for the formation of stones in the gall bladder. Bile is a complex fluid in which cholesterol is held in solution by proper proportions of salts and certain fatty substances. If the composition of the bile is abnormal, cholesterol will come out of solution, producing cholesterol gallstones, the most common type.

By feeding patients a naturally occurring bile acid, chenodeoxycholic acid, for prolonged periods, Dr. Rudy Danziger of the Mayo Clinic in Rochester, Minn., decreased the size of some cholesterol gallstones and made some others disappear completely. These results suggest that there are better ways to get rid of gallstones than by removing the gall bladder.

Cancer detection. In the mid-1960s, Dr. Phillip Gold of McGill University in Montreal, Canada, found a carcinoembryonic antigen (CEA) in the blood of some patients with cancer of the colon. As the test became more sensitive, CEA could be detected in the blood of 96 per cent of patients with this type of cancer.

In August, 1971, Dr. Paul LoGerfo of the Columbia University department of surgery in New York City reported that the antigen was found not only in cancers of the colon but also in cancers of the breast, lung, and prostate—particularly in the more advanced cases. Other cancers showed smaller amounts of the material. The antigen was also found in a few normal individuals and patients with noncancerous diseases.

Research is now being carried out in many laboratories to see if a diagnostic

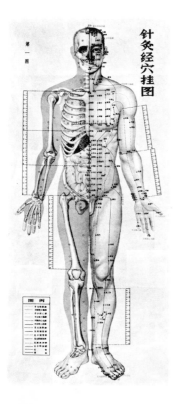

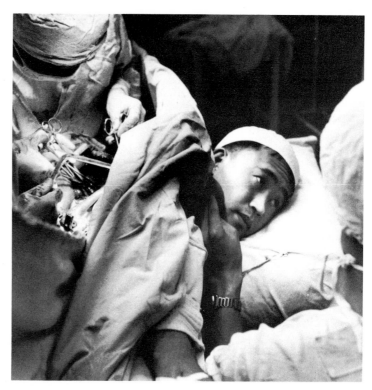

Medicine

Continued

Chinese chart, *above*, shows points along the body's energy meridians where acupuncture needles can be placed. Tuberculosis patient under acupuncture anesthesia, *above right*, remained fully conscious as doctors removed his right lung and a rib.

test based on this antigen can be developed for certain types of malignant disease. It is already clear that the test will be helpful to doctors in checking to see if tumors are recurring in patients who have had them removed by surgery.

Control of tumor growth. It is known that rapidly growing tumors are supplied by a large number of blood vessels from the organ that gave rise to the tumor and sometimes also from surrounding tissues. In April, Dr. M. Judah Folkman, surgeon in chief at Children's Hospital Medical Center in Boston, reported to the New York Cancer Society that small clumps of tumor cells first grow through nourishment by the diffusion of essential materials from their immediate environment. If blood vessels do not grow into them, the tumors become no larger than pinheads.

All tumors produce a tumor angiogenesis factor (TAF), which causes a rapid growth of blood vessels from surrounding tissues. After these blood vessels enter the tumor, it grows rapidly. Folkman suggested that if TAF can

be blocked, tumor growth could be halted before it damages its host. Through this approach, called "anti-angiogenesis," it may be possible to develop antibodies that will inactivate or destroy TAF. This method should prove to be particularly helpful in the treatment of brain tumors, which have many blood vessels. Surgery might be employed to remove the major foci of a tumor and anti-angiogenesis therapy used to control what remains.

Support for a failing liver. Dr. George Abouna, department of surgery, Medical College of Georgia, in August, reported the dramatic recovery of three patients who had suffered severe liver damage from viral hepatitis.

Unlike acute and chronic kidney failure, in which an "artificial kidney" is available to maintain life, there had been no comparable device for a failing liver. Patients with massive destruction of their livers, most commonly caused by viral hepatitis, have generally died after coma, hemorrhage, or infection.

Each of Dr. Abouna's patients was connected to an isolated baboon liver

Medicine

Continued

or human liver and the patient's blood was pumped through the outside liver before it returned to his body. Not only did the extra liver remove toxic products from the blood, but it also produced the factors that are necessary for normal blood coagulation, thus avoiding the catastrophic hemorrhages that could occur during such a procedure. The extracorporeal liver perfusions—passing the patient's blood through another liver outside his body—lasted from 20 to 50 hours.

Dr. Abouna showed that extracorporeal liver perfusion was superior to simply exchanging the patient's blood by multiple transfusions, a procedure that produces only brief periods of slight improvement. The livers of all his patients showed signs of regeneration.

Although it is apparent that life may be maintained by this type of perfusion, the livers of some patients are so badly damaged that they will not regenerate. The only hope for these patients is transplantation of another human liver, which is now reported to have a 28 per cent, one-year survival rate.

Cardiovascular surgery. The use of segments of veins as a by-pass around an area of obstruction in the coronary artery remained the most popular treatment for atherosclerotic disease and the pains of angina pectoris. This procedure was performed approximately 20,000 times during 1971. In spite of some question of overuse of the operation, hospitals are reporting excellent results in relief of angina. An estimated 85 per cent of these grafts continue to carry blood for more than a year. However, it will be several years before surgeons know if this procedure really prolongs life longer than alternative therapies.

Heart transplantation was carried out infrequently and generally in only a few specialized centers in 1971 and 1972. Dr. Norman Shumway, of Stanford University, one of the pioneers in this field, reported a 45 per cent, one-year survival rate and a 35 per cent, two-year survival rate after transplantation. During the past two years, the number of survivors has increased. Nevertheless, a major effort is still being made to perfect an implantable artificial heart.

Coral skeleton, *below right*, is so much like human bone, *below*, that it is being used to cast prosthetic devices that are implanted in humans.

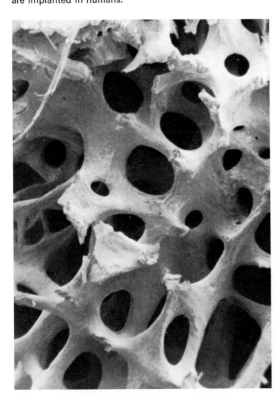

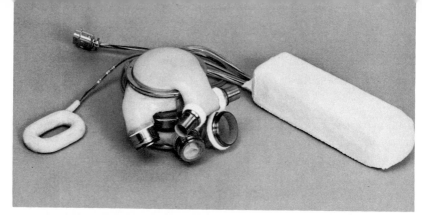

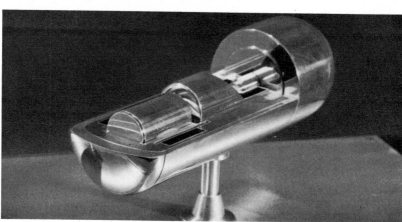

The first totally implantable artificial heart system, *above right*, includes, from left to right, pick-up coil for recharging batteries, motor-driven heart, and battery pack and control computer. A nuclear-powered engine, *right*, also implantable, can be used as an alternate source of power.

Medicine

Continued

Orthopedic surgery. In addition to producing much discomfort, fractured bones, and particularly those that heal slowly, cost Americans millions of dollars annually in time lost from work. In October, 1971, surgeons at the University of Pennsylvania reported that they had used electricity to heal an ankle fracture that had not responded to other treatment. Another successful test of this method was detailed by Dr. Leroy S. Lavine, orthopedic surgeon at Downstate University in Brooklyn. Lavine reported in March that a pseudarthrosis (false joint) of the tibia (shinbone) of a 14-year-old boy fused after a small electric current was applied to both sides of the bone. The current was supplied for a period of four months by flashlight batteries attached to buried metal electrodes.

Lavine believes that *piezoelectricity* (electricity produced by pressure) effect is involved in the healing process. In this phenomenon, mechanical forces can be converted to electrical impulses by crystalline solids such as bone. Electron microscopy study of this process suggested that the electric current had caused the fibrous tissue cells at the fracture site to make bone. Doctors hope to use this method for a wide range of treatments for nonjoining fractures without need for electrodes.

Acupuncture. Interest in the ancient art of healing with needles became widespread in the medical community in 1971 and 1972. The first use of acupuncture for anesthesia in the United States was reported by the Downstate Medical Center in Brooklyn, N.Y., on May 15, 1972. Anesthesiologist John W. C. Fox used four fine-gauge, 1¾-inch-long needles on a 23-year-old medical student undergoing an operation for removal of a small growth on his tonsils. The needles were inserted and rotated at the prescribed rate of 100 times per minute in the webbing between the thumb and forefinger of each hand and in the skin between the second and third toes of each foot. Successful acupuncture anesthesia was also reported in two operations in late May at the Albert Einstein College of Medicine.   [John Bryan Price, Jr.]

Meteorology

National Oceanic and Atmospheric Administration (NOAA) scientists reported in May, 1972, that the carbon dioxide (CO_2) content of the atmosphere may increase 20 per cent by the end of this century. An increase in CO_2 levels in the atmosphere could cause the earth's climate to warm up, because CO_2 traps the infrared radiation that is given off by the earth.

Human activities may play a major role in changing the climate by affecting the chemical and physical properties of the atmosphere. For example, burning such fuels as coal and oil adds large amounts of CO_2, and carbon monoxide (CO) to the atmosphere.

However, human activities may also increase the concentration of dust and other particles in the atmosphere. And this could reduce the intensity of solar radiation reaching the earth, causing the climate to grow cooler. In addition, the condensation trails left by jet airplanes flying at high altitudes may artificially make the atmosphere cloudier. This also could decrease the solar radiation that reaches the earth.

Scientists can only guess at the ultimate effects that these and other man-made atmospheric changes will have on the climate. Unfortunately, they know relatively little about trends in CO_2, particulate content, cloudiness, or other atmospheric properties.

To begin gathering such information, the United States and several other nations agreed to establish a network of remote sampling stations that will monitor global trends in atmospheric gases and particles. The proposal was endorsed in June, 1972, by the United Nations (UN) Conference on the Human Environment, meeting in Stockholm, Sweden. The UN World Meteorological Organization will coordinate the network of 110 monitoring stations.

Ten of these will be so-called baseline stations located in isolated, relatively pollution-free areas of the world. They will provide a standard for determining the seriousness of pollution levels. All of the stations will measure levels of CO, CO_2, and particles, as well as ozone, sulfur compounds, radioactivity, and solar radiation.

France's meteorological satellite, Eole, was launched by NASA to track up to 500 weather balloons drifting in the Southern Hemisphere. It transmits data on pressure, temperature, and winds through a tracking station to the French National Center for Space Studies for analysis. The data is then forwarded to NASA.

Satellite Eole

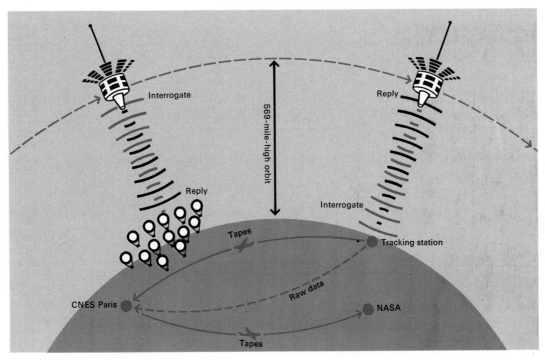

331

Meteorology

Continued

The United States plans to operate four of the base-line stations. Mauna Loa Observatory, atop a dormant volcano in Hawaii, and a station in Antarctica are already in operation. The other two will be at Point Barrow, Alaska, and in American Samoa.

Climate and cities. In 1971 and 1972, scientists began analyzing data from the Metropolitan Meteorological Experiment (Metromex), a cooperative project of United States government and university scientists who are trying to determine the influence of cities and industrial plants on local weather and climate. Whether or not cities affect the amount of local precipitation remains a controversial point. It is well known that cities alter the local climate in many ways, such as artificially heating the air, obstructing wind, and emitting gaseous and particulate pollutants. Particles provide a nucleus around which water vapor condenses and may in this way affect cloud formation and precipitation.

Since 1970, Metromex scientists have been collecting air samples, analyzing particles, measuring wind patterns, and making detailed maps of precipitation around several cities. Some investigators claim that cities measurably affect the distribution and amounts of precipitation, but the data are still inconclusive.

Great Lakes project. The International Field Year for the Great Lakes (IFYGL) began in April, 1972. The deteriorating condition of the Great Lakes spurred this major Canadian-U.S. effort to better understand the chemical, biological, and hydrological processes affecting these valuable water resources. The meteorological phase of the project is focusing on evaporation and precipitation. The IFYGL is a concentrated, yearlong study of Lake Ontario, using towers, buoys, ships, aircraft, and special weather radar devices. It is part of a four-year, $15-million program in which about 600 scientists and technicians will participate. The goal of the project is to provide a scientific basis for predicting future lake levels and lake quality, and for assessing the effects that changes in lake use will have on the environment.

Drawing by Chon Day; © 1972 The New Yorker Magazine Inc.

Meteorology
Continued

Numerical weather prediction. Computers continued to assume an increasing role in weather forecasting, and the mathematical models of the atmosphere used for weather prediction grew increasingly complex.

The U.S. National Meteorological Center (NMC) introduced a limited-area fine-mesh grid for North America in 1971. Numerical weather forecasts are computed for grids of a certain size superimposed on maps of the earth. The larger the grid size, the less precise the forecast. The limited-area fine-mesh model cut the standard grid size used for North America in half.

A similar fine-mesh prediction model went into operation in Great Britain. With it, the British Meteorological Office hopes to forecast the details of rainfall over the British Isles.

Weather modification. On Sept. 26 and 28, 1971, U.S. scientists of the Navy, Air Force, and NOAA criss-crossed Hurricane Ginger in airplanes and dropped more than 250 burning canisters full of silver iodide into its clouds. This was the most recent hurricane seeding of Project Stormfury, an experiment to determine if man can control hurricanes by seeding the clouds.

In addition to inducing precipitation, cloud seeding may be able to modify the dynamics of circulation systems such as hurricanes. Computer simulations have indicated that seeding the clouds outside the hurricane eyewall – the ring of strongest wind and heaviest rain – may reduce the maximum wind speeds.

Unfortunately, Hurricane Ginger did not have a well-defined eyewall on the days that it came within range of Stormfury aircraft. So it was not a fully satisfactory target for the project. To find more and better targets for experimentation, Stormfury scientists have proposed moving the project to the western Pacific Ocean, where they can seed the more abundant typhoons. However, the U.S. government decided not to move Stormfury to a new site in the Pacific in 1972.

Federal support for research on artificial weather modification increased to $25 million in fiscal 1972, from $15-million in 1971.   [Jerome Spar]

Microbiology

In 1971 and 1972, scientists reported experiments that proved what several earlier reports had only suggested – the genes of bacteria, one of the simplest forms of life, can be incorporated into the cells of complex plants and animals. The new work is particularly exciting because it proves that the incorporated bacterial genes continue to function in their new homes. In one experiment, the bacterial genes corrected a genetic defect in human cells. The experiments also lent some support to a view of cell evolution known as the capture theory. Part of this theory is that certain gene-containing parts of modern complex cells were once bacteria. See THE CURIOUS ANCESTRY OF OUR CELLS.

In December, 1971, Belgian biologists Lucien Ledoux and Raoul Huart of the Center for the Study of Nuclear Energy and Nichel Jacobs of Vrije University in Brussels reported that the deoxyribonucleic acid (DNA) – genetic material – of two different bacteria could be incorporated into germinating seeds of the common weed mouse-ear cress (*Arabidopsis thaliana*), which is related to mustard weed. More startling yet, after the mature plant had flowered, the bacterial DNA was found in its seeds.

First, the scientists chemically sterilized seeds and planted them in sterile glass containers. Meanwhile, the bacteria – *Streptomyces coelicolor* and the common intestinal bacterium *Escherichia coli* – were grown in a medium that contained the radioactive chemical thymidine, which was incorporated into their DNA, making it radioactive. Next, they extracted radioactive DNA from the bacterial cells, purified it, and added it to the containers of germinating seeds.

The plants that grew from the seeds were harvested at different stages of growth, and their DNA was isolated and purified. Then, the DNA was centrifuged. Bacterial DNA is heavier than plant DNA and, consequently, if it is centrifuged with plant DNA, it spins out further in the container that is holding them.

Ledoux, Huart, and Jacobs found in the centrifuged material some radioactive DNA too heavy to be pure plant DNA and too light to be pure bacterial

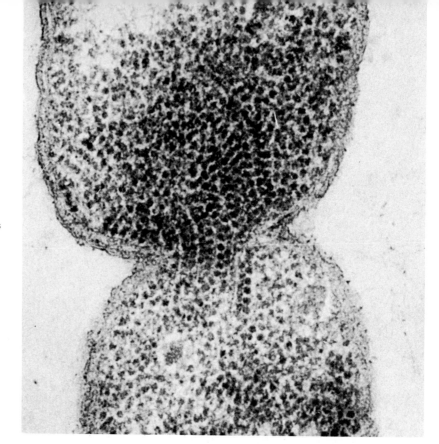

As bacterium reproduces by division, polysomes – chains of beadlike structures that help to produce proteins – form at the dividing point. The French scientists who took the electron photomicrograph think the polysomes might be producing proteins of the new membrane that is forming between the two daughter cells.

Microbiology

Continued

DNA. They separated this DNA, by treatment with ultrasonic-sound vibration, into molecules of DNA and centrifuged once again. This time, the scientists got two distinct types of DNA. One was heavy and radioactive and behaved exactly as the bacterial DNA should. The other was relatively light, not radioactive, and behaved like plant DNA. The only plausible explanation for these observations was that the bacterial DNA had been integrated into the plant DNA.

By checking progressively older plants with a radiation-detection device, the scientists determined where the bacterial DNA moves as a plant matures. The bacterial DNA first localizes in the early leaves of the plant. When flowering begins, however, it accumulates in the flowering buds. Later, it is found in the embryos of the seeds that developed from the flowers.

It is surprising that bacterial DNA is taken up by a plant. It is astounding that it is transferred to the next generation. If these findings are verified – and especially if the bacterial genes function

in the plant cells – current ideas about evolution and many basic principles of genetics may have to be re-evaluated.

A plant tumor commonly called crown gall is produced when the bacterium *Agrobacterium tumefaciens* infects wounds of certain plants. This plant tumor is comparable in many respects to cancerous animal tumors. Consequently, studies of crown gall are important, and scientists at several laboratories have tried to determine how the bacterium causes the tumors to develop in plants. Experiments by two research groups threw new light on this subject during the past year.

Several years ago, it was discovered that DNA of the bacterium, but apparently not the bacterium itself, was present in the developing tumor tissue. Until the results of two research groups were published in 1971, almost nothing was known of how the bacterial DNA gets into the tissue and whether the DNA functions there.

Bacteriologists David Yajko and George Hegeman at the University of California, Berkeley, grew *A. tume-*

faciens in a medium containing radioactive thymidine, which became incorporated into the DNA of the bacteria. Then they incubated the bacteria for short periods of time with slices of carrot root tissue. After the tissue was washed, to remove any bacteria on the surface, it was placed in a radioactivity-detecting apparatus. As the scientists expected, the device gave a positive reading, proving that bacterial DNA was present.

Next, Yajko and Hegeman grew the bacteria in a medium that would cause certain of the bacterial constituents, but not their DNA, to become radioactive. After these bacteria were grown with carrot root tissue, the researchers found no radioactivity in the tissue. This proved conclusively that the bacterial DNA that enters the plant tissue does not get in with intact bacteria.

Yajko and Hegeman also proved that the bacteria attach themselves to the plant tissue to transfer their DNA. The scientists grew the bacteria that contained radioactive DNA with carrot root tissue in the presence of the enzyme deoxyribonuclease. This enzyme completely breaks down any DNA not enclosed by a cell membrane. The amount of radioactive bacterial DNA taken up by the plant tissue was the same in the presence of the enzyme as it was without the enzyme. This disproves an earlier hypothesis that the DNA is released from dead, disintegrated, bacterial cells before it is taken up by the plant. Just how the DNA leaves the intact bacterial cells and enters the plant cells remains to be discovered. Undoubtedly, several research groups soon will be working on this problem.

Another research team of molecular biologists, led by M. Stroun at the University of Geneva in Switzerland, reported that the DNA of *A. tumefaciens* is reproduced in tomato plant tissue as the plant tumor is formed. The scientists infected wounded plants with the bacteria and then grew the plants for seven hours while they were exposed to radioactive thymidine. DNA synthesized during this time contained the radioactive thymidine. The researchers then extracted, purified, and centrifuged the DNA in the developing tumor tissue. Three classes of radioactive DNA were found. One had the density of plant DNA, another the density of bac-

terial DNA, and the third had a density intermediate between the two. Stroun and his associates disrupted and recentrifuged the intermediate DNA, and detected radioactive plant DNA and radioactive bacterial DNA. This proves that both plant and bacterial DNA were synthesized in the plant during the seven hours after infection.

Stroun and his colleagues went on to demonstrate that the bacterial DNA was transcribed in the plant tissue. In transcription, the genetic code of DNA is copied onto molecules of ribonucleic acid (RNA), which are then used as a template, or pattern, for protein synthesis in the cytoplasm of a cell. Transcription is catalyzed by an enzyme called RNA polymerase.

The antibiotic rifamycin inhibits the RNA polymerase of bacteria. When Stroun's group treated wounded tomato plants with this antibiotic soon after infecting them with *A. tumefaciens*, no tumor developed. The only plausible explanation is that the bacterial DNA in the plant must be transcribed into RNA to produce the plant tumor.

The scientists then actually proved directly that RNA made from bacterial DNA is produced in the wounded and infected plant. First, they grew the infected plant with radioactive uridine for seven hours. This chemical is incorporated into newly synthesized RNA. Next, the RNA in the developing tumor tissue was isolated and purified by several complex procedures.

The scientists used a very sensitive technique called RNA-DNA hybridization to determine if any of this RNA had been made using bacterial DNA as a template. This technique is based on the fact that RNA will stick to the DNA that acted as a template for its production. Stroun and his co-workers simply mixed DNA isolated from bacteria with the radioactive RNA purified from the infected plant. Some of the RNA hybridized with the DNA, showing that bacterial RNA had been produced in the plant tissue.

Transfer of nitrogenase genes. Another recent example of transfer of bacterial genes may have significant practical implications. This transfer is not from bacteria to higher life forms, but from one kind of bacteria to another kind of bacteria.

The genes transferred were those responsible for the synthesis of the enzyme nitrogenase. This enzyme converts the inert atmospheric gas nitrogen into the biologically useful form of ammonia. Nitrogen is required for life, yet only a limited range of bacteria and the closely related blue-green algae possess the enzyme that allows them to tap the plentiful supply in the atmosphere. Bacteriologists R. A. Dixon and John Postgate of the University of Sussex in England reported in May, 1972, the transfer of the genes for nitrogenase from one bacterium to another, which previously had not been able to utilize nitrogen gas.

The scientists mixed populations of *Klebsiella pneumoniae*, which produces nitrogenase, with *E. coli*, which does not. The *K. pneumoniae* cells they used were susceptible to the antibiotic streptomycin and the *E. coli* cells were from a strain that is resistant to it.

They placed the mixed culture in a medium containing streptomycin and readily available sources of food except nitrogen, which was supplied as the gas.

The streptomycin killed the *K. pneumoniae*, but some *E. coli* survived and grew. They must have acquired genes needed to produce nitrogenase.

The significance of this work is not that genetic information can be transferred among bacteria. Gene transfer in bacteria has been known for years. But the fact that genetic manipulation has been used to transfer a potentially useful function from one bacterium to another is new and exciting.

It does not stretch the imagination too far to hope that the genes for nitrogenase might be introduced into higher plants. Although this goal may be a long way off, it is certainly worth pursuing. Such plants, able to obtain the nitrogen they need from the atmosphere, would completely revolutionize agriculture, particularly in underdeveloped countries that are now unable to produce enough food for their people. It would no longer be necessary to fertilize soil with commercial fertilizers, and, as a side benefit, that would also end several serious ecological problems that such practices often create. [Jerald C. Ensign]

Molecular Biology

See Biochemistry,
Chemistry,
Genetics,
Microbiology

Neurology

Much neurological research reported in 1971 and 1972 reflected increased interest in how man and other mammals can recover abilities lost because of injury to the central nervous system—the brain and spinal cord. Most other organs of the body show a remarkable ability to heal and to resume their functions after being damaged. However, serious injury to the central nervous system is usually permanent and causes lifelong abnormalities. The only exception to this occurs when the injury is inflicted when the animal or person is either in the fetal stage or the first few days after birth.

Neurons, or individual nerve cells, in the central nervous system transmit information to other neurons at sites called synapses. Neurons involved in similar responses are clustered together in the central nervous system. Their axons, the fibers along which they transmit their impulses, travel together in a bundle that ends at a distant group of neurons with which the axons form synapses. When injury severs or seriously damages a bundle of axons, the response they control is lost.

Regrowth of cut axons. Recovery of abilities lost after damage to the central nervous system depends on the regrowth of cut axons and their accurate reconnection to the same neurons with which they formerly formed synapses. Fishes and amphibians can make this reconnection, but man and other mammals cannot.

In August of 1971, neurobiologist Anders Björklund and his associates at the University of Lund in Sweden reported experiments with rats in which, after axons in the rat's brain were cut, they regrew vigorously for a time. The workers also noted that the newly formed axon sprouts grew in the right direction—into the site of injury. This confirmed earlier reports that some axons in mammals can regenerate. However, regenerating axons in the central nervous system of mammals rarely, if ever, grow all the way back to their original synapse.

The glial scar. A series of studies still underway at the University of Florida in Gainesville adds new details and confirms old ones about why regenerating mammalian neurons do not grow back

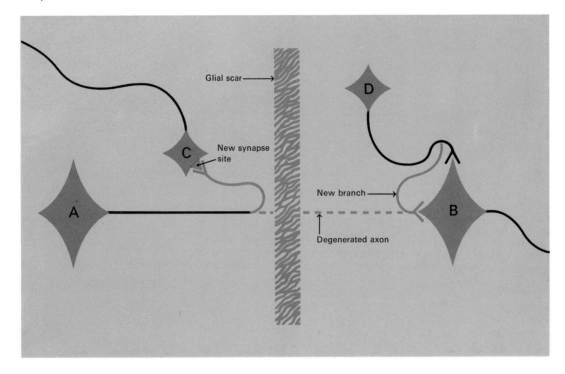

Glial scar

New synapse site

D

C

New branch

A

Degenerated axon

B

Neurology

Continued

The axon of neuron A, in a mammal, normally synapsed on (connected to) neuron B. When the axon was cut, a glial scar formed. The regenerating axon, unable to penetrate the scar, synapsed at a new site that formed on nearby neuron C. The vacated synapse site on neuron B was occupied by a new axon branch from neuron D. With A's axon synapsed, and its original site occupied, A's axon would not reconnect to B even if the scar were to be removed.

to their original points of synapse. One of the first responses to neural injury occurs at the damaged site. The glial cells, which surround the neurons of the central nervous system, begin to multiply rapidly and form a dense scar. Experiments have shown that severed axons in goldfish regrow through the glial scar, but regenerating axons in mammals cannot penetrate the scar.

Other events occurring in the region of injury have been studied by neurobiologist Gerald Bernstein and his coworkers in Gainesville. When they looked at neurons near the injury, they noticed the development of new sites at which synapses could be formed.

In the rat, the regenerating axons blocked by the glial scar formed synapses with these local neurons. On the other hand, in the goldfish, the regenerating axons easily penetrated the glial scar and connected with their original terminal neurons. However, if a plastic barrier was used to block the regenerating goldfish axons, some of them, like those of the rat, synapsed incorrectly on the changed local neurons. Once these

local contacts were completed, even if the plastic barrier was removed, the axons did not grow toward the neurons on which they originally synapsed. This "contact inhibition" could be overcome in the goldfish only if the regenerated goldfish axons were recut after the barrier was taken away. Then, the axons would regrow across the glial scar and connect again with their original terminal neurons.

Earlier studies in which the glial scar was softened or not allowed to form showed that even when healing mammalian axons are not hindered by it, they fail to find their way back to their original synapses. This failure is probably related to the chemistry of the nerve cells involved.

Chemical failures. The chemistry of both the damaged nerve cells and those with which they originally synapsed plays a role in guiding growing axons of damaged neurons. For example, at about the same time that an injured neuron's axon begins to grow, the rest of the cell begins to attempt to meet its regenerating axon's demands for nu-

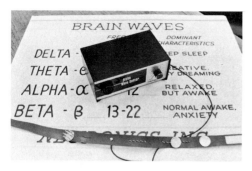

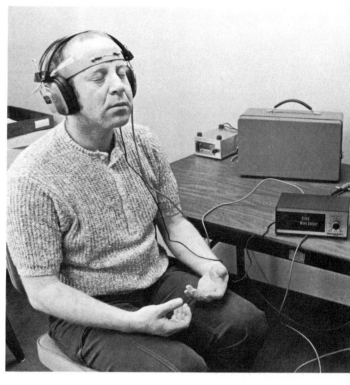

Devices, *above,* that measure the alpha waves produced irregularly by the brain have found their way out of the laboratory and into the hands of laymen. A fad developed in which people trained themselves to produce alpha waves at will, *right.* A beep in the earphones tells the man when he is creating the waves. The "alpha state" supposedly is unmatched for relaxation.

Neurology

Continued

trients, especially protein. Successful regeneration almost certainly depends on maintaining increased protein synthesis within the neuron and delivering sufficient protein to the growing axon tip. Experiments have revealed increased rates of protein transport along the axon toward the regenerating tip of damaged neurons in goldfish. Damaged mammalian axons, however, do not seem to generate a similar increase. Thus, failure to meet protein demands of the regenerating axon may be one of the reasons that mammalian nerves fail to regenerate in the central nervous system.

In 1971, neurologist Robert Moore, working with Björklund and his associates, confirmed work suggesting that at least part of the problem in mammals may be the failure of a chemical signal sent from the neuron on which the injured axon normally synapsed. In the original work, done in 1969, neurobiologist Geoffrey Raisman of the University of Oxford, England, studied two bundles of axons–the fornix and the median forebrain bundle. Both normally synapse with the same terminal neurons in

the rat's brain. Using the electron microscope, Raisman could distinguish between the synapses of axons of the two bundles. He found that when either bundle of axons was cut, the intact axons of the other bundle developed new axon branches after the ends of the disconnected axons had degenerated. These healthy axon branches then moved in and occupied the vacant synaptic sites.

Moore and his co-workers checked the same two bundles chemically. Synapses formed by axons in the median forebrain bundle contain the substance noradrenalin whereas those formed by axons of the fornix do not. When the scientists destroyed the axons in the fornix, the number of nerve terminals containing noradrenalin increased. After they cut the median forebrain bundle, however, they found that those containing noradrenalin decreased substantially. In mammals, cut axons may fail to reconnect to their correct synapses because the signal for reconnection ceases when the original synaptic sites are occupied by branches of adjacent intact neurons. [Thomas H. Meikle, Jr.]

Oceanography

Scientists made several new discoveries about the ocean floor in 1971 and 1972, and launched new expeditions to further investigate seabed areas. The United States and other nations also began joint studies of two little-known aspects of the ocean, Arctic ice floes and mid-ocean circulation. The location of large cold-water eddies provided scientists with an unusual opportunity to study this ocean phenomenon.

Seabed discoveries. A five-month study, completed in November, 1971, provided new details about the northeast Pacific Ocean floor. Canadian, Japanese, and U.S. scientists aboard the National Oceanic and Atmospheric Administration (NOAA) research ship *Surveyor* found the southern edge of a huge sediment plain, perhaps larger than the central United States. Complex deep-sea channels up to 600 feet deep and 6 miles wide run through the sediment plain.

The *Surveyor* expedition revealed that two undersea mountains off the Oregon-Washington coast, Bear and Cobb seamounts, are still growing because of volcanic action or a general uplifting of the area. Someday they may become islands. The scientists also found a deep, sediment-filled trench that suggests offshore oil deposits may be present.

New seabed studies. In March, 1972, a crew of scientists aboard another NOAA research ship, *Oceanographer*, embarked on the first cruise of a three-year program to study the floor of the northwest Pacific Ocean. The study encompasses a little-known area of the seabed, 2,000 miles long and 240 miles wide, running from the Hawaiian Islands to the Philippine Sea.

The scientists suspect they will find the world's oldest oceanic crust, unchanged since the late Paleozoic Era about 250 million years ago. They will study the geologic features and history of the crust in that area and try to determine whether the Pacific Plate, thought to be the largest of the enormous tectonic plates that form the earth's crust, is gradually slipping under the Philippine Sea Plate. See EARTH'S HEAT ENGINES.

Beginning in late March, French, English, Venezuelan, and U.S. scientists aboard the NOAA ship *Discoverer* spent three months in the Atlantic Ocean studying the area between the West Indies and the undersea mountain range known as the Mid-Atlantic Ridge. They were trying to learn more details about how the Atlantic Ocean Basin developed. In addition to the North Atlantic Plate, South Atlantic Plate, and African Plate, scientists believe this area contains a number of smaller tectonic plates whose movements figured in the formation of the basin. They hope that analysis of the data collected on this expedition will enable them to apply knowledge about moving plates to regional as well as worldwide geological phenomena.

A detailed six-month study of the North Pacific seabed, extending 800 miles off the North American coast from California to British Columbia, began in April, 1972. The expedition's scientists, using *Surveyor*, will map the topography of the area in detail and measure the earth's magnetic field and the pull of gravity there.

Arctic ice floes. The discovery of oil on the North Slope of Alaska has, in part, inspired the United States, Canada, and Japan to undertake an extensive study of ice movement in the Arctic Ocean. The project, known as the Arctic Ice Dynamics Joint Experiment (AIDJEX), will take more than two years to complete. The first personnel and equipment were flown into the project's main camp, an ice floe almost 300 miles north of Point Barrow, Alaska, in March. In addition to the main camp, there are two smaller camps 60 miles apart.

The AIDJEX scientists are studying forces, such as wind, ocean current, ice stresses, and tidal forces, that control ice movement. They are particularly interested in pressure ridges that project into the air above the ice and into the water below. By reacting to wind and current, these pressure ridges may control the motion of an ice floe in much the same way that sails and keels control sailboats. Information gathered by the AIDJEX scientists might enable them to predict or, eventually, even to control ice movement. This could aid ships in navigating through Arctic waters.

The scientists will use buoys to collect weather data and current meters to take water-flow readings. Ice-floe movements will be tracked by acoustic bea-

Oceanography

Continued

The catamaran *Lulu* hovers over *Edalhab II,* a movable underwater habitat used for the Florida Aquanaut Research Expedition (FLARE), *above. Lulu* moved *Edalhab* to sites off the Miami, Fla., coast from January to mid-April, 1972. FLARE diver-scientists study the ecology and geology of reefs, *above center,* and observe how fish react to traps, *above right.*

cons, laser beams, and satellites. A Japanese experiment is designed to measure strains within the ice. Both divers and remote-controlled submersibles will explore the underside of the ice.

Cold-water eddies. Oceanographers from NOAA and the University of Rhode Island studied a large Gulf Stream cold-water eddy for more than a year, the longest time that such an eddy has been under observation. The cold-water eddy, about 100 miles in diameter, was first sighted by an environmental satellite in April, 1971. At that time, it was off Cape Hatteras, N.C.

Like other cold-water eddies, it was formed when the Gulf Stream made one of its unpredictable fluctuations off course. Part of this fluctuation, or meander, broke off from the main body of warm Gulf Stream water. The warm water of the meander then surrounded colder water outside the Gulf Stream, forming the eddy in the center.

Unlike other cold-water eddies, which head out to sea, this almost circular body of water, rotating counterclockwise, moved about 1 mile a day in a

relatively straight line down the East Coast. By April, 1972, it was about 120 miles east of Cape Kennedy, Fla., in the area of the Atlantic known as the Sargasso Sea. Oceanographers determined that the surface temperature of the eddy was from seven to eight degrees colder than that of the Sargasso Sea. They also believe that the eddy might have been more than a mile deep.

This eddy disappeared in April, 1972, and oceanographers think that it was absorbed by the Gulf Stream. Meanwhile, another eddy that is perhaps larger and colder than the first was discovered. The NOAA scientists plan to use the *Researcher* for an extensive study of the second eddy in late 1972. These eddies have raised interesting questions, such as whether a countercurrent runs southward under or near the Gulf Stream, and whether eddies usually are absorbed into the Gulf Stream, restoring energy to it.

Ocean circulation. In May, 1972, the research vessel *Knorr* of the Woods Hole (Mass.) Oceanographic Institute sailed for an area in the Sargasso Sea near

Oceanography

Continued

Bermuda to prepare for the Mid-Ocean Dynamics Experiment (MODE), scheduled for March through June, 1973. The *Knorr* placed instruments on a network of buoys, which will help MODE scientists study ocean circulation.

So little is known about mid-ocean dynamics that oceanographers are not even certain that a regular, identifiable pattern of circulation exists. Using buoys, current meters, and sophisticated instruments aboard three U.S. ships and a British ship, the MODE scientists will attempt to make a three-dimensional map showing levels of temperature, salinity, and the rate at which currents flow in this section of the Atlantic Ocean.

This information will help theoretical oceanographers construct the mathematical models that aid them in predicting the behavior of major ocean currents. Scientists must understand ocean circulation more fully in order to estimate how various contaminants and other materials will be distributed in parts of the ocean and to study the ocean's effect on weather and climate.

To illustrate the difficult task of theoretical oceanographers in studying oceans, an analysis of data collected in 1966, 1967, and 1970 by scientists from Liverpool University in England showed that different masses of ocean water mix in a very complex manner. The scientists studied the area where the warm, salty Mediterranean Sea meets the cooler, less salty Atlantic Ocean at the Strait of Gibraltar. Instead of uniformly mixing together, the water from these two bodies forms a pattern similar to steps in a staircase, with each level of water having a different concentration of heat and salt.

Ocean pollution. In August, 1971, Woods Hole oceanographers reported finding DDT and PCBs (polychlorinated biphenyls) in samples of marine organisms taken from the Atlantic Ocean. Concentrations of these pollutants were about the same in marine animals that dwell in the upper ocean layer and in those that feed in the upper layer at night but return to the depths during the day. This work seems to support that of other scientists who sug-

gest that extremely large quantities of DDT travel through the atmosphere to the ocean and finally settle in the oceanic abyss. See NEW PERIL TO OUR OCEANS.

Central to the findings of the Woods Hole study is the fact that DDT does not readily dissolve in water but easily dissolves in fats and oils, such as the fatty tissue in living organisms. Following normal biological pathways, DDT residues would be expected to concentrate heavily in aquatic organisms that are toward the end of the food chain. However, this does not seem the case.

The study showed that position in the food chain has no apparent relation to concentrations of DDT in fish and shellfish. These aquatic animals can absorb DDT in two ways: Indirectly, by consuming other fish; and directly, by taking DDT up from the water through gills or other body openings. Plankton, at the beginning of the food chain, contain the highest concentrations of DDT, while higher aquatic animals toward the end of the chain, such as tuna, contain low concentrations. So, it appears that fish largely absorb DDT directly from the water, and this keeps them in equilibrium with DDT concentrations in the water. Also, if DDT concentrations were magnified along the aquatic food chain, massive extinctions of higher aquatic species could be expected, but these have not occurred.

However, so little is known about how marine organisms interact with DDT and PCBs and the complexity of the aquatic food chain, that scientists fear there may be as yet unknown, indirect effects from these pollutants.

In September, 1971, National Aeronautics and Space Administration (NASA) officials announced that a device for monitoring plant life in the ocean had been successfully tested. The device, which measures chlorophyll levels in lakes and seas, can detect the effects of pollution on plant life – for example, when phosphate concentrations cause the rapid growth of algae. Although development of the device is still far from completed, NASA officials hope to eventually develop a system of satellites equipped with this device that can monitor the oceans on a worldwide basis. [Darlene R. Stille]

Alcoa Seaprobe, equipped with remote scanning, coring, and retrieval devices, can find mineral deposits on the ocean floor and locate and recover ships and other objects that sank in deep water.

Physics

Atomic and Molecular Physics.

Many experiments in this field measure the properties of single particles – isolated atoms, molecules, or ions. To study intrinsic atomic or molecular properties, physicists must complete their observations during the brief period between events that alter the state of the particle, usually collisions with other particles or with the walls of the container. In most experiments, the atoms' velocity, and therefore their energy, is not well defined, because the time for study is so short.

Professor Hans G. Dehmelt and his students at the University of Washington have developed a technique for overcoming these difficulties. Using strong electric and magnetic fields, they store ions at constant energy in a small volume of space, free from collisions for as long as five hours.

These trapping techniques were used in 1972 at the Joint Institute for Astrophysics in Boulder, Colo., where Gordon H. Dunn and F. L. Walls studied the recombination of ions and electrons. They used a combination of magnets and charged electrodes to hold positive ions such as N_2^+ or H_2O^+ in space. The magnetic field keeps the ions moving in tight spirals about the magnetic field lines. The electrostatic field keeps their spiral paths from shifting along the magnetic field lines.

The electrodes also serve as the detectors. Electrical noise originates in this trap, its amount remains relatively constant at different frequencies. However, there is always a sharp dip in this noise at one frequency, the frequency at which the ions spiral, for the particular ions in the trap. The frequency of this dip thus identifies the species of particles in the trap, and the depth of the dip gives the number of particles.

Dunn and Walls shot electrons having identical energies into the trap to neutralize the ions. By observing the changes in the dip, they determined how fast the electrons and ions recombined over a range of electron energies. Because the energy of both a trapped particle and the combining electron is well specified, the reaction rates can be studied very precisely.

Lamb shift. A new Lamb shift determination by Daniel E. Murnick and Marvin Leventhal of Bell Telephone Laboratories and H. W. Kugel of Rutgers University was reported in December, 1971. The Lamb shift, named for its discoverer, Willis E. Lamb, Jr., of Yale University, refers to the very small difference between the energy levels of two excited states, the $2S_{1/2}$ and $2P_{1/2}$ levels, in atomic hydrogen.

Such a shift also occurs in heavier atoms that have been stripped of all but one of their electrons and are therefore hydrogenlike. The heavier the ion, the greater the shift. The magnitude of the shift is a test of the value calculated using quantum electrodynamics.

The experimenters measured the Lamb shift in carbon atoms that had lost five electrons (C^{5+}). A 25- to 35-million electron volt beam of carbon nuclei (C^{6+}) is fired from a Van de Graaff accelerator into argon gas. Some of the carbon nuclei pick up an electron and become C^{5+} in the excited $2S_{1/2}$ state. The ion in this state has a very long lifetime unless it is converted into the $2P_{1/2}$ state, in which case it rapidly decays to the ground state, $1S_{1/2}$, with the emission of an X-ray photon.

The $2S_{1/2}$ to $2P_{1/2}$ conversion is triggered by passing the high velocity (6-million meters per second) ions through a 100,000-volt per centimeter electric field. The rate of this transition is measured by how far downstream the X rays are emitted. The Lamb shift is calculated from the observed decay time. The value of the Lamb shift derived from this experiment is 780.1 $\pm$ 8.0 gigahertz (GHz), which agrees with the calculated value of 783.7 $\pm$ 0.25 GHz.

A new technique for observing the Lamb shift in atomic hydrogen was also announced early in 1972, by T. W. Hansch, I. S. Shahin, and Arthur L. Schawlow at Stanford University. In the past, the width of spectrum lines limited the accuracy of the Lamb shift measurement in hydrogen.

The Doppler effect produces most of the broadened hydrogen lines. This effect describes the change in frequency of the radiation produced by the motion of the emitting hydrogen atoms. Atoms moving toward the observer have a higher frequency, those moving away a lower frequency than stationary atoms. Similarly, in absorption, atoms moving toward and away from the source of radiation appear to have absorption lines

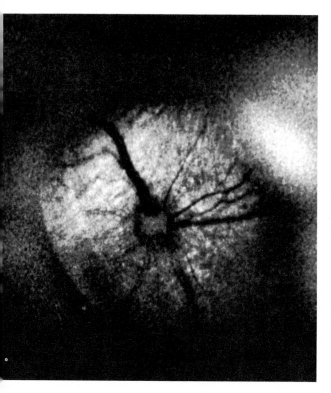
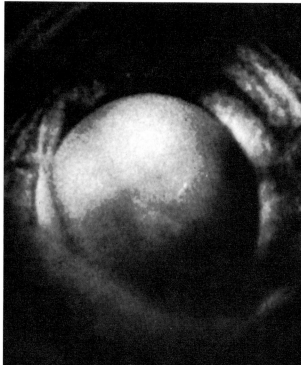

Physics

Continued

An image of the eye of a cat was frozen in three dimensions on a single laser-produced hologram. Thus a single hologram permits study of the blood vessels of the retina, *above*, or the iris, *above right*.

Starlike glass sphere one-thousandth of an inch in diameter is held aloft in air by laser light in an experiment by Bell Laboratory scientists.

at lower and higher frequencies. The large thermal motions in the gas broadens the lines.

In hydrogen, the decay from the $2P_{1/2}$ state to the ground state is in the visible light range, at 6563 angstroms (A). The line width is about 0.1A. Even at liquid helium temperatures it is still great.

The Stanford physicists completely resolved this line by using laser-saturation spectroscopy. The technique allowed them to look at only those atoms that are moving neither toward nor away from the observer. The Stanford experiments used a continually tunable laser producing 80 pulses per second as the source of light. The laser beam is split into two beams. One, a strong saturating beam of about 100 milliwatts (mw) was periodically interrupted by a shutter. The second, a probe beam, has a power of about 5 mw.

If the laser beam has a slightly higher frequency than the center of the absorption line, its light will be absorbed by atoms that are moving away from the saturating light source at just the right velocity to compensate, by Doppler

shift, for this frequency mismatch. The same will be true of the probe beam and the atoms moving away from it. Because the two beams pass through the hydrogen in different directions, each will interact with a different population of atoms.

If the laser frequency coincides exactly with one of the natural frequencies of the hydrogen atom, then the two beams will interact with the same atoms, those with zero velocity relative to the two beams. The strong saturating beam will be absorbed by some of these atoms, and they will not be available to absorb the probing light. This decreased absorption, recurring periodically only at the frequency with which the saturating beam is interrupted, identifies the exact wave length of the hydrogen line.

The reported line widths, 0.0036A to 0.0043A, are about 10 times narrower than those obtained in the best previous experiments. Although this resolution is not competitive with other methods of obtaining the Lamb shift, it can be used to measure other atomic constants more accurately.   [Karl G. Kessler]

Elementary Particles. "Things are turning out simpler than we had any right to expect." These words from Richard P. Feynman of the California Institute of Technology (Caltech) are typical of the optimistic mood of many particle physicists in 1972. For the last decade, each new discovery in particle physics has made the exotic subnuclear world seem more complex. Scores of new particles were discovered, each with its own peculiar properties.

Theories invented to account for these facts also grew more and more complex, a disastrous situation for a field whose only purpose is to discover simple laws that underlie the nature of matter. But now, it seems, particle physics may have turned a corner.

The most dramatic progress has come from "photographing" the inner workings of protons and neutrons. These particles are studied by bombarding them with electrons. The angle through which each electron is deflected, and the energy it loses in the process, are measured. These measurements can be used to reconstruct a picture of the target particle. Because of the wavelike properties of the electrons, the greater the energy, the finer the detail that can be observed. Much recent work has focused on deep inelastic scattering, in which very high energy electrons are used. The experimenter looks for those electrons that emerge with an energy much smaller than when they went in.

Scale symmetry. Physicists feared that such experiments would be hard to interpret. The energy the electron leaves behind goes into producing new, unstable particles whose unknown structure masks that of the proton or neutron. Early experiments at energies of a few billion electron volts (GeV) bore out these fears. Electrons usually lost exactly the energy required to produce the known unstable particles.

In 1972, a team effort by physicists from Stanford University and Massachusetts Institute of Technology working at the 2-mile-long Stanford Linear Accelerator (SLAC) carried this work to higher electron energies and greater fractions of energy lost. There, the data smoothed out, giving a continuous distribution of electron energies.

SLAC theorist James Bjorken discovered the data were simple in a startlingly new way. They had a peculiar symmetry called scale invariance.

Stated simply, scale invariance means that a law of physics is unchanged if you change the units of distance and time. Most of our ordinary physical laws do not behave this way. Newton's law of gravity, for example, contains a constant, g, whose value depends on the units of distance and time; Bjorken's formula contains only quantities that do not. When the electron-scattering measurements are expressed in terms of these quantities, they become independent of the energy of the bombarding electrons.

It is too early to be sure just how significant scale invariance may be. The inverse-square laws, which appear in all four fundamental forces now known, become scale invariant at speeds approaching that of light. At the very least, Bjorken's discovery may imply that the inverse-square laws remain valid at very short distances.

This is important because many physicists have speculated that part of the trouble with understanding elementary forces might be that space itself loses its smoothness at small distances, and force laws based on distance break down. At the other extreme, scale invariance might prove to be a fundamental law, a deep truth about the nature of space and time as significant as Einstein's theories of relativity. The next step is to determine whether this simple property holds for other types of particle reactions. See Close-Up.

Are partons quarks? Other features of electron collisions are being interpreted by Feynman's "parton" method. Feynman assumes the proton or neutron is made up of subparticles, or partons, and interprets the electron-collision data to find the properties of the partons. If the electron hits a parton hard enough, the forces binding it to its companions become negligible.

The first efforts in this direction have already yielded a dividend. The partons appear to be spinning with an angular momentum one-half the fundamental Planck constant, $\hbar$. Thus, the new results hint that partons could be Caltech theorist Murray Gell-Mann's hypothetical "quarks," particles with spin $\frac{1}{2}\hbar$ and charge either $\frac{1}{3}$ or $\frac{2}{3}$ as large as that of the proton and electron.

A New
High for
High Energy
Physics

In physics, new frontiers open whenever a new scale of energy is reached. Atomic and molecular physics occurs at energies of a few electron volts, nuclear physics at millions of electron volts (MeV), and elementary particle physics at billions of electron volts (GeV). In each case, physicists encounter a rich structure–composed of multiple energy levels, different types of matter, and detailed dynamics.

Results from deep inelastic electron-proton scattering experiments using the Stanford Linear Accelerator in 1971 and 1972 strongly indicate that we are now in a transition region (see PHYSICS, ELEMENTARY PARTICLES). The familiar GeV energy scale is no longer significant, but a new, higher energy scale has not been reached.

All physical bodies interact only through four forces–strong, electromagnetic, weak, and gravitational. I believe that at this new scale the weak forces may become comparable in strength to electromagnetic forces. The possibility that these two forces may become unified is profoundly significant.

Two important parameters characterize every force: Its range and its coupling constant or strength. Consider two protons one centimeter (cm) apart. There are only electromagnetic and gravitational forces between them. The electromagnetic force is about 10^{36} times stronger.

When we move the protons closer, to less than 10^{-13} cm, the dominant force becomes the strong force, which is about 137 times stronger than the electromagnetic force. Strong forces bind all nuclei together. Physicists define their coupling constant as 1, so the corresponding coupling constant of the electromagnetic forces is $1/137$.

At still smaller distances, the weak forces come into play. They are responsible for beta decay and the decay of almost all particles that have lifetimes varying from about 10^{-6} second to about 10^{-11} second. The precise range of the weak forces is not known, however. Intensive experimental studies show that the range should be less than 10^{-14} cm. Equally strong theoretical studies indicate that the range must be greater than 5×10^{-17} cm.

According to quantum mechanics, the range of a force and the energy transfer needed to probe it are closely related. The range of the weak forces–smaller than 10^{-14} cm but larger than 5×10^{-17} cm–implies that the new basic energy scale should be larger than 2 GeV, but smaller than 300 GeV.

What lies in this new energy region? An attractive idea is to assume that the weak forces, like all other forces, have a carrier. The photon carries the electromagnetic forces, the graviton carries the gravitational forces, and the pion carries the strong forces. If the carrier of the weak forces, called the "intermediate boson," exists, the new basic energy scale may be represented simply by its mass–somewhere between 2 GeV and 300 GeV.

We cannot know its exact value until future experimental results become available. Nevertheless, much circumstantial evidence suggests common origins and interactions for the weak forces and the electromagnetic forces. Thus, these two interactions may have the same coupling constant–$1/137$. This would make the range of the weak forces about 5×10^{-16} cm and the new basic energy scale about 37 GeV.

This is more energy than can be transferred in a particle collision at the new National Accelerator Laboratory near Batavia, Ill. But it lies just barely within the capability of the new intersecting storage rings at CERN in Geneva, Switzerland. Intensive experiments are now being carried out at CERN under the direction of physicists Leon Lederman from Columbia University and Rodney Cool from Rockefeller University.

In the 1860s, two other physical forces, electric and magnetic, were unified into the electromagnetic force. This unification opened the door to a world dominated by the technology of the electromagnetic interaction.

Electromagnetism also formed the basis of Einstein's special theory of relativity, which predicted the relationship between energy and mass that led us into the nuclear age. There seems to be a close parallel between that unification and the present possible one of the weak and electromagnetic forces. If this new unification is indeed achieved, there is no reason why it cannot have at least as great an impact on our civilization.　　　　[Tsung-Dao Lee]

Physicists at the National Accelerator Laboratory are led in a traditional Chianti toast by Director Robert W. Wilson, as the NAL produced its first 200-GeV proton beam, March 1, 1972.

Physics

Continued

The quark theory assumes that all known particles are ultimately built up from three subparticles. This accounts for all particles discovered to date. Whatever success the theory has had, however, has been clouded by the failure to observe a "free" quark in 10 years of determined searching.

If the electron-scattering data continue to suggest that partons are quarks, the problem is even more perplexing because Feynman's calculations imply that it should not be hard to knock a parton loose. Furthermore, the quark theory is of little use in predicting the forces that act when two particles collide at high energy. Any of the known particles may transmit forces between other particles, and it is difficult to predict which of more than 200 known particles are the most important in any given collision.

Most successful theories of collisions have been inspired by the "bootstrap" hypothesis of theorist Geoffrey Chew of the University of California, Berkeley. It states that all particles are equally fundamental (or nonfundamental) and

that there is no basic building block like the quark. In bootstrap calculations, the forces acting in particle collisions are described by mathematical functions that vary smoothly with particle energy and momentum and do not require knowing how the force is transmitted. Then, these formulas are used to predict when particles will combine to produce composite states. Collisions of one kind of particles suggest the existence of other kinds, and the process is repeated until one gets back to the original particles. This is a herculean labor, requiring a veritable army of theorists and experimenters.

A new duality. The bootstrap game never predicts the existence of a particle that is not also predicted by the quark model. Two theories that start from seemingly contradictory assumptions lead to the same results.

Physicists anticipated that this might happen. As early as 1967, when the bootstrap data was still in a primitive state, David Horn, Richard Dolen, and Christoph Schmid, then at Caltech, devised a test of consistency between the

Physics

bootstrap and quark approaches. They predicted the force between two particles by adding up contributions from the known composite states these particles can form, a method suggested by the quark theory. The result seems to agree with that obtained from the bootstrap formulas.

The implications are uncanny. If the quark theory is correct, the particles formed from quarks are engaged in a vast conspiracy to make the bootstrap look good. If the bootstrap is right, the only way for nature to be consistent is to behave as if quarks existed, even though the bootstrap theory rules them out.

In the past, when rival theories led to the same results, physicists often found a deeper theory that removed their apparent contradictions and made this agreement natural. The present situation resembles that in the early decades of this century, when studies of atoms were turning up simple regularities that made no sense in terms of existing physical laws. But the quantum theory, a revolution in scientific and philosophical thought, emerged from that puzzle. Obviously, it is too early to be sure that we are on the verge of a comparable breakthrough, but there are signs that some simple insight into the nature of matter lurks just beyond the horizon.

NAL reaches 200 GeV. Not all the good news in 1972 came from the theorists. At 1:08 P.M. on March 1, the giant proton synchrotron at the National Accelerator Laboratory (NAL) near Chicago succeeded in accelerating protons up to its design energy of 200 GeV. Much work remains to be done before the 6-mile ring reaches its full potential as a research tool, but a few experiments have already begun. Appropriately, the first was an American-Soviet collaboration. The Russian members of the team, from the Serpukhov Laboratory, brought with them equipment that can study proton-proton collisions inside the accelerator by squirting a thin jet of hydrogen gas in the path of the proton beam. This permitted simple experiments to begin before the machine builders could magnetically turn the beam out of the accelerator. This wedding of a Russian technique and a U.S. accelerator marks a new high point in East-West scientific cooperation. [Robert H. March]

Nuclear Physics. Prior to 1972, nuclear physicists and chemists had accepted as a basic fact that the periodic table of the naturally occurring elements extended from hydrogen, with its single-proton nucleus, to uranium, with 92 protons in its nucleus. Heavier elements unquestionably were produced in the initial phases of the creation of the universe—and later by nuclear physicists. But none of these had a radioactive lifetime comparable to the 4.5-billion-year age of the earth. Therefore, even if once present, they had all long since decayed and disappeared. This "fact" is no longer true.

Darlene Hoffman of the Los Alamos Scientific Laboratory and her collaborators at the General Electric Corporation KAPL laboratories in Schenectady, N.Y., demonstrated during 1972 that plutonium (with 94 protons in its nucleus) does occur in nature. It is an extremely rare component of a rare earth mineral ore, bastnasite, from the Mountain Pass Mine in southern California. Apart from its importance as a new extension of the periodic table, this discovery supports the theory that the heavy elements on earth were formed in the supernova explosion of a giant star somewhere near our galaxy about 5-billion years ago.

To higher energies. Large proton accelerators, which have been fully occupied in elementary particle research, are now being used to probe the nucleus. Higher proton energy results in a correspondingly shorter proton wave length and thus makes it possible for physicists to observe much finer detail in the deep nuclear interior.

Beyond this, high-energy protons can be used to produce a galaxy of new short-lived particles, the mesons and hyperons. These, in turn, can be used as new probes of nuclear structure.

High-energy neutron transfer. Much of our knowledge of the motions of neutrons and protons in nuclei—particularly those near the nuclear surface—has come from collisions between a projectile and target nucleus in which a single neutron or proton was transferred into, or out of, a particular orbit within the nucleus.

One of the most useful projectiles has been the deuteron, a rather loosely bound neutron-proton pair. If neutrons

The Los Alamos Meson Physics Facility provides a new tool for nuclear researchers. Experiments at the $57-million, half-mile-long accelerator complex began in July, 1972.

Physics

Continued

are transferred to target nuclei, for example, the angular distribution of the liberated protons provides a characteristic signature for the nuclear orbits into which the neutrons are captured. Such studies normally involve deuterons in the 10- to 20-million electron volt (MeV) energy range.

In 1972, a group of Swedish physicists at the University of Uppsala led by Bo Hoistad worked at much higher energies, using 185-MeV proton projectiles, and discovered a remarkably analogous behavior. Here, the proton can be considered as a bound neutron-pion (π^+-meson) pair. Using a carbon 12 target, these workers found that the outgoing groups of pions corresponded in energy to the formation by neutron transfer of several low-lying quantum states of the nucleus.

Even more surprising, however, the angular distributions of the pions were very much like those of the emerging protons from the similar carbon 12 reaction carried out with low-energy deuterons. This has opened up an entirely new kind of nuclear-reaction study in-

volving pions and, hopefully, many heavier strange particles.

Heavy ions (normally any nuclear species heavier than helium) are extensively used both as low- and as high-energy projectiles. They permit large clusters of neutrons and protons to be transferred between projectile and target, giving access to very high spin situations, and—in marked contrast to the high-energy proton work—focus on the surfaces of nuclei rather than on their interiors.

The new results of cluster-transfer studies include the discovery that special structures exist at much higher energy in nuclei than had been anticipated. Physicists have long asked what happens to a nucleus as more and more energy is added to it. Do the neutrons and protons, like molecules in a heated water droplet, simply increase their average energy until some evaporate? This behavior has been observed—but is there a more sophisticated mechanism involved? The answer is yes.

Adriano Gobbi and his collaborators at Yale University and Rolf Siemssen and his collaborators at Argonne National Laboratory near Chicago have demonstrated that, under some conditions, the nucleus uses most of its increased energy to create especially tight clusters of neutrons and protons. The remaining energy then appears as the simple motions of these tightly bound clusters about one another.

In silicon 28, for example, it has been found that at 40 MeV of excitation the nucleus appears to have the structure of two carbon 12 clusters bound together by a helium 4 cluster. This is far above the added energy at which any sharp excited states had ever been anticipated in nuclei. This helium 4 cluster (an alpha particle) moves in an orbit about both of the carbon 12 clusters in a manner very reminiscent of the electrons binding the two protons in a hydrogen molecule.

Heavy ions allow the study of excited nuclei spinning at least 10 times faster than any seen before. When a parameter as fundamental as angular momentum in a natural system can be increased by a factor of 10, surprises can be anticipated. Andrew Sunyar and his collaborators at Brookhaven National Laboratory on Long Island found that in dysprosium 156, as the nucleus is spun faster and faster, an abrupt and striking charge occurs in its moment of inertia, which indicates an abrupt change in the nuclear shape. The reason for this is as yet unknown.

New accelerators. The 800-MeV proton linear accelerator, the heart of the Los Alamos Meson Physics Facility (LAMPF), became operational in July, 1972. It gives the United States an unmatched facility for high-energy nuclear science. Smaller but similar facilities, based on high-energy proton cyclotrons, are nearing completion in Vancouver, Canada, and in Zurich, Switzerland.

The University of California's heavy ion linear accelerator, (Hilac), at Berkeley is being rebuilt as a Super-Hilac and by late 1972 will provide a very powerful facility for such studies. The Oak Ridge, Argonne, and Los Alamos national laboratories have developed designs for a more powerful national facility in heavy-ion science. A decision on one of these designs as a national facility to complement Super-Hilac may be reached in 1973.   [D. Allan Bromley]

Plasma Physics. A subtle but distinct change has occurred in choosing the most promising approach for obtaining unlimited energy from controlled thermonuclear fusion. Until 1972, the only serious candidates were various schemes for magnetically confining a hot plasma long enough to let the energy-releasing nuclear reactions occur. In the estimation of most physicists, the tokamaks—doughnut-shaped plasma devices heated by an electric current in the plasma—ranked highest among these schemes. Their promise had first been demonstrated in 1969 by a Russian team led by Lev A. Artsimovich. Much of the American fusion effort had switched to tokamak research in response to the Russian experiments.

In the past year, however, a new concept for a fusion reactor has generated much excitement. The idea is to use an extremely short, ultrapowerful, finely focused laser pulse to vaporize, ionize, and heat a deuterium- and tritium-containing ice pellet. The pellet must be brought to such a high temperature so rapidly that no magnetic confinement of the resulting plasma is needed. Enough nuclear reactions will occur before the fuel is scattered by the ensuing microexplosion. The inability of physicists to confine plasma adequately with magnetic fields has been the major stumbling block to producing fusion power. As a result, it is understandable that a scheme which avoids this step altogether is tempting.

Research into the possibility of laser fusion has been proceeding in various laboratories since the mid-1960s, notably at the United Aircraft Research Laboratories in East Hartford, Conn., under Alan F. Haught; at the University of Rochester under Moshe J. Lubin; and at the Lebedev Laboratories in Moscow under Nobel laureate Nikolai G. Basov. But until 1972, most physicists expected that lasers would only be able to produce and possibly heat a plasma that would be confined in a magnetic trap.

Physicists estimated that producing a fusion plasma without magnetic confinement would require laser pulses from 10,000 to 1 million times more powerful than the 1,000-joule "monsters" now available. The pulse had to be delivered to a fuel pellet during 1 to

100 picoseconds (1 picosecond $= 10^{-12}$ second). This is from 10 to 1,000 times shorter than is currently possible.

However, physicists now recognize that by using an advantageous laser pulse shape and fuel pellet configuration, they can reduce these forbidding numbers by a thousandfold or more. The energy that can be deposited in a pellet is proportional to the square of its density and to the fourth power of the fraction of the laser light that it absorbs. Therefore, relatively small improvements in these quantities can produce very large gains.

Physicists can, for example, shape the pulse in such a way that the first portion of it to strike the pellet will vaporize and ionize it and produce the kind of highly absorbing plasma that can fully absorb the main portion of the pulse a few picoseconds later. In addition, they can illuminate the pellet equally from all sides so as to implode it, and thereby increase its density.

This effort can be greatly aided by good pellet design. For example, with a properly designed hollow pellet, the laser burst may raise the density of the solid 1,000 times. It is curious that the success of laser fusion today appears to depend on the ability to produce a laser implosion much as the success of fission bombs 30 years ago depended on producing a chemical implosion.

New estimates. In the light of this new understanding, estimates for the laser power needed to fuse enough plasma to produce more energy than it consumes have been lowered by a factor of a thousand—to from 10,000 to 1,000,000 joules. Much of the research in this area is classified as secret, both in the United States and in Russia, because of its potential use as a "clean" trigger for hydrogen bombs. As a result, its present status is not really known.

Based on what has been published, however, Basov's program appears to be the most advanced. He has produced 1,000-joule pulses lasting as short a time as 2 nanoseconds (1 nanosecond $= 10^{-9}$ second), and he is expected to test a 10,000-joule laser soon. Lubin's is the most advanced program in the United States. He has built a 500-joule laser

Physics

with pulses lasting several hundred picoseconds, and he is planning a symmetric configuration of six lasers that will deliver 5,000 joules within 100 picoseconds or less. In addition, two American groups—at the Los Alamos and the Lawrence-Livermore Laboratories of the Atomic Energy Commission—have been conducting studies.

Electron beams. Closely allied with the idea of laser fusion is the concept of using an extremely intense beam of very energetic electrons to deposit the required energy in the fuel pellet. The advantage is that existing electron-beam generators can be easily scaled up more economically and to higher efficiencies than can existing lasers producing the same energy.

For example, the most powerful electron generator, the Aurora generator at the Harry Diamond Laboratories in Washington, D.C., produces a 1.7-million ampere, 15-million electron-volt pulse lasting about 80 nanoseconds. Energy densities in such beams are up to 2 joules per cubic centimeter (cm^3). This is 40 times higher than in the most energetic laser beams today, and compares favorably with the energy density of 2.5 joules per cm^3 needed in a fusion plasma.

Previous research on intense, energetic electron beams has shown that they are the most effective means of heating an existing plasma in a very short time. For example, a 1-million ampere beam of electrons having an energy of several million electron-volts can heat a plasma 1/10 the density needed for fusion to 50,000 electron-volts—hot enough for fusion—in several hundred nanoseconds.

For pellet fusion, however, a Soviet group under Evgeny K. Zavoisky and L. I. Rudakov of the Kurchatov Institute in Moscow has calculated that a 10-million-volt, 10-million-ampere electron beam 0.02 cm in diameter, lasting 1 nanosecond, would be required. This is far beyond present capabilities. All present electron beams have a pulse duration of more than 10 nanoseconds, and focusing them, which is so essential to successful pellet fusion, is more difficult than focusing lasers. Therefore, a choice between electron beams and lasers cannot yet be made by fusion researchers. [Ernest P. Gray]

Solid State Physics. Until 1972, physicists had never observed any material in which the nuclei magnetically aligned themselves in an ordered way. Now, in experiments carried out at ultralow temperatures, they have found signs that this effect occurs.

Every electron and most atomic nuclei behave like tiny magnets, having a spin and magnetic moment. Physicists know that electrons interact magnetically with each other. In some solids, they align themselves very regularly. An iron magnet is one example of electron magnetic order.

Each nucleus has its own characteristic magnetic strength, which is about 2,000 times weaker than an electron. Thus, in most materials, at everyday temperatures, the nuclei contribute little to the total magnetism.

Nuclei usually do not have a regular alignment, but point in completely random directions. When a solid, liquid, or gas sample is placed in a magnetic field, however, it becomes slightly magnetized. Both the electron and nuclear spins partially align with the field. In this slightly polarized state, extremely sensitive techniques can be used to measure nuclear magnetism.

Physicists expose the polarized sample to a radio-frequency signal and measure the nuclear absorption of radio energy. This technique is called nuclear magnetic resonance (NMR). The strength of the interaction between the nuclei and the applied magnetic field determines the frequency at which nuclei can absorb the radio energy. By measuring this NMR frequency, physicists can determine the characteristic magnetism of the nuclei in the sample.

Even in the absence of an applied magnetic field, electrons in some crystals spontaneously align their spins parallel to each other. In iron, for instance, this magnetic order persists until it is destroyed by thermal vibrations at about 1043° Kelvin (K.). (Kelvin temperatures are degrees centigrade above absolute zero.) In substances called ferromagnets, the spins all tend to point in the same direction. In other substances called antiferromagnets, the directions of the spins alternate regularly.

Only at extremely low temperatures, close to absolute zero, can physicists expect the nuclear magnets to line up

A new type of electron microscope in the laboratory of Humberto Fernandez-Moran at the University of Chicago has resistance-free magnet windings to focus the electrons.

Physics

Continued

in such a regular manner. Nuclear alignment would be ideally studied in a material that has no interfering electronic magnetism.

Such a material is helium 3, the rare natural isotope of helium that can be obtained in relative abundance from the radioactive decay of hydrogen 3, or tritium. An individual He-3 atom has two electrons and possesses no electronic magnetism because the electron spins are paired off. The nucleus, however, is magnetic. The more abundant isotope He-4 has no nuclear magnetism. The four nucleon spins are paired off.

For years, physicists thought that the He-3 nuclear spins in solid He-3 would align themselves antiparallel to each other at temperatures below about $0.002°$ K., producing a nuclear antiferromagnetic state. But no clear observation of this had been reported.

In 1950, the Russian theorist I. Pomeranchuk proposed cooling a mixture of solid and liquid He-3 by compressing it. At ultralow temperatures, squeezing such a mixture would not increase its temperature, it would drive it down.

Thus, physicists could study the magnetic properties of both the solid and the liquid states simultaneously while they were cooling. His proposal was verified by experiments done in both Russia and the United States.

In very precise experiments still in process in mid-1972, Cornell University physicists Douglas Osheroff, Robert Richardson, and David Lee squeezed He-3 according to Pomeranchuk's suggestion. They found sudden, surprising changes in the pressure of their He-3 sample at two definite temperatures, $0.0027°$ and $0.0020°$ K.

Throughout their experiment, the researchers studied the NMR signal from both solid and liquid He-3. A shift in the NMR frequency occurred at the same temperature that the pressure first changed, but only in the liquid portion of their sample. Thus, this transition does not appear to be the nuclear antiferromagnetic transition predicted for the solid.

However, this transition does indicate that some kind of change takes place in the arrangement of the nuclear spins in the liquid. These observations, especially the sudden change at the lower temperature, require more study before the magnetic alignments can be adequately described.

The Cornell experiments have stimulated much interest in the fundamental ordering of nuclear magnets. Several groups of experimenters in the United States, Russia, and Great Britain are studying this phenomenon, not only in He-3 but also in metals.

Hexagonal superconductors. In 1972, Bernd T. Matthias of the University of California, San Diego, and Bell Telephone Laboratories and his Bell Labs colleagues reported they had discovered a new group of superconducting compounds with a hexagonal structure that has an unusually high transition temperature, about 13° K. These compounds all contain lithium, titanium, and sulfur.

Physicists continually search for superconducting materials that maintain their lack of electrical resistance until they reach relatively high temperatures. Most previous success in finding such high transition temperatures had been with compounds having a particular cubic crystal structure—that of the niobium-tin alloy Nb_3Sn. For example, a record transition temperature of about 21° K. is observed in such a structure, made up of the element niobium along with a mixture of aluminum and germanium to replace the tin.

Hexagonal crystals had been previously studied and their transition temperatures had been found to be much lower. The Bell Labs researchers' success in obtaining the high transition temperature was to ensure that the hexagonal structure of the sample does not change after it is made. If this compound is allowed to stand for even a few hours at room temperature, its crystal structure begins to change and its transition temperature drops. Even though solid transformations are known to exist at normal temperatures, scientists had not sufficiently realized their importance in superconductors.

Since superconducting compounds usually have been studied long after they have been prepared, there is hope for finding higher transition temperatures for other unstable structures.

Phonon generators. In 1971, Bell Labs researchers Venkatesh Narayanamurti and Robert C. Dynes developed a simple method of generating high-frequency phonons by heating a thin superconductor. The phonon, a quantum unit of vibrational energy, is often used to describe the conduction of heat in a crystal.

Heating the end of an ordinary bar of metal creates a large number of phonons at many different frequencies. The phonons diffuse toward the cooler end and give up some energy to the bar as they move along it, creating a temperature distribution.

Physicists have discovered that phonons have unusual properties. For instance, under special conditions achievable only in very pure crystals and at low temperatures, they can move like projectiles within the material, passing by millions of atoms without losing their energy. Scientists can easily create low-frequency phonons by using quartz transducers. Although extensive studies have been made of these, little is known about the properties of high-frequency phonons.

The new phonon generator relies on the unique properties of a superconductor. The electrons in a superconductor are paired off, with opposite spin directions, and they move about in a highly ordered way. If these pairs are broken apart by raising the energy of the superconductor, the electrons will try to pair again. In so doing, they fall to a lower energy state.

By heating one side of their thin piece of superconducting tin (at 1.3° K.) with a pulse of heat, Narayanamurti and Dynes broke up some of the pairs of electrons. As the pulse ended, the electrons, in re-establishing their stable superconducting state required by the low temperature, gave off energy in the form of phonons with a specific high frequency. By applying a magnetic field to the superconductor, the physicists could also tune the phonon energy.

No other source of well-defined phonons of such a high energy has been reported. Such a tunable phonon generator should open a whole new field of phonon experimentation in both solid state and low temperature physics. It is being explored at several laboratories in the U. S.   [Joseph I. Budnick]

Psychology

Controversy over how the control of human behavior might alleviate society's major problems was renewed in September, 1971, with the publication of *Beyond Freedom and Dignity* by B. F. Skinner of Harvard University. In this best-selling book, Skinner points out the directions that a "technology of behavior" should take in helping man control his behavior.

Skinner developed this technology by studying the rates at which animals, such as pigeons and rats, respond to stimuli and the procedures by which the rates could be changed with rewards and punishments. These procedures are called schedules of reinforcement. The simplest of these is known as a continuous reinforcement schedule. In this schedule, the response to be learned is reinforced every time it occurs.

Behaviorism in education. Behaviorist techniques are currently being applied in education. This has required changing the curriculum and the style of teaching to maximize the reinforcements students receive. The most important reinforcements are intrinsic rewards, such as the personal satisfaction the student gets from both learning and making correct responses. Extrinsic rewards include being told that a response is correct, but there are also the more contrived rewards, such as good grades, membership in honor groups, and diplomas.

One example of a behavioral educational technique is programmed instruction, which includes the use of teaching machines and programmed textbooks. In programmed instruction, materials are presented to the student in an increasing order of difficulty. By reading small items of information and answering short questions based on them and on what he has already learned, the student usually rapidly masters the material. The questions are designed to make correct answers almost certain, and the student is told immediately if his answer is correct. This provides a continuous reinforcement schedule.

The problem with programmed instruction is that it is usually geared to the slowest learners. As a result, the student acquires information at a slow rate.

A chimpanzee quickly learns to reach through a slot and match an object he can see, *below right*, with one he can feel but not see, *below*. This newly found ability to link information from two senses is similar to man's ability to read by relating written words to sounds.

Psychology

Continued

In the mid-1960s Fred S. Keller, now a professor of psychology at Western Michigan University, developed a different behavioristic teaching method. His system divides course materials into small units. Each student works on a unit at his own pace and takes a short essay-type examination when he believes he is ready for it. The examination requires full mastery of the unit. If the student fails to pass the exam, he must take it again before going on to the next unit in the program.

The precision teaching approach is a modification of the Keller plan, and many variations of it can be used. Created in the late 1960s by Ogden R. Lindsley at the University of Kansas, it emphasizes the importance of keeping a precise record of the student's behavior. For example, the arithmetic teacher keeps a record of the number of arithmetic problems solved. This technique does not use extrinsic rewards or punishments. Students are tested at periodic intervals. To get an observable record of achievement, teachers record the number of correct responses and the rates at which correct and incorrect responses are made. Changes in response rate can be associated with–and, therefore, controlled more effectively through–changes in the student's environment, teacher, or curriculum.

Defining improvement in terms of the rate with which correct responses are made allows the teacher or administrator to evaluate changes of curricula and teaching methods almost immediately. Furthermore, because improvement is defined as an acceleration in the rate of making correct responses, the measure is essentially independent of original abilities. This plan, or modifications of it, is being tried in many high schools, and in more than 100 colleges around the country.

In 1972, psychologist James Kulik of the University of Michigan's Center for Research on Learning and Teaching reviewed reports from both instructors and students using this technique. He reported that students taking courses under a Keller-type plan spend an abnormally large amount of time on their studies. The drop-out rate among these students is rather high, however, and most of the drop-outs say that the course is too demanding. Grades tend to be high, because a student who finishes the course has mastered the required materials.

On final examinations, students taught by Keller-type methods do as well or better than those taught in conventional ways, but the higher drop-out rate under the Keller-type methods makes an exact comparison difficult. Most students who finish the Keller-type course prefer it to courses taught in a conventional manner.

Blessing or danger. Applying behavioristic principles to education in the form of programmed instruction or a Keller-type plan will have increasing impact. One reason for this is that it is a more economical method of teaching, serving many students with a minimum number of teachers. Furthermore, while these methods are designed to handle a large number of students, they also provide for an individual approach to learning, placing the emphasis on the individual student and his rate of learning. By freeing teachers from many of the routine aspects of teaching, behaviorist methods also make more time available for special work with individuals or small groups of students.

On the other hand, there are dangers in the wholesale acceptance of a Keller-type plan for education. The technique's long-term effects are still unknown. In addition, there are serious questions about what the goals of education should be, especially at high school and college levels. Are the right goals adequately fulfilled by simply acquiring high rates of correct responses? Do the new methods create the desire to learn and develop the ability to think creatively? Do they engender a lasting motivation to evaluate and solve the problems of society and the individual?

Beyond the classroom. Charting behavior with an eye toward controlling it can easily be extended beyond the classroom to include such activities as smoking, drinking, and eating between meals. More complicated behaviors, such as feelings, can also be observed and counted. For example, two people living together can record how often they make each other feel happy or unhappy and the reasons. When the chart is posted in the home, they become aware of how certain actions affect each other.   [Robert L. Isaacson]

Science Support

Radical protest in the scientific community lessened in 1971 and 1972, and federal funding of science and technology increased. No single issue dominated public attitudes toward science. Instead, attention focused on a broad spectrum of issues, ranging from international space cooperation to ethics and science.

Unemployment was still a serious problem. An estimated 65,000 scientists and engineers were unemployed in their fields in 1971 and the first half of 1972. Chemists in particular fared poorly. The American Chemical Society's annual survey, which had found 2.7 per cent of its members unemployed in March, 1971, showed that 3 per cent were out of work in March, 1972. Chemists under 25 years of age had a 24 per cent unemployment rate; women chemists, a 7.3 per cent unemployment rate.

Graduate school enrollments in all fields of science and engineering fell off in the face of the bleak job market. A National Science Foundation (NSF) survey, conducted in June, 1971, found that enrollments in physics departments at 10 leading U.S. institutions of higher education had declined about 34 per cent. Overall graduate enrollments in science and engineering had dropped 9 per cent. As a result, fewer teachers were needed in graduate programs, creating a tight academic job market.

Reordering priorities. The federal government began to shift the emphasis away from defense and aerospace spending to such domestic problems as health, environment, and transportation. As an indication of the new official attitude toward science and technology, President Richard M. Nixon said in his State of the Union message to Congress on Jan. 20, 1972, "A nation that can send three people across 240,000 miles of space to the moon should also be able to send 240,000 people across a city to work." In an attempt to correct the situation, the National Aeronautics and Space Administration (NASA) and the Department of Transportation planned joint efforts to develop new modes of mass transportation.

In October, 1971, President Nixon dedicated the Army Biological Defense Research Center at Fort Detrick, Md.,

President Nixon meets with the three members of the new President's Cancer Panel. They are, from left to right, Chairman Benno C. Schmidt; R. Lee Clark, a cancer surgeon and research administrator; and Robert A. Good, a noted immunologist.

formerly the principal U.S. germ-warfare laboratory, as a cancer-research center. The defense center had become obsolete when Nixon renounced most biological weapons in 1969.

In January, 1972, the President signed legislation authorizing $1.6 billion to be spent over the next three years in finding a cure for cancer. The law includes making the National Cancer Institute an independent agency within the National Institutes of Health (NIH) in Bethesda, Md. The cancer program had created controversy in Congress between those who wanted the institute completely separated from the NIH and those who favored keeping it under NIH auspices. A three-man board, the President's Cancer Panel, will oversee the cancer bureau's activities. Its annual budget will be presented to the President. See ESCALATING THE WAR ON CANCER.

Funds for science. The proposed budget for fiscal 1973 (July 1, 1972, to June 30, 1973) added $700 million for research on special civilian problems, such as urban transportation. It offered an increase in authorized funds for research and development (R&D) from $16.4 billion for 1972 to $17.8 billion for 1973. Domestic R&D funding rose by 15 per cent. By contrast, defense spending was up only 9 per cent.

The 1973 budget proposed increases in authorized funds for ongoing programs: $133 million for environmental research; $93 million for the cancer-cure program; $88 million for energy research; $43 million for finding ways to reduce the hazards of natural disasters; $10 million to fight drug abuse; and $40 million for emergency health-care systems.

A NSF report issued in April, 1972, showed that between 1969 and 1971 all research and development spending had risen 4 per cent. But inflation increased by 3 per cent over the same period, making the real rise only 1 per cent. Also, research spending, which had been 3 per cent of the gross national product during the mid-1960s, fell to 2.5 per cent by 1972.

New NASA projects. The President substantially altered the future of

"Good evening. I'm George Graham, Harvard '71. I have an A.B. cum laude in physics, and I can recommend the roast duck unreservedly."

Drawing by Weber; © 1971 The New Yorker Magazine, Inc.

A paper airplane homes in on Senator Hubert Humphrey as protestors attempt to disrupt his speech at the annual meeting of the American Association for the Advancement of Science in December, 1971.

Science Support

Continued

NASA in May by agreeing to a joint U.S.-Russian manned mission in 1975. American astronauts will dock an Apollo command module with a Russian Soyuz spacecraft in earth orbit. See SPACE EXPLORATION.

James C. Fletcher, administrator of NASA, said that the planned $250-million mission will create 4,400 jobs in the depressed aerospace industry. It may keep from 1,000 to 2,000 persons who are currently employed in the Apollo program from losing their jobs.

On Jan. 5, 1972, President Nixon approved the development of a reusable space shuttle. It will take six years to design and build the shuttle at an estimated cost of $5.5 billion. Fletcher said the new project will create 50,000 aerospace industry jobs. However, some congressional opponents criticized the space-shuttle program as a waste of federal funds that could be used to solve domestic problems.

Changes in the AEC. On July 21, 1971, Glenn T. Seaborg, who had headed the Atomic Energy Commission (AEC) for 10 years, announced

his resignation. He was succeeded by James R. Schlesinger, Jr., a former Office of Management and Budget official, on August 8. Schlesinger was soon confronted with environmentalist attacks on the AEC.

On July 23, a three-judge panel, reviewing a proposed nuclear reactor to be built at Calvert Cliffs, Md., had decided that the AEC had not fulfilled its responsibilities under the National Environmental Policy Act of 1970. The court ordered the AEC to adopt stricter licensing procedures for protection of the environment. After he took office, Schlesinger was urged by the nuclear industry to appeal the decision. However, Schlesinger refused, and the AEC instituted the stringent new procedures.

In a tough speech to the industry in October, Schlesinger announced that the old policy of promoting and protecting the nuclear industry was at an end. "It is not the responsibility of the AEC to solve industry's problems, which may crop up in the course of commercial exploitation," Schlesinger said. "The AEC's role is a more limited one, primarily to perform as a referee serving the public interest."

Basic versus applied research. The Nixon Administration's new interest in science appeared to be directed toward applications of technology that could produce concrete results. For example, the Administration felt that science and technology ought to be able to do something to improve a faltering U.S. world-trade position.

In April, 1972, Secretary of Commerce Peter G. Peterson implied that our difficulties in the competitive world market might be the delayed result of a slowdown in research and development during the 1960s. He called for a change in U.S. patent policies, saying that most government-owned patents are not being used in commercial production. Peterson recommended instituting an incentive system to reward holders of government research and development contracts for finding ways of applying the technology they develop.

This emphasis on applied research caused some scientists to fear that basic research might be neglected. H. Guyford

Stever, who became head of the NSF on February 1, told Congress that basic research "provides the scientific underpinning" needed for the continued application of science and technology to the needs of the nation.

An increase in funds for basic research allayed the scientists' fears. The proposed 1973 budget for the NSF, the main federal supporter of basic research at universities, was increased to $653-million from $622 million in fiscal 1972. The AEC budget was increased to pay part of the cost of opening the National Accelerator Laboratory near Batavia, Ill., on March 1.

Science and ethics. Potential harm to society from new developments in science and technology continued to concern individual scientists. A group of scientists at a conference sponsored by the New York Academy of Sciences in December advocated establishing a system for warning the public about potentially harmful scientific advances and urged other scientists to speak out on related ethical questions. A bill to set up an Office of Technology Assessment to

Federal Spending for R & D

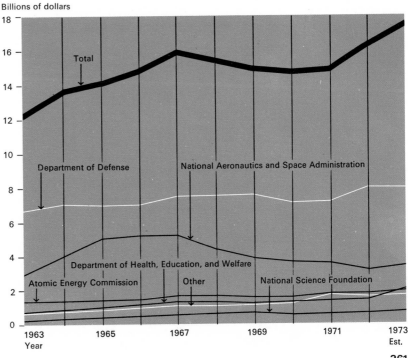

Billions of dollars

study possible dangerous side effects of new technology was passed by the House of Representatives in February.

The involvement of scientists in public debates over various national problems raised ethical questions. The new era of relevance for science was criticized by some spokesmen, who feared that the best minds might be drawn away from basic research.

Philip Handler, president of the National Academy of Sciences, warned that the real job of science was not to solve immediate problems but to remain a decade ahead of society. Others favored scientists becoming involved with national needs. George B. Kistiakowsky, former science adviser to Presidents Dwight D. Eisenhower, John F. Kennedy, and Lyndon B. Johnson, urged greater political involvement.

There were only two major protests against the scientific establishment. In December, 1971, during disturbances at the annual meeting of the American Association for the Advancement of Science in Philadelphia, someone hurled a tomato at Senator Hubert H. Humphrey when he attempted to address a symposium. On Dec. 7, 1971, a bomb exploded at the Stanford Linear Accelerator Center near Palo Alto, Calif.

A book for laymen, *The Limits to Growth,* prepared by Dennis L. Meadows, created controversy when it was published in March. It describes a computer model of the future, forecasting disastrous results for the world if growth continues in any or all of five areas—population, food production, industrialization, pollution, and consumption of nonrenewable natural resources.

Public officials praised the book and called for immediate action on its conclusions. However, many economists and scientists were skeptical about the study, particularly because the supporting data had not been published in scientific journals. They criticized the book for not giving a sound scientific basis for its conclusions. The controversy illustrated a major problem of "relevant" research: Is the public entitled to see startling conclusions before experts have had a chance to review the supporting evidence?   [Deborah Shapley]

Space
Exploration

An era in the U.S. space effort began drawing to a close in 1972 as the $25-billion Apollo moon exploration program neared completion. The last moon mission, Apollo 17, was scheduled for December. In the years ahead, most of the U.S. space program's money and energy will be directed toward developing the space shuttle, a reusable vehicle that can shuttle from earth to orbit and back. The first space shuttle operations are planned for the late 1970s.

One of the Apollo program's architects, Wernher von Braun, announced on May 26, 1972, that he was retiring as planning chief for the National Aeronautics and Space Administration (NASA) after almost 25 years of leadership in the U.S. space program. The German-born rocket engineer, who translated his vision of a system for transporting men to the moon into the Saturn 5 rocket, decided at the age of 60 to carve out a new career for himself in private industry.

U.S.-Russian mission. An era of international cooperation in space began on May 24, 1972, when President Richard M. Nixon signed an agreement for a joint space venture with the Russians. Plans for the mission, scheduled for 1975, call for two or three astronauts in an Apollo-type spacecraft to rendezvous and dock in space with two cosmonauts in a Soyuz spacecraft.

The two vehicles will link up about 145 miles above the earth using a U.S.-built docking mechanism. Crewmen will pass through an airlock as they move from one craft to the other. The airlock is required because the American craft uses pure oxygen at 5 pounds per square inch pressure for its breathing atmosphere, while the Russian craft uses a combination of nitrogen and oxygen at 14.7-pounds pressure.

Exact details of the joint mission are still to be worked out, but one objective is to develop an emergency rescue system for stranded spacemen. The crews will test an orbital rendezvous system, suitable for both Russian and U.S. craft, and docking equipment that will fit space vehicles of both nations. The international crew will be together for about three days. Then, the U.S. space-

Pioneer 10, launched on March 2, 1972, has journeyed farther out into the solar system than any other craft. It will pass Jupiter in 1973 and continue on a never-ending journey into interstellar space.

Space Exploration

Continued

men will remain in orbit for several more days gathering scientific data on the earth's resources.

The U.S. astronauts will receive part of their training in Russia; the cosmonauts, in the United States. Russian and U.S. engineers have been meeting since October, 1970, to work out the technical details of the docking system.

Meanwhile, three U.S. astronauts were already learning Russian. It is almost certain that two of them, if not all three, will represent the United States in the first international manned space flight. They are John L. Swigert, Jr., Apollo 13 lunar module pilot; Thomas P. Stafford, Apollo 10 pilot; and Donald K. Slayton, the only one of the original seven astronauts who has never flown in space. He was grounded in 1962 by a heart condition but received medical clearance in 1972.

U.S. moon flights. Manned flights to the moon proceeded at a slower pace in 1971 and 1972 than in the early phases of the Apollo lunar-exploration program. However, from the standpoint of useful information gained, the Apollo

15 and 16 missions were by far the most important to date.

The lunar-landing crews of Apollo 15 and 16 had a great advantage over the earlier moon explorers. The Lunar Rover, a battery-powered, jeeplike machine, gave them greater mobility. Without surface transportation, previous crews had been limited to traveling short distances on foot. Rover enabled the later explorers to travel miles from their landing sites and sample a wider variety of lunar soil and terrain types.

Apollo 15. The target area for Apollo 15, the Hadley-Apennine region, was chosen to exploit this new mobility. This region, located in the lunar uplands and farther north than any previous landing site, included the high Apennine Mountains, the deep chasm of Hadley Rille, and several other unusual geological features. In preparing to explore this scientifically fascinating region, astronauts David R. Scott and James B. Irwin received unusually intensive training in practical geology.

The crew of Apollo 15 also included Alfred M. Worden, who piloted the

A Message
To the Stars

Mankind's first serious attempt to communicate with extraterrestrial civilizations occurred on Mar. 2, 1972, when the Pioneer 10 spacecraft was launched. Pioneer 10 is the first spacecraft designed to pass through the asteroid belt between the orbits of Mars and Jupiter and then examine the environment of Jupiter. It is also the first man-made object destined to leave our solar system, and it carries a message for any civilization that may exist among the stars.

Although the spacecraft is the speediest object ever launched by mankind, space is very empty and the distances between the stars are immense. If Pioneer 10 were headed for the nearest star, about 4.3 light-years away, it would take about 80,000 years to get there.

But it is not directed toward the nearest star. Instead, it will travel toward a point between the constellations Taurus and Orion. Pioneer 10 will not enter the planetary system of any other star for at least the next 10 billion years, even assuming that all the stars in the Milky Way have planetary systems. However, while on this journey through inter-

stellar space, it is possible that the spacecraft will be encountered by extraterrestrial beings who have developed interstellar space flight.

My artist wife Linda, astronomer Frank Drake, and I, all of Cornell University, were able to convince the National Aeronautics and Space Administration that the first man-made object to leave our solar system should carry a plaque that intends to communicate the place, time, and something about the nature of the spacecraft's builders.

So we designed a message which was etched on a 6 x 9-inch gold-anodized aluminum plate. The plate is attached to the antenna supports of Pioneer 10. We expect that the very small amount of erosion due to micrometeorites in interstellar space will allow this message to remain intact for tens of millions of years and probably much longer. This is the longest expected lifetime of any human artifact.

We used binary notation for numerical information on the plaque, because this is the simplest system of numeration. There are only two symbols in

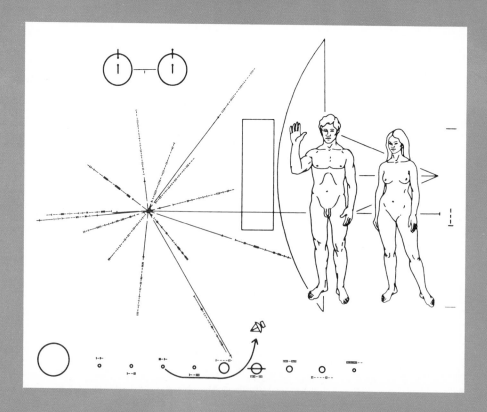

binary notation, "1" and "0," represented on the plaque as "I" and "−."

At the top left of the plaque is a schematic sketch of the two energy states of the hydrogen atom, with parallel and antiparallel proton and electron spins. When the atom spontaneously changes between these states, it emits a radio photon having a wave length of 21 centimeters (cm), about 8 inches, and frequency of 1,420 megahertz. Since there is characteristic distance and characteristic time associated with the transition, it provides a universal standard for measuring distance (centimeters) and time (megahertz) that should be recognizable to scientists of any civilization advanced enough to have developed the technology for interstellar flight. The binary number I beneath this sketch represents these standards of time and distance.

On the right of the plaque is a scale drawing of two human beings. The binary number I − − −, representing the decimal number 8, is placed between two marks indicating the actual height of the humans. The woman is represented as 64 inches tall. A sketch of the Pioneer 10 spacecraft, drawn to the same scale, appears behind the man and the woman. Any civilization that acquires the plaque will also acquire the spacecraft, and they should be able to measure the craft and determine that the dimension indicated on the plaque is 8 times the 21-cm standard unit of length. This will confirm that the symbol at the upper left of the plaque represents the transition between the energy states of the hydrogen atom.

Binary numbers are also shown in the radial pattern on the diagram. The binary numbers in this case represent time rather than distance. The pattern and the binary numbers show the position and period of 14 pulsars with respect to our solar system.

Pulsars are rapidly rotating neutron stars that were produced by catastrophic stellar explosions. They are natural and regular sources of cosmic radio emission, galactic clocks that are running down at generally known rates.

We believe that a scientifically advanced civilization will readily understand that the radial pattern represents 14 pulsars as seen from the place where the craft was launched. Whoever finds the message must ask themselves where and when it was possible to see the 14 pulsars in such relative positions. They will find it could have been only in a small volume of the Milky Way Galaxy and during a single year in the history of the Galaxy. There are perhaps a thousand stars in that small galactic volume, but it is likely that only one has the arrangement of planets with the relative distances shown by the drawing and binary numbers at the bottom of the plaque. The drawing of the planets also shows the trajectory of the Pioneer 10 spacecraft launched from earth to pass by Jupiter.

Assuming that an advanced civilization has kept detailed records of pulsars, they will be able to determine the solar system from which Pioneer 10 came and their equivalent of the exact year in which it was launched.

While we believe the content of the message to this point will be clear to an advanced extraterrestrial civilization, they may have difficulty understanding what is most obvious to the average man on earth – the sketch of the human beings. Extraterrestrial beings that are the product of billions of years of independent biological evolution may not resemble humans at all. The creatures shown on the plaque will probably be deeply mysterious to them.

The man and woman are not holding hands or otherwise in contact because the recipients might think the drawing represents one creature, not two. Only the man has a hand raised in greeting, so the recipients will not think our arms are permanently bent at the elbow.

The Pioneer 10 plaque has generally been praised, but it has also provoked some criticism. Some persons were concerned that mankind may be endangered because we announced the position of the earth. However, the earth's location has been announced for the past 50 years by radiobroadcasts.

The sketch of the humans was the greatest point of controversy. Some believed it was too explicit; others, not explicit enough. However, composing the Pioneer 10 message has forced us for the first time to ask seriously: What pictorial representation of mankind is true, appropriate, and understandable in a first contact with an extraterrestrial civilization? [Carl Sagan]

command and service module. Although he did not land on the moon, scientists said that Worden's tasks in lunar orbit were as important to the mission as those of Scott and Irwin on the moon. See GEOCHEMISTRY.

Apollo 15 was launched on July 26, 1971, and the lunar module landed at Hadley-Apennine on July 30. After opening the hatch at the top of the lunar module, Scott observed all around him "a gently rolling terrain [with] tops of the mountains . . . rounded off [and] no sharp, jagged peaks." Later, when he and Irwin drove to the foothills of Mount Hadley and then to the edge of the sinuous Hadley Rille in their television-equipped Rover, they encountered some of the most spectacular scenery ever beheld by intrepid explorers anywhere.

Meanwhile, orbiting overhead, Worden made visual observations of the surface that gave scientists new clues about the formation and development of the moon. He also collected vast amounts of information with the scientific gear in his charge.

On three separate trips across Hadley Plain, Scott and Irwin traveled more than 17 miles and collected about 170 pounds of samples. These samples included the first deep drill cores ever taken on the lunar surface. The astronauts stayed on the moon a record 67 hours, spending 18½ hours outside the lunar module. The Apollo 15 trio returned to earth on the afternoon of August 7, bringing back enough material to greatly increase the harvest of new knowledge about the moon.

Apollo 16. On April 16, 1972, astronauts John W. Young, Thomas K. Mattingly, and Charles M. Duke, Jr., left Cape Kennedy, Fla., on a mission that essentially duplicated the objectives of Apollo 15, but on a different part of the moon. The target of Apollo 16 was in the Descartes Highlands. This region has much in common with the back side of the moon, which is never seen from the earth. The Descartes Highlands are also far different from the lowland areas visited by the crews of Apollo 11 and 12 in 1969.

The flight was marked by several crises, such as paint peeling off the lunar module, and the crew was forced to cut short their mission by a full day because of difficulties with the backup control system. Still, Apollo 16 achieved all its major objectives, and the astronauts brought back a record 210 pounds of rock and soil samples.

Young and Duke traveled about 17 miles in their Lunar Rover visiting some of the most rugged terrain ever viewed at close hand by moon explorers. Their 20¼ hours outside the lunar module was the most spent by any landing crew.

About a month after the Apollo 16 astronauts left the moon, a large meteorite hit close to the Apollo 16 landing site. Quake detectors left by the astronauts picked up seismic signals from deep inside the moon—as much as 600 miles down. One geologist commented that since such a cosmic event could be expected only a few times each century, scientists were fortunate to record a meteor impact of such substantial size in such a short time.

Apollo 17's December, 1972, lunar landing, the last in the program, is planned for the area near the Taurus Mountains and Littrow Crater, a region that contains both highlands and lowlands. If present theories are correct, the astronauts can expect to find volcanic rocks that were spewed out from deep within the moon. The astronauts will also set up a device capable of detecting water up to 6 miles beneath the lunar surface. Another Apollo 17 experiment will study the earth's tidal pull in relation to the moon.

Navy Commander Ronald E. Evans will pilot the Apollo 17 command module. Navy Captain Eugene A. Cernan and Harrison H. Schmitt, a civilian geologist, will land on the moon.

Beyond Apollo. The accent began to shift from Apollo to Skylab, a series of three earth-orbital missions of much longer duration than any flown up to this time. Operating in a specially modified third-stage Saturn 5 rocket casing, the Skylab crews will live and work under zero-gravity conditions for 28 days on the first mission and 56 days on both the second and third. The flights will begin early in 1973 and will probably end early in 1974.

Named to command the first Skylab flight was Navy Captain Charles Conrad, Jr., veteran of two Gemini missions and commander of the Apollo 12 lunar

Space Exploration

Continued

flight. Joseph P. Kerwin, the first physician named to a U.S. space crew, was designated science pilot; Navy Commander Paul J. Weitz, pilot.

Skylab 2's crew will be Navy Captain Alan L. Bean, veteran of Apollo 12; Owen K. Garriott, electrical engineer; and Marine Major Jack R. Lousma. Skylab 3 will have an all-rookie crew: Marine Lieutenant Colonel Gerald P. Carr; Edward G. Gibson, engineer-physicist; and Air Force Lieutenant Colonel William R. Pogue.

The space shuttle was authorized by President Nixon on Jan. 5, 1972. In March, NASA officials decided that the shuttle will use solid-fuel rockets and have a reusable booster stage. The booster rockets will detach at an altitude of about 25 miles and fall into the ocean, where they will be recovered.

The orbiter part of the shuttle, resembling a delta-wing airplane and designed to hold four men, will be able to return to earth on its own power and land on conventional airport runways. The shuttle equipment is intended to last for more than 100 missions. Flight testing on the shuttle is scheduled to begin in 1976. A complete program should be in operation by the 1980s.

Russian activity. In 1971, the Russians made 83 successful launchings to earth orbit or beyond, compared with 31 for the United States. However, most of Russia's space activity was confined to the Cosmos satellite category, which includes military satellites.

In late 1971 and early 1972, there were no manned Russian missions, following the tragedy of June, 1971, which claimed the lives of three Soyuz 11 cosmonauts. This absence of missions surprised many Western observers, because the cause of the disaster, an improperly sealed spacecraft hatch, did not seem particularly difficult to correct. Also, the Russians did not follow up on the brief Salyut flight of early 1971, which had been announced as the first step toward establishing a manned Soviet space station.

However, the Russians did launch three unmanned lunar probes: Luna 18, on Sept. 2, 1971; Luna 19, on Sept. 28, 1971; and Luna 20, on Feb. 14, 1972.

The Russian Mars 3 landing capsule, shown before it was launched, transmitted for a short time from the Martian surface on Dec. 2, 1971.

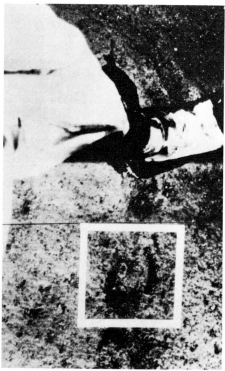

Space Exploration

Continued

Luna 20, Russia's unmanned moon probe, returned to the earth on February 25, carrying samples of the moon's soil in one of its cylinders, *above*. The spacecraft removed the samples from the lunar surface, *above right*, with an automatic drill.

The first of these crashed on landing. The second went into orbit with no apparent attempt to land. Luna 20 landed on the moon, then automatically removed a few ounces of soil from the surface and brought it back to earth.

Journeys to Mars. The big U.S. achievements in unmanned space flight were the extensive exploration of Mars at close range and the start of exploration beyond Mars. Mariner 9, a television-equipped spacecraft, was launched toward Mars on May 30, 1971. It went into Mars orbit on November 13 at the height of the greatest dust storm ever observed from the earth.

For several weeks, photographs returned to the earth showed nothing but the tops of impenetrable dust clouds. Then, gradually, the high winds that had whipped Mars's rare atmosphere into a frenzy died down.

By mid-February, data from Mariner 9 had enabled scientists to map about 85 per cent of the planet's surface in great detail. The probing telescopic eyes of Mariner's television cameras revealed Mars as a planet scarred by long, canyonlike valleys and studded with ancient volcanoes, some of them far larger than any found on the earth. Instruments aboard Mariner 9 also sent back a great deal of data on temperature and atmospheric conditions.

The unfolding picture of the planet made scientists revise their theories about Mars almost daily and gave some of them new hope that the planet might hold some form of life. One emerging theory suggested that Mars, instead of being a "dead" world, may actually be one where life is beginning to develop. Viking, a new type of U.S. spacecraft, will attempt to probe this mystery in 1976. See Mars Is an Active Planet.

During the worst part of the dust storm on Mars, Russia's Mars 2 and 3 spacecraft made the first landings of man-made objects on that planet. Mars 2 landed on November 22, but returned no data. Mars 3 landed on December 2, and sent signals for about 20 seconds before transmission stopped. Both craft touched down during the fierce dust storm, so weather conditions probably caused the failures.

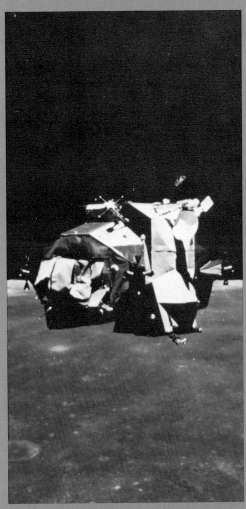

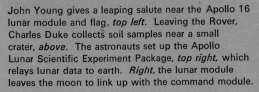

John Young gives a leaping salute near the Apollo 16
lunar module and flag, *top left*. Leaving the Rover,
Charles Duke collects soil samples near a small
crater, *above*. The astronauts set up the Apollo
Lunar Scientific Experiment Package, *top right,* which
relays lunar data to earth. *Right,* the lunar module
leaves the moon to link up with the command module.

To Venus and Jupiter. The Russians also launched another spacecraft, Venera 8, toward the planet Venus on March 27, 1972, carrying an improved landing package to drop on the planet's surface. Temperatures and atmospheric pressures on the Venusian surface are so high that none of the earlier Russian landers survived for as long as half an hour. The United States has not yet attempted a landing on Venus.

Meanwhile, a little U.S. spacecraft, Pioneer 10, was crossing the asteroid belt far out beyond Mars en route to Jupiter, the largest planet in the solar system. This vast body may be an "almost-star," because it radiates more energy than it receives from the sun.

Pioneer 10 was launched March 2, 1972, the first spacecraft ever targeted for a destination farther away than Mars. It will fly past Jupiter on Dec. 3, 1973, taking pictures as it goes. Its speed increased by a "slingshot" effect due to the giant planet's gravity, the spacecraft, powered by atomic batteries, will continue on an unending journey into interstellar space. Pioneer

10 will be tracked until about 1980, when it will pass the planet Uranus nearly 2 billion miles from the earth.

On the slim chance that some interstellar civilization may stumble upon the little wanderer thousands or even millions of years from now, an imaginative group of scientists headed by astronomer Carl Sagan and his artist wife designed a plaque that should explain to any chance extraterrestrial discoverer where the spacecraft came from and what manner of creatures built it. See Close-Up.

Early in 1972, NASA abandoned the unmanned Grand Tour missions, scheduled for the late 1970s. The Grand Tour would have explored the outer reaches of the solar system during a rare aligning configuration of the planets Saturn, Uranus, Neptune, and Pluto. Heavy funding needed for the higher-priority space shuttle program was the main reason for canceling the Grand Tour missions. As a substitute, NASA proposed sending two Mariner-type spacecraft on fly-by missions to Jupiter and Saturn in 1977.   [William Hines]

Transportation

In October, 1971, construction started in Morgantown, W.Va., on what will be one of the most advanced urban rapid-transit systems ever built. This Personal Rapid Transit (PRT) system is expected to be ready for routine service by late 1976. The 3.2-mile system will connect Morgantown and West Virginia University's three campuses.

The PRT's 15-foot-long vehicles will be driverless. Each will carry 8 seated and 7 standing passengers at speeds of up to 30 miles per hour (mph). The air-conditioned and heated vehicles have 60-horsepower electric motors that receive power from a third rail. They will run on rubber tires and be guided by wheels that extend out from both sides of the car and run against the sides of a U-shaped elevated guideway.

The system is designed for use in congested areas. During peak demands, the vehicles will operate on scheduled service and on an as-called basis at all other times. In the latter case, a passenger pushes a button in the boarding station to indicate his destination. A central computer then assigns the first available

vehicle. The passenger should not have to wait more than 30 seconds during scheduled service or more than 2 minutes during on-demand service.

Transpo 72. Activity in developing other PRT systems for use in crowded urban areas increased dramatically in 1972. Four such systems were exhibited at the International Transportation Exposition at Dulles International Airport near Washington, D.C., from May 27 to June 4. They were developed by the Dashaveyor Company, a subsidiary of Bendix Corporation; Ford Motor Company; Monocab, Incorporated, a subsidiary of Rohr Industries; and Transportation Technology, Incorporated (TTI), an affiliate of the Otis Elevator Company.

The Ford and Dashaveyor systems are 12-seat, rubber-tired vehicles that run on a guideway. They are operated automatically, either individually or while linked in trains, and can be manufactured in a variety of sizes. Monocab utilizes six-passenger, completely automatic vehicles. The vehicles hang from an overhead monorail.

Personalized rapid transit systems use small cars
that operate as computer-controlled taxis without
drivers. Electric-powered vehicles travel on an
overhead monorail in one system, *left*, and on
a concrete-walled guideway in another, *above*.
A third, *below*, offers the most flexibility. In
this system, the cars float on cushions of air
and are able to move in any horizontal direction.

The Little Engine That Might

In the 1920s, German inventor Felix Wankel had an idea for an engine with a piston that spins like a turbine instead of thrashing up and down as in the conventional auto engine. Wankel did not obtain financial backing to develop the engine until 1951. Then NSU Motorenwerke AG, a small German firm that is now a subsidiary of Volkswagen, Inc., thought the engine could be used for small cars and motorcycles. So they patented it.

Today, many auto engineers and executives regard the Wankel engine as the engine of the future, the eventual replacement for the conventional piston engine. More than a score of companies, including General Motors Corporation (GM) and Ford Motor Company are paying NSU for licenses to manufacture the Wankel engine.

The engine is simpler in design and has far fewer parts than a reciprocating engine. There are no valves, rocker arms, camshafts, push rods, connecting rods, valve springs, nor valve lifters. Because of its small size, light weight, and less complex design, it should be more reliable and easier to manufacture than conventional engines.

The Wankel is a four-cycle engine with a triangular-shaped piston, or rotor. As the rotor spins, it picks up gas, which is compressed, ignited, and exhausted all in one revolution. The different steps occur at the same time at different faces of the three-sided rotor. The Wankel rotor has three power strokes on each revolution. This equals two cylinders in a standard piston engine. Most Wankels have two rotors on the same drive shaft.

In early Wankel engines, the seal at the three edges of the rotor wore rapidly. This reduced efficiency and gave the Wankel the reputation of being a "dirty" engine because of incomplete combustion. However, the rotor seals have been improved, and engineers say they are no longer a problem.

The engine is still inherently more polluting than a conventional engine because the irregular shape of its rotor traps pockets of unburned gas that are swept out the exhaust port. However, in the mid-1970s, U.S. antipollution regu-

As the Wankel engine rotor spins around its axis it sweeps gasoline and air into the chamber (intake), and then squeezes them against the side (compression). When the spark plug fires (ignition), the gases ignite, expand, and force the rotor to sweep burned gases out the exhaust port (exhaust).

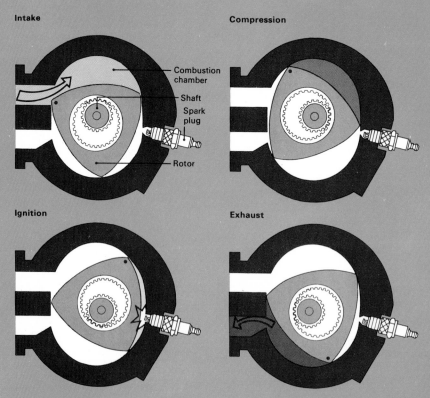

Intake

Combustion chamber
Shaft
Spark plug
Rotor

Compression

Ignition

Exhaust

lations for cars will require the use of thermal reactors and catalytic converters to clean exhaust systems even of conventional engines. Wankel advocates believe it will be less expensive to add the equipment to rotary engines.

GM acquired its Wankel license in 1970, when these new rules were adopted. The firm agreed to pay $50 million over a five-year period, after which it would have worldwide rights to build Wankels for all vehicles except airplanes. GM maintained in June, 1972, that it still had not decided on whether to build the engine. However, federal environmental officials said GM planned to build a limited number of Wankel-powered vehicles in 1975. Some industry analysts expect a few Wankel-powered Chevrolet Vega GT's, GM's small car, to be offered in late 1974. It would have front-wheel drive, considered ideal for the small rotary engine.

Henry Ford II, board chairman of Ford, admits that GM is well ahead, but maintains it will be "the very last part of this decade" before rotary engines are produced in any quantity for autos.

However, Ford's German subsidiary has a license, and Chrysler Corporation was negotiating for a license in 1972. Firms that supply machinery to the automobile industry are developing equipment to mass-produce rotary engines.

Yet, in 1972 the Wankel rotary engine was used in only a tiny fraction of the 25 million motor vehicles built in the world. Japan's Toyo Kogyo is the only auto firm selling a Wankel-engine automobile, the Mazda, in volume. It sold 11,623 of the cars in the United States in 1971 and projections for 1972 were 60,000.

Volkswagen is expected to build a Wankel car in the late 1970s. And Daimler-Benz, the German maker of Mercedes-Benz cars, has built a three-rotor Wankel capable of 190 miles per hour and is expected to produce Wankels in 1975.   [Robert W. Irvin]

Transportation

Continued

TTI's system has the most innovative suspension and propulsion. Instead of wheels, its vehicles travel on a cushion of air, which provides a smooth, nearly frictionless ride. Electromagnetic propulsion is by linear-induction motor. Thus, the vehicles move without contacting any part of the guideway. The air cushion also permits the cars to move in a horizontal direction. Vehicles can stop at a station and move sideways into a docking area to load and unload passengers, keeping the main guideway free for express traffic.

Tracked air cushion vehicles (TACV), designed for high speed ground transportation between cities, began to be implemented during the year. The U.S. Department of Transportation (DOT) completed new tracks at its test center near Pueblo, Colo., and contracts were awarded to Rohr Industries, Incorporated, of Chula Vista, Calif., for construction of a 150-mph TACV and to Grumman Aerospace Corporation of Bethpage, Long Island, for a 300-mph TACV. The 150-mph test vehicle is sponsored by DOT's Urban Mass Transportation Administration (UMTA). It will be tested at Pueblo during the first half of 1973, and will be installed in a passenger-carrying demonstration line at a yet to be selected location. The UMTA vehicle will be propelled by a linear-induction motor and will ride on a thin cushion of air.

The 300-mph vehicle was being developed by the Federal Railroad Administration for testing at Pueblo in August, 1972. Three jet engines will propel the 51-foot-long prototype during testing. They will be replaced by a linear-induction motor when the vehicle goes into commercial use. Then, residual thrust from turbofans used to provide pressure for the air cushion will also help to propel the vehicle.

Auto-Train, a privately sponsored rail passenger service, caught the public's eye in December, 1971. It carries about 100 automobiles and their passengers between Lorton, Va., and Sanford, Fla. The autos are loaded into double-decked cars, and the passengers make the 15-hour overnight trip in reconditioned lounge or Pullman cars. In

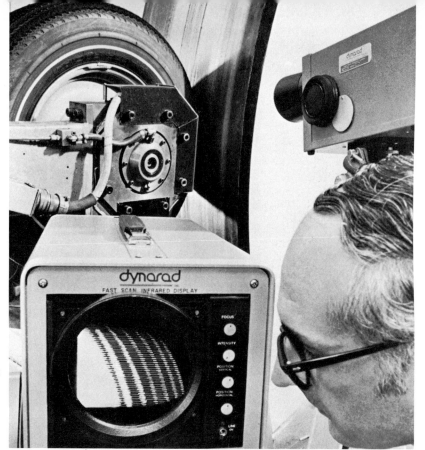

Heat picture of a moving tire appears on a cathode ray tube during safety tests conducted at a B. F. Goodrich research center. This equipment gives heat reading at 600,000 points on the tire every second.

Transportation

Continued

the first month of service, the trains were completely booked up.

Leadership in Europe. Rail passenger service in Europe continued to receive major attention. Every country was steadily improving its domestic service and installing new equipment. France, West Germany, and Great Britain were building experimental trains that will travel more than 300 mph using linear-induction motors.

In contrast, rail passenger systems were virtually neglected in the United States. Out of nearly $25 billion of public money spent on intercity transportation in the U.S. in 1971, the railroads got only $6.5 million, or 0.26 per cent.

Exclusive busways. Of all short-term efforts to reduce urban traffic congestion in the United States, few have been as successful as the use of exclusive traffic lanes for buses. This concept is now in operation or planning for the Shirley Highway from northern Virginia to Washington, D.C., for the Lincoln Tunnel from New Jersey to Manhattan, and for routes in Los Angeles, Oakland, Pittsburgh, and Seattle.

Exclusive busways are promising means for developing corridors that can later be upgraded to full rapid-transit service using automated, guided vehicles. They also represent the only form of mass transit that can now be financed by the otherwise sacrosanct Highway Trust Funds.

The first steam-powered bus in the United States was exhibited in Washington, D.C., on Nov. 17, 1971. It is one of several that began tests in California during the year as part of a $1.6-million grant from UMTA. The otherwise conventional bus has a three-cylinder, 203-horsepower steam engine that uses steam at 850° F. and 800 pounds per square inch of pressure. The bus began carrying commuters in Oakland, Calif., on Jan. 24, 1972.

In February, San Francisco began tests on a steam turbine bus designed by the Lear Motors Corporation, and the Southern California Rapid Transit District of Los Angeles began testing a six-cylinder, reciprocating-steam-engine bus built by Steam Power Systems, Incorporated.   [James P. Romualdi]

Zoology

Bumblebees living in a northern climate have a problem with body temperature regulation. Even in the summer, it can get chilly when skies are overcast, yet bumblebees must continue to forage for food. Their flight muscles will not work properly unless kept warm, so they must generate enough heat to overcome the cooling effects of the environment. Biologist Bernd Heinrich of the University of California at Berkeley reported, in January, 1972, how they do this.

Heinrich observed some of his bumblebees flying at temperatures as low as 5° C. (41° F.). By inserting tiny thermometers into the thorax of these bees, he was able to measure their body temperatures under a variety of environmental conditions.

He found that instead of generating heat by beating their wings, the bees take in enough sugar in the nectar to sustain a high metabolic rate. This makes it possible for them to maintain their body temperature high enough to permit muscle contraction and flight.

Cold feet in the Arctic. Keeping warm is especially important among such animals as foxes and wolves that must walk on extremely cold ground. Biologists Robert E. Henshaw, Larry S. Underwood, and Timothy M. Casey of Pennsylvania State University completed a study of these animals at the Naval Arctic Research Laboratory in Barrow, Alaska. In March, 1972, they reported that arctic foxes and gray wolves keep their feet from freezing by holding the foot temperature slightly above the freezing point – even when plunged into a mixture of ethanol, ethylene glycol, water, and dry ice at a temperature of −38° C. (−36.4° F.).

The animals achieve this by virtue of a rich network of blood vessels just beneath the skin in the foot pad. Warm blood from the body is selectively circulated through these capillaries continuously. This limits heat exchange to the surface of the pad, which is in contact with the cold environment. Although surface tissues cool to a temperature just above freezing, this mechanism does not allow them to freeze completely, which would be disastrous. Thus, these animals have adapted to Arctic life to

Sightings of the nearly extinct California condor, *below right,* plus four huge feathers, *below,* found near its tracks, *bottom,* show that a colony of these big scavenger birds was living in northwest Mexico in June, 1971. The world population is only about 60 birds.

Zoology

avoid tissue death by freezing – something man and other creatures of temperate or tropical climates cannot do.

Bats in belfries. Some animals such as bats withstand cold temperatures by going into hibernation. Their body temperature drops to just a few degrees above freezing during the winter months. Under natural conditions, bats are never pregnant during hibernation.

However, P. A. Racey of the Zoological Society of London set about to find out how forced hibernation affects the course of gestation. He collected bats in the belfries of old English churches during the spring, several weeks before they were due to give birth. Some of them were kept at 5 or 10° C. (41 or 50° F.) without food, and were thus forced into hibernation. In March, 1972, he reported that body temperatures of the hibernating group dropped, the metabolic rate slowed down, and the development of the offspring was retarded accordingly. These animals gave birth about two weeks later than normal.

The other bats were kept in the cold, but had access to food, and they re-mained active. Their young were born only a few days late.

Conversely, if bats are kept at extra-warm temperatures, they will give birth several days earlier than normal. This suggests that a speeded up metabolism accelerates the rate of fetal development.

Pit vipers and boa constrictors possess sensory organs that are highly sensitive to infrared waves, the heat given off by warm objects. This enables them to detect warm-blooded prey even in the dark. Although man has known for 40 years about the existence of these organs, how they act has remained unknown. That is, are the snake organs stimulated by specific wave lengths of infrared, or by a spectrum of wave lengths according to how much the receptor tissue is heated?

To resolve this problem, researchers John F. Harris and R. Igor Gamow of the University of Colorado used a carbon dioxide laser setup to give brief pulses of far-infrared stimulation. They implanted electrodes into the brains of the snakes so they could measure the evoked potential from the brain when

"I think I know what's causing your migraine."
Drawing by Richter; ©1972 The New Yorker Magazine, Inc.

Penguin's backpack transmits cardiovascular data to the University of Washington scientists stationed in Antarctica. The scientists are studying the biology of animals that live near the South Pole.

Zoology

Continued

the snakes were exposed to the infrared rays. The results of their tests, reported in June, 1971, indicate that snakes react to all wave lengths of infrared and that the sense organs therefore operate on thermal principles.

Salt glands for sea life. Sea snakes are the only reptiles that can spend their entire lives at sea, some of them even giving birth in the ocean. Because their body fluids are about one-third the concentration of seawater and their kidneys cannot excrete urine in a concentration greater than that of the body fluids, there must be some way the snake gets rid of the excess salt it takes in with its food. It was reported in July, 1971, that, unlike other reptiles which secrete concentrated salts from modified tear glands or nasal glands, the sea snake does the job by means of modified salivary glands under its tongue.

These investigations were carried out on specimens of the yellow-bellied sea snake captured by biologists William A. Dunson, Randall K. Packer, and Margaret K. Dunson on the research vessel *Alpha Helix*, off the coast of Costa Rica.

They collected the fluid with tubes introduced into the duct of the salivary glands. When the snakes were given injections of sodium chloride, the concentration and rate of flow of the fluids secreted by the glands increased markedly. In this way, the sea snake avoids becoming saturated with salt.

Mercury in fish. The oceans contain many other dissolved substances, not the least important of which is mercury. Analyses reported in March, 1972, have shown that some of the tuna and swordfish on the market have been contaminated with high concentrations of mercury. Observers have assumed that even the oceans are being polluted by the toxic by-products of man's industries.

A team of chemists headed by George E. Miller of the University of California at Irvine has analyzed the mercury content of museum specimens. Seven tuna caught between 1878 and 1909 and preserved in the Smithsonian Institution, plus a swordfish from the Museum of the California Academy of Sciences caught in 1946, were analyzed. The chemists found that these fishes have

Zoology

Continued

Felix, an overweight orang-utan housed at Brookfield Zoo west of Chicago, *opposite page,* confounds the experts by using his knuckles as aids in walking. Unlike chimpanzees and gorillas, orang-utans had never been known to walk in this way.

about the same concentrations of mercury as have been detected in fresh and canned specimens. It seems clear that certain species of fish concentrate more mercury than others, and that if we want to eat them we must also consume the mercury that they contain.

Grafts of tissues and organs from one animal to another are rejected because the host organism produces antibodies against them. Biologists Alan E. Beer and Rupert E. Billingham at the University of Texas Southwestern Medical School wondered why the embryos and fetuses of mammals are not similarly rejected from the inside wall of the mother's uterus during pregnancy.

Suspecting that there might be something about the uterus that permits foreign tissue grafts to survive, they transplanted segments of a rat's tail skin into the animal's uterus. They reported in 1972 that when such a rat was given a simultaneous injection of estrogen to simulate pregnancy, the grafts of skin became attached to the inner lining of the uterus. However, even with the estrogen, skin grafts from other rats were

soon rejected from the uterus, just as they would have been elsewhere on the body. This proved that the lining of the uterus is not a privileged site for foreign-tissue grafts, and that the survival of the embryo must be attributed to some special feature in the placenta.

They also discovered that grafts of the female rat's own tail skin were able to implant on the uterine wall in much the same way that the early embryo attaches itself. The tissues merge together and blood vessel connections are made.

If the rat is given repeated injections of estrogen, the cylinder of tail skin not only survives but its epidermis, or outer layer of cells, grows out onto the lining of the uterus and replaces the tissues that normally grow there. Other studies have shown that even suspensions of dissociated, or separated, epidermal cells, if injected into a uterus, will likewise implant on the uterine wall, where they form small nests of tissue.

Red blood cells undergo a constant turnover so that the production of new cells carefully balances the death of old ones. Biologist Joseph A. Grasso at

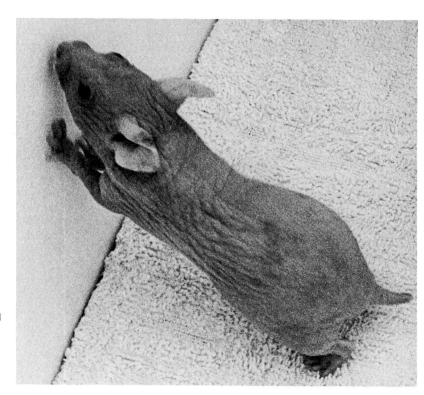

Hairless hamster, bred in England in 1972, may be useful in the study of skin tumors because it need not be shaved before a chemical carcinogen is applied.

Zoology

Continued

Milton, *above right,* and Matilda, musk oxen from the San Francisco Zoo, went to China as gifts from the United States. In exchange, China sent two pandas, Ling-Ling, a female, *above,* and a male, Hsing-Hsing.

Boston University has been studying red blood cell regeneration in salamanders. In February, 1972, he reported that if the drug acetylphenylhydrazine is injected into these amphibians, virtually all of their red blood cells are wiped out within three days. Nevertheless, the salamanders can survive quite well even though rendered totally anemic. Their circulating blood remains completely devoid of the cells for the first week or two, but then large cells appear in the plasma. These proliferate and give rise to immature red blood cells. During the next month, these cells multiply and become fully mature red blood cells.

Marks of crayons. The reproductive cycle of the female red deer has been studied by biologists Fiona Guinness, Gerald A. Lincoln, and Roger V. Short of the University of Cambridge in England. Their studies, reported in 1971, were carried out on wild deer on Rum Island, some 16 miles off the western coast of Scotland. Eleven young females were captured and raised in captivity. During the fall of their second year, they became sexually mature. The object of the study was to find out how frequently these females came into heat, or estrus. In a wild population, such deer would be expected to become pregnant the first time they came into heat. But this could be prevented in this captive population, and the estrous cycles could be timed.

One cannot determine from outward appearances when a female is in heat, but male deer know. A stag, sterilized by vasectomy, was added to the herd, and a harness with a crayon mounted in it was strapped to his chest. Whenever a female came into heat the male would mount her and leave a colored mark on her rump. Because the male was sterile, the female did not become pregnant and later came into heat again.

The researchers learned that these animals commence their reproductive activity each year in October, and they remain fertile sometimes as late as March. Between these dates, they go through a series of cycles, each one of which lasts an average of 18 days, with a minimum length of 15 days and a maximum of 20.   [Richard J. Goss]

380

Men of Science

The scope of scientific endeavor ranges from seeking the laws of nature to harnessing those laws for man's purposes. This section, which recognizes outstanding scientists and engineers, features two whose work represents each end of this research scale.

382 **Humberto Fernandez-Moran** by Richard S. Lewis
While probing the world of inner space to determine the cooperative nature of life at the atomic level, this multidisciplinary scientist envisions a better role for science in his native Venezuela.

398 **Christopher Kraft** by William J. Cromie
Adapting his talents to a newborn era, this space age engineer directed the manned space flights from their inception through the spectacular journeys that transported men to the moon.

414 **Awards and Prizes**
A list of the winners of major scientific awards over the past year and summaries of the work that earned them special recognition.

423 **Deaths**
Brief biographical notes on world-famous scientists who died between June 1, 1971, and June 1, 1972.

Humberto Fernandez-Moran

By Richard S. Lewis

Physician, inventor, and one of the world's great electron microscopists, he has developed the tools to search for the ultimate structure of life

On the campus of the University of Chicago, across Ellis Avenue from the famous Henry Moore statue *Nuclear Energy*, stands an unimposing complex of buildings called the Research Institutes. In the basement, a door curiously marked "Clean Room" opens into a multi-million-dollar laboratory and an international staff of scientists and assistants headed by a man who has been variously called a wizard, a poet, an artist, and one of the great electron microscopists of our age. The man is Humberto Fernandez-Moran, the A. N. Pritzker Professor of Biophysics in the Pritzker School of Medicine.

Physician, biophysicist, and inventor, Fernandez-Moran is known principally for having developed two great tools for scientific research. One is the superconducting electron microscope, through which he can see many of the structural patterns of matter and life at the molecular level. The other is the diamond knife, a "scalpel" with an edge so sharp that it can literally slice through individual molecules.

In his superclean laboratory, where visitors must don nylon smocks and plastic overshoes to avoid tracking in dust, Fernandez-Moran's far-ranging and energetic mind keeps both visitors and staff dazzled. His laboratory is the only one of its kind in the world. The research there ranges from studying three-dimensional laser holography and the phenomenon of superconductivity to the visualization of the deoxyribonucleic acid (DNA) molecule and the fine structures of moon rocks returned by the crew of Apollo 15 from the Hadley Rille. But most of the time Fernandez-Moran pursues his primary interest—the examination of the fine structure of nerve membranes. Under the superconducting electron microscope, he has identified membrane substructures—incredibly tiny biological units that form the building blocks of the brain and central nervous system. At the age of 48, he believes his greatest work is still ahead of him. It is a search for the fundamental structure and order of living systems. He regards this search as the key to understanding life.

Fernandez-Moran is a compact man with short dark hair, luminous dark eyes, and a nose that was slightly flattened years ago in a student boxing bout in Germany. His manner still has the gallantry of the Venezuelan gentry from which he came.

Born in Maracaibo, Venezuela, in 1924, Fernandez-Moran had earned a B.A. degree from Schulgemeinde Wickersdorf at Sallfeld, Germany, by the time he was 15 and an M.D. from the University of Munich at 21. He survived the chaos of Nazi Germany because of his neutral Venezuelan passport. Returning to Venezuela in 1945, he entered the University of Caracas to study tropical medicine and received his second medical degree at 22. Fascinated by diseases of the central nervous system and the brain, he interned in neurology and neuropathology at the George Washington University Medical School in Washington, D.C., in 1946.

He then went to Stockholm, Sweden, to begin his lifelong investigation of the brain. From 1947 to 1949, he was a research fellow at the Nobel Institute of Physics. The director, Karl Manne Siegbahn, inspired Fernandez-Moran to develop the diamond knife, a tool that would enable electron microscopists to probe the nature of living material at the molecular level.

The sharpest and most sophisticated cutting instrument in the world today, the diamond knife is about a tenth of an inch long and has a cutting edge from 20 to 50 angstroms thick. (An angstrom [A] is 1/100-millionth of a centimeter, about the diameter of an atom.) The width of a red blood cell, by comparison, is 80,000 A. This instrument, widely used today in surgery as well as in biological research, is regarded by some surgeons as the most important advance in surgical tools since the steel scalpel. For example, it has been used extensively and successfully in eye cataract operations. It is also an important tool in industry and research. It can slice any material, including metal. Fernandez-Moran uses it today not only to slice thin sections of brain and other nerve tis-

The author:
Richard S. Lewis
is the editor of the
Bulletin of the Atomic
Scientists. He is a
past contributor to
Science Year and was
science editor of the
Chicago Sun-Times.

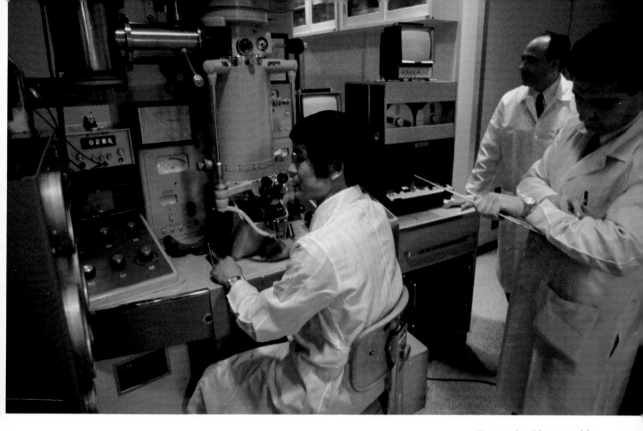

sue for examination under his superconducting electron microscope, but also to cut ultrathin sections of moon rocks so scientists can analyze their fine, crystalline structure.

The diamond knife earned Fernandez-Moran the prestigious John Scott Medal awarded by the city of Philadelphia in 1967, putting him in the company of former medal winners Marie Curie and Jonas Salk.

In Stockholm, the young scientist continued his postgraduate work and research at the Institute for Cell Research and Genetics of the Karolinska Institute, where he received an M.S. in cell biology in 1951. The following year, at the age of 28, he received a Ph.D. in biophysics from the University of Stockholm. During his studies in Stockholm, he also was a clinical assistant and resident at the Serafimerlasarettet Hospital. In addition, he served as the Venezuelan scientific and cultural attaché to Sweden, Norway, and Denmark from 1947 to 1954. Along the way, he somehow found time to meet, court, and marry a tall, blonde girl named Anna Browallius, sister of Irya Browallius, one of Sweden's best-known writers.

Fernandez-Moran returned to Venezuela in 1954, planning to fashion a career that would combine medicine, biology, and the education of the scientists that Latin America needed. The government quickly authorized him to develop a center for research in neurology and brain physiology—a center that ultimately would cost $50 million and become a magnet for researchers from all over the world. The Instituto Venezolano de Neurologia y Investigaciones Cerebrales (Venezuelan Institute of Neurological and Brain Research) was chartered on April

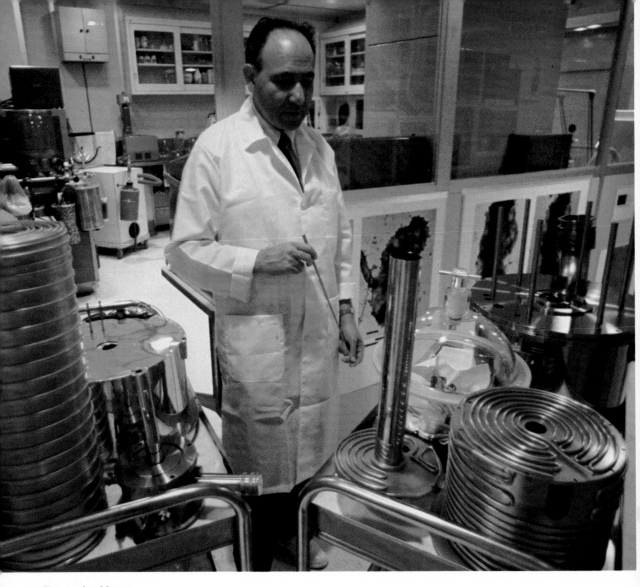

Fernandez-Moran
inspects some of
the components of a
new type of high field
superconducting lens
system that he and his
colleagues developed.

29, 1954. At the end of 1955, a $3-million medical research building was dedicated atop Altos Pipe (Pipe Mountain), near Caracas.

In 1955, Fernandez-Moran headed a Venezuelan delegation to the first United Nations Conference on the Peaceful Uses of Atomic Energy, held in Geneva, Switzerland. In 1957, he led another delegation to the first Inter-American Symposium on Nuclear Energy at the U.S. Atomic Energy Commission's Brookhaven National Laboratory in Upton, N.Y. Attracted by the fundamental nature of nuclear physics, Fernandez-Moran, on his return, persuaded the Venezuelan government to invest in an atomic reactor for his institute.

Then, in January, 1958, came a shattering blow. A military junta ousted the regime of Colonel Marcos Pérez-Jiminéz. For a single day during the coup, when all other officials in the deposed government had fled, Fernandez-Moran stayed at his post as minister of education and then turned over the government, as head of state, to the new regime. Then he, too, was persuaded by the new regime to leave the country with his family. His departure, however, was not without grace. He was given a lifetime diplomatic passport in recognition of his scientific and educational contributions to his Venezuelan homeland.

The Venezuelan scientist was cut off from his institute just four years after he had founded it. Afterwards, the institute's academic nature was altered from specializing in brain research to become a broader research and training center in science and applied technology. That was not the kind of institution the "Wizard of Altos Pipe," as Fernandez-Moran has been called, had in mind when he painstakingly planned it. He wanted to build a great world center of neuromental physiology on the Altos Pipe. But that was not to be, and that is his personal tragedy.

Fernandez-Moran became, in effect, a political exile. He went to the Massachusetts General Hospital in Boston, where he organized the Mixter Laboratories for Electronic Microscopy. He served for four years as an associate in neurosurgery and was also a visiting lecturer in biology at the Massachusetts Institute of Technology and a research associate in neuropathology at Harvard University.

In Boston, Fernandez-Moran began to focus his creative energy on improving the resolving power of the electron microscope. But it was not until he came to the University of Chicago in 1962 that he made a dramatic improvement.

The electron microscope uses a beam of electrons to illuminate an object instead of the conventional beam of light of optical microscopes. Magnetic fields focus the electron beam instead of the optical instrument's glass lenses.

Optical microscopes are limited in their ability to magnify an object by the size, or wave length, of light waves. Even the shortest of light waves is hundreds or thousands of times larger than some of the objects researchers want to look at. The most powerful light microscopes can magnify an object only 2,000 times, and can resolve objects no smaller than 2,000 A wide.

The electron microscope, invented about 40 years ago, pierced the light wave "curtain," and made it theoretically possible to see atoms. Electrons have wave characteristics similar to those of light, but their wave lengths are several hundred thousand

In the ultrahigh-vacuum bell jar, a thin film of metal is evaporated onto a glass slide. The film is used to convert black and white electron micrographs to color.

times shorter. A beam of electrons, generated outside the microscope, is injected into the instrument from the top and directed by an electromagnetic lens onto a thin specimen. As the particles of the beam pass through the specimen (which is the reason it has to be sliced very thin), they are scattered, but they are then focused by other magnetic lenses to produce an image of the specimen. Near the bottom of the microscope, the focused beam strikes a fluorescent screen, which transforms the electronic image into a visual one, seen through a small window.

A major problem in all electron microscopes has been the "thermal noise," or heat, generated by the electric current in the electromagnets that focus the image. Heat tends to jiggle the lenses and distort the image. Fernandez-Moran virtually eliminated this problem by immersing the windings of the electromagnetic lenses in liquid helium at temperatures of 4.2° Kelvin, just above absolute zero. At this low temperature, a phenomenon called superconductivity appears in the windings. All resistance to the flow of electricity through them vanishes.

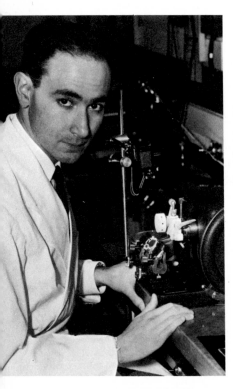

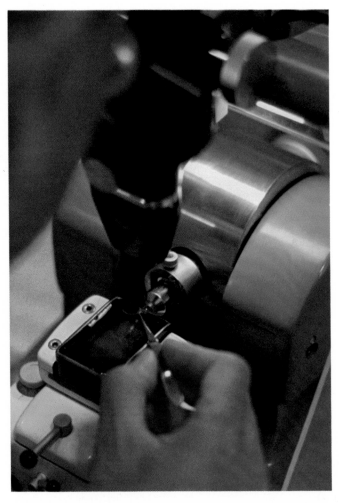

In his early 20s, Fernandez-Moran invented the diamond-knife ultramicrotome, *above.* A modern version, *right,* prepares ultrathin slices of brain tissue for viewing under the electron microscope.

A section of the electron microscope laboratory can be seen through the lenses of the laser optical bench. The bench is used to reconstruct 3-dimensional images from holograms taken by electron microscopy.

Thus, no heat is produced and current continues to flow indefinitely, even after the power is turned off. The heatless flow of electricity sustains a constant magnetic field and produces an undistorted image.

The superconducting electron microscope magnets at the University of Chicago are cooled by a huge refrigerator that occupies five stories of the Research Institutes. Liquid helium is a superfluid; it travels virtually without resistance through the more than 40 feet of chilled pipe. Because of this, the superconducting, superfluid microscope system will continue to produce undistorted images for nearly 30 minutes after the electricity to the helium pumps and magnetic lenses is turned off.

Fernandez-Moran has spent 12 years improving this superfluid helium system. His reward has been the ability to see crystalline lattices as small as 2.06 A wide and, in biological specimens, structures 3 A wide. In addition to eliminating thermal noise, he has reduced building vibration, another barrier to high-power microscopy, by mounting the nine electron microscopes in his basement laboratory on individual concrete blocks. The blocks are supported by springs and insulated from floor motion by shock pads.

Designing electron microscopes is the work of an electronic engineer or a physicist, rather than of a physician, but Fernandez-Moran's motives arise from his medical training. "I was a practicing physician who was depressed by the utter futility of seeing patients die of brain tumors despite all our efforts. I turned to basic research with the desire to learn more about these tumors," he said. With the electron microscope, Fer-

The components of "the power plant of the cell," the mitochondrion, were isolated and identified using the electron microscope. A model of a mitochondrion appears to the right of its textbook illustrations.

nandez-Moran was able to see the structure of the nerve fiber in the brain for the first time.

Under the microscope, a filigree structure of the nerve's myelin sheath, which encases the nerve fibers, reveals a grouping of molecules much like that in a crystal. This paracrystalline order suggests to Fernandez-Moran that the brain itself is like a liquid crystal. "The discovery of the paracrystalline character of the myelin sheath was so important to me," he says, "since it meant that a major constituent of the brain was invested with exquisite regularity and elaborate, coherent, and meaningful design."

The electron microscope also revealed fibers that Fernandez-Moran postulates act as wave guides for electromagnetic energy in the infrared or ultrahigh-frequency ranges generated in the crystalline brain tissue structure. By this mechanism, the brain transmits an orderly beam of electromagnetic energy, much like a laser beam. Such laserlike energy, he believes, can be used to retrieve stored information (memory) from any part of the brain. Remembered images can be seen again and again in the "mind's eye" much like a hologram, the stored three-dimensional image produced by intersecting laser beams.

In developing the tools to examine the ultimate structure of life—the diamond knife to pare thin sections of membrane, and an electron microscope chilled by a superfluid helium system to magnify them millions of times—Fernandez-Moran has built a new concept of the way living systems work. He believes these systems have a moving order, a "momentum order," through which energy flows and in which the chemical components function cooperatively. Possessing momentum order, the whole cell becomes greater than the sum of its chemical components. It becomes alive.

On this aspect of the Venezuelan scientist's work, the Austrian-born biochemist Hans Selye observed: "It is true that the further you take things apart optically, as Fernandez-Moran does, or chemically, the better you may understand life. But life is a correlation of parts which, in themselves, are not life. As [Albert] Szent-Gyorgyi has said, in the study of life you dive from the highest levels to the lower until life fades out and you have atoms and molecules."

Under the microscope, Fernandez-Moran could see how the components of living systems are arranged in a static order—life frozen under an electron beam. This shows him how the parts, such as the cell's mitochondria that energize living systems, are arranged, but not how they cooperate as moving parts. This is somewhat like trying to diagnose an automobile's mechanical problem without starting the engine. Fernandez-Moran wants to see life at the molecular level, in its running state, with its engine turned on.

To understand life, he realized, he had to understand momentum order, which must be the underlying principle of all cooperative phenomena in nature. All living things are examples of cooperative phenomena—components that work together to develop and sustain the whole being. Such an investigation can take a lifetime or more. He has decided to devote the rest of his productive years to it. "The best part of my coming years I want to devote to nailing down momentum order and the nature of cooperativity," he says. "The real thrust of the next decade in biology is to explore cooperativity in the biological domain."

Cooperative phenomena, Fernandez-Moran believes, may also be seen in the functioning of human societies. One can visualize momentum order throughout all aspects of life, in fact—in the reproduction and genetic differentiation of cells, in the functioning of the brain for the storage (memory) and retrieval (recollection) of information, and in the rhythmic flow of a ballet troupe on the stage. Perhaps this principle divides life from nonlife. The ability to perceive momentum order in natural systems, in living cells, in nerve fibers, and membranes may reveal the way to cure the neuromental diseases of mankind.

Fernandez-Moran is convinced, too, that successful human societies exhibit a trait similar to the cooperativity that guides the flow of life at the molecular level in biological systems. Society in the United States, in particular, has exhibited such cooperativity throughout its short history in developing so rapidly, he observed.

After a long day in the laboratory, Fernandez-Moran meets his wife, Anna, for a leisurely stroll across the University of Chicago campus.

"It may be regarded as the typical and most successful example of what I might call Homo-Cooperativity Phenomena," he said. "It is precisely this rare quality of being able to elicit the best efforts from individuals of all nationalities without distorting them in the process that accounts for the phenomenal success of the American team effort, ranging from the economic to the scientific and engineering cooperative projects. It explains why the United States now controls two-thirds of the world's capital, using only 5 per cent of the world's qualified labor force to do so. Nowhere are conditions more favorable for attainment of true cooperative phenomena at present than between the Americas."

Perhaps Fernandez-Moran applies biological principles to social organization and functioning in the hope of unifying the two worlds. In this way, he became, as he puts it, "a wanderer between two worlds." His worlds are those of science in the developed nations and education in the underdeveloped, "handicapped" countries. He still lives in both worlds, his intellect at work in his Chicago laboratory, his heart on Pipe Mountain.

Emotionally, Fernandez-Moran cannot abandon hope that someday he may again play a role in Venezuela's scientific development. Latin America harbors a virtually untapped source of singularly talented human beings, he told the American-Venezuelan Alumni Association in November, 1971, in Caracas and Maracaibo. If he has any political ambitions, they are tied to "a vital urgency to ... overcome the present wide discrepancy in the scientific and technological levels of development between the Latin American countries and the ... more developed countries in North America, Europe, and Eurasia."

Not only is he one of the most distinguished scientists of a country where science and politics mix, but he is also a member of an economic élite in a land that has been ruled by one élitist group or another since it won independence from Spain. His father, Luis, was an official in the Bolívar-Maracaibo oil district. He is proud, too, of his Basque and Sephardic Jewish heritage and its admixture with Carib Indian blood. It represents to him an élite of exiles and suffering. Among his ancestors were two generals, Jose Trinidad Moran and Jose Rafael Urdaneta, who both fought for liberation 150 years ago under Simón Bolívar. Fernandez-Moran is imbued

On the way to their lakefront apartment, the Fernandez-Morans pass the newly constructed Joseph Regenstein Library.

At home, *above,* they are joined by daughters Veronica, left, and Brigida and the family cat, Bubbles. *Left,* Fernandez-Moran examines a rare book from his priceless collection.

with Bolívar's dream of Pan American union. "The Americas represent an indivisible entity. It will either survive as the backbone of this planet in a complete and voluntary cooperation between north and south, or it will disintegrate piecemeal," he said.

Despite his intense patriotism, Fernandez-Moran is not a political activist, but a dreamer of a Messianic deliverance of his people from the handicaps of poverty and limited education. Only in science is he an activist. But he is, as he says, "a prisoner of my dreams." Would he re-enter politics to realize them? Selye dismisses the idea. "Nobody as interested in science as he is could possibly be a politician," said Selye. "Yet, his patriotism is remarkable. Not many scientists have it."

Among Fernandez-Moran's prized possessions are documents and awards from his diplomatic and scientific careers, *above,* and a German telescope, *below,* made more than 100 years ago.

Looking to the future of his scientific work at the University of Chicago Fernandez-Moran stands in front of a reminder of his past— an antique map of his Venezuelan homeland.

Fernandez-Moran admits that his hope of establishing a great brain research center for Venezuela may have been quixotic. He tends to view himself as fated to challenge the unknowable and seek the impossible. On many occasions, he has met with reverses. But this quality of accepting reverses with grace and persistence is what Selye says he admires most in the Venezuelan.

It might appear that the life style of Fernandez-Moran is confusingly varied, almost chaotic. But he has been "inoculated against chaos," he explains, in the manner that children of the very poor in his native Venezuela were immunized as infants against poliomyelitis—by exposure to it in the squalor into which they were born. His varied interests are, in fact, related quite logically in his far-ranging investigations.

To the Venezuelan scientist, the microscopic world "is the key to the future." If, for example, the electron microscope could be used to make small objects appear larger, it could also be applied to make large objects look smaller. It could demagnify—reduce sheets of data to microscopic size, for storage on microfilm. Perhaps this is what the brain does with data perceived by the senses from the environment. The brain may encode certain electromagnetic radiations in the spectrum of visible light, certain oscillations in the atmosphere it interprets as sound, certain molecules of gas it senses as odors, and certain tactile impressions produced by the sense of touch. These are stored somewhere in the billions of brain cells, retrievable on demand. In the brain, which is millions of times more efficient in storing information than a computer, a staggering amount of information is reduced to a submolecular imprinting and stored.

Using his superconducting electron microscope, Fernandez-Moran could make a start toward realizing, in a synthetic information storage system, the storage potential of the brain. He and his laboratory team could reduce the printed page of a book 1 million times in a single imprinting, mount the print on an electron microscope slide, and store it for retrieval by remagnification. An entire book could be reduced in this way to the size of a dot, hardly visible to the naked eye. "Such techniques" he says, "make it feasible to think of storing the entire Library of Congress miniaturized on a sheet the size of typewriter paper."

His ultimate goal in information retrieval is to duplicate the fine structure of the nervous system, down to the molecular level. The ultimate in the condensation of data is the DNA molecule, which contains the blueprint for the cell and the whole organism. These spiral structures can be seen with the superconducting electron microscope. Fernandez-Moran believes that he can use the diamond knife and the electron microscope to practice a rudimentary form of genetic engineering. The knife makes it possible to slice the DNA molecule and restructure it, thus reshaping the entity that evolves from it. Such microsurgery may one day free mankind from its burden of hereditary diseases.

He thinks eventually it should be possible to synthesize the components of nerve cells—to make prosthetic neural circuits that would repair or restore nerve, motor (muscle), and sensory damage.

It may also be feasible to hook up the human sensory nervous system to a computer so that a person could communicate directly with the machine. An airplane or spacecraft pilot might be connected, through sensory organs, perhaps in the fingertips, to his guidance computer to learn the relative position and attitude of his vehicle in space.

The visions of Fernandez-Moran are far-ranging ones, enhanced by an enormous body of specialized knowledge and uninhibited by fear of failure. He has been described as a "renaissance man" because of the diversity of his interests, because he is something of a classical scholar as well as a scientist and linguist. But the Wizard of Altos Pipe is essentially a modern scientist, working at the molecular pole of biology.

Selye dedicated a book, *In Vivo* (1967), to Fernandez-Moran, "As a token of my great admiration for his work on the finest particles of life." This is a mark of esteem, indeed, for *In Vivo* is a defense of supramolecular biology, the biology of the whole human being, as opposed to Fernandez-Moran's focus on the microscopic details of cell structure.

"We stand at opposite poles and that is the cause of our mutual attraction," said Selye. "He is quite prepared to lose perspective to get to the finest detail, while I am willing to sacrifice detail to reach the broadest perspective." Between these two great scientists ranges the whole spectrum of biological research.

One day, Selye relates, he visited Fernandez-Moran's laboratory. "I began to realize the grandeur of his scientific contribution. There was the latest model of his famous diamond knife, with which he could physically cut glycogen molecules into smaller sugars. Then there flashed the terrifying thought through my obsolete mind: Imagine this great genius using all his enormous intellect and knowledge to build an instrument with which to restrict his visual field 2 million times!"

But it is at such resolutions, in his tiny, submolecular, restricted world, that Fernandez-Moran believes the tools he has fashioned will make it possible to perceive the ultimate structure and order of life.

Christopher Kraft

By William J. Cromie

**His decision-making skill and coolness
in crises helped to land men on the moon**

Chris Kraft was sitting at the management console in the Manned Spacecraft Center near Houston when the message came in saying things had begun to go awry. It was April 20, 1972, and the Apollo 16 astronauts had separated their lunar module, Orion, from the command module and were preparing to make man's fifth landing on the moon. Suddenly, Thomas K. Mattingly, piloting the mother ship, discovered that something was wrong with the backup control system for the engine that would bring the astronauts back to earth. He reported the problem to mission control, and Chris Kraft, new director of the Manned Spacecraft Center and veteran flight director, swung into action.

Kraft knew that it would be dangerous for the astronauts to stay in lunar orbit without a backup control system to get home in case the main system failed. Should both control systems fail, there

The Manned Spacecraft Center's director sits
behind models of Skylab and the space shuttle.

At a control console directing one of the earliest launches, Kraft sports a winged cap such as that worn by the mythical Roman god Mercury, for whom the first manned space program was named.

was only one way to save the astronauts—use the powerful rockets on the lunar module to steer the spacecraft back to earth. This would mean abandoning plans to descend to the moon. "It looked like we would have to scrub the landing," Kraft says. He had about 10 hours before a definite decision had to be made.

After determining exactly what the problem was, Kraft had engineers and technicians run tests and gather data on the engine from all over the United States. In only two hours, he and his men pulled all this information together and decided the backup system could do its job. So Kraft let the Apollo 16 landing proceed.

Crises such as this are nothing new to Kraft. Making crucial decisions and finding solutions to life-or-death problems have become routine for him. This has been part of his work since the very beginning of the U.S. space effort. From an engineering point of view, he has had more to do with getting manned missions into space and keeping them there than anyone else.

Yet Chris Kraft almost wound up working for an automobile manufacturer, trying to build a better car engine. When he entered college,

The author:
William J. Cromie is president and editor of Universal Science News, Incorporated. He also wrote "Mars Is an Active Planet" for this edition.

he didn't know what he wanted to do. It was a combination of inspiring teachers, his own adaptability, and happenstance that enabled him to find his unusual place in the world.

Christopher Columbus Kraft, Jr., was born in Phoebus, Va., on Feb. 28, 1924. His father had been born in New York City three days before the 400th anniversary of Columbus' arrival in the Americas. "That," says Kraft, "prompted my grandmother to name her son Christopher Columbus. When I came along, I became 'Junior'."

Before Chris, Jr., "came along," his family moved from New York City to Phoebus. The whole family lived in a modest duplex next to a saloon that his grandmother ran. His mother was a nurse, and his father was a finance clerk at the soldier's home in Kecoughtan, Va.

When he was 3 years old, young Chris fell into a fire and severely burned his hand. Despite a series of operations, he still cannot completely open that hand. It kept him from becoming a Navy pilot in World War II, but it did not prevent him from playing baseball.

Both his father and mother were ardent baseball fans and New York Yankee rooters. Occasionally, his father would take the whole family to Washington, D.C., to see a Yankee game. On one trip, when he was 9 years old, Chris bought a baseball and had Babe Ruth and Lou Gehrig autograph it. He now keeps the ball in a plastic case in his den, and it is one of the first things that he shows to visitors.

Phoebus is near Hampton, Va., where the National Advisory Committee for Aeronautics (NACA) did flight research and tested aircraft at the Langley Aeronautical Laboratory. Chris never even suspected it at the time, but Langley was to become the key to his whole career. "I didn't spend much time at Langley as a boy," Kraft says, "but I had an

Kraft met President John F. Kennedy, who was touring Mission Control escorted by astronaut John Glenn, after Glenn's historic orbital flight in 1962.

intuitive interest in airplanes. I built model airplanes with engines, and thought I'd like to design engines for a living."

Phoebus had no senior high school so Kraft went to Hampton to complete those grades. At Hampton High School he met a woman who, by his own evaluation, had the greatest effect on his choice of a career. She was a math teacher, Mrs. Marguerite Stevens. "That woman planted math in my brain," he says. "She had a greater impact on my choice of becoming an engineer than anybody."

Kraft decided to study mechanical engineering in college. In September, 1941, he entered Virginia Polytechnic Institute in Blacksburg, Va. It was then a military school. Freshmen, known as "rats," had to put up with traditional harassment from upperclassmen and were assigned menial tasks. Kraft adapted easily to the system and confined his dissent to throwing upperclassmen's mattresses out the window into the rain and snow. He joined the freshman baseball team and eventually became the center fielder on the varsity.

Kraft feels this experience with the "rat" system and athletics benefited him later in life. "As you gain rank in the cadet corps, you have to face all kinds of human problems that you normally wouldn't as a regular university undergraduate. You force yourself to become a leader whether you like it or not.

"As for the benefit of athletics, controlling a space flight is wholly a matter of teamwork. You depend on other people's knowledge and inputs for the information you need to make decisions. I don't know where you would get better training for this than having done it as part of an athletic team."

Kraft's whole life began to change in his sophomore year, when he took an elective course in aerodynamics. It was taught by Professor Lee Seltzer, head of the aeronautical engineering department. "I was becoming more and more disenchanted with mechanical engineering," Kraft recalls. "Spending three or four hours a day designing a punch press was not very exciting when I saw seniors in aeronautical engineering designing airplanes. Seltzer and others were able to bring out in me a fascination for aerodynamics. What they were teaching sounded like something I wanted to do as a matter of desire, more than as a way of making a dollar to survive."

In December, 1944, Kraft graduated from Virginia Polytechnic with the rank of cadet captain, a B-plus scholastic average, and a .325 batting average. World War II was drawing to a close, and there were plenty of jobs for aeronautical engineers at nearby Langley Aeronautical Laboratory. So he joined NACA and was assigned to the stability and control branch of the flight research division. This was an important step in his career, since space flight later proved to be a natural extension of his work at Langley.

Kraft worked as a flight-test engineer, measuring the flying qualities of airplanes to determine how their performances might be improved. He greatly admired the head of the stability and control branch at

Kraft confers with astronaut Edgar D. Mitchell, left, and director of flight crew operations Donald K. (Deke) Slayton, right, and studies mounds of computer data before deciding it is safe to go ahead with the Apollo 16 moon landing.

Langley, William Hewitt Phillips. When Phillips became interested in the stresses on airplanes and the passenger discomfort caused by air turbulence, he asked Kraft to work on the problem with him.

Kraft designed what he called a "gust alleviation device," then took charge of modifying an old twin-engine Beechcraft C-45 to test it. To his enormous satisfaction, he found that it smoothed out 90 per cent of the plane's motion caused by turbulent air. Unfortunately, the system was too costly, complicated, and heavy to be practical, and the airlines turned it down. Nevertheless, Kraft says, "This was a tremendous experience for me. I learned the world of managing a project—bringing it from the thought stage to a test phase to prove its worth. It is the achievement I'm most proud of in aeronautics."

In 1950, when he was 26 years old, Kraft eloped with Elizabeth (Betty) Anne Turnbull. They had known each other since their junior year at Hampton High. Mrs. Turnbull did not want her daughter to get married, although, Kraft says, "She didn't have anything in particular against me." They were secretly wed at an Episcopal Church.

The Krafts' first child, a son named Gordon, was born in 1952. That same year they built their first house in a peach orchard in East Hampton. Three years later, their daughter, Kristi-Anne, was born. The fol-

With television screen
showing the Apollo
16 astronauts safely
on the moon, Kraft can
unwind with colleagues
at Mission Control.

lowing year, 1956, was a low point in the aeronautical engineer's life. "Like most people, I came to a point in my career when I wondered whether to stay with the work I was doing or go into something else," he remembers. And then, during the 1956 Christmas holidays, Kraft's father died of a heart attack.

About a week after his father's funeral, Kraft complained of stomach pains, and the doctor told him he had developed an ulcer. "I decided to change my way of living after that," he says. "It was a big struggle, but I got around to where I didn't let things bother me." Kraft's new way of looking at things proved invaluable to the U.S. space program.

At about the same time that Kraft was treating his ulcer, the Russians launched Sputnik I, the first artificial earth satellite. The date was Oct. 4, 1957, and after that, NACA became the nucleus of the U.S. effort to catch up in space. Men who worked with rockets, supersonic aircraft, and telemetry were recruited for the new space effort. At that time, Kraft was involved with telemetry, the radio transmission of data from instruments aboard aircraft. When Kraft was asked to join the space team, he says, "It took me less than 10 seconds to say 'yes'."

A Space Task Group of about 35 people was formed to put together a U.S. space program. In October, 1958, NACA became the National Aeronautics and Space Administration (NASA), and project Mercury —the U.S. effort to build and fly a manned spacecraft—began. "We had to write specifications for the first manned spacecraft, and then we had to select someone to build it," Kraft recalls. "None of us had ever done anything like that before. It was kind of interesting," he adds with characteristic understatement.

Kraft's world at NACA had been flight control. "I wrote out the maneuvers the pilot should make—the speeds and distances he should fly," Kraft says. "During the flight, I was either in the plane or on the ground talking to the pilot by radio. Things never went exactly as planned, and there were always immediate decisions to make." This is where Kraft fit naturally into the space program—as the man in overall charge of directing a flight.

"In the beginning, we didn't even know we would require a flight director, but gradually the need became obvious," he explains. "At first, flights were short and you only needed a few people on the ground to control the various aspects of a mission. But, as missions grew longer and more complex, the number of people needed to keep track of all the systems increased. Each flight required shifts of flight controllers, backed by hundreds of technicians, all under the supervision of a flight director." The facilities expanded from a few concrete block houses at Cape Kennedy to the elaborate Mission Control Center near Houston, and Kraft played a major role in every stage of this development. "We sort of grew with the system," he says.

The system had many growing pains. The first Atlas-launched flight, aimed at qualifying the Mercury spacecraft, blew up after less than a minute in the air. The first Redstone rocket with a spacecraft on top of it rose only about an inch above the earth before falling back onto the launch pad. A safety officer had to destroy Mercury-Atlas 3 by radio command 40 seconds after lift-off because its guidance system failed. Just 13 days before this, on April 12, 1961, the Russians had electrified the world with the orbital flight of cosmonaut Yuri A. Gagarin. "That was really a blow," says Kraft. "We had all been working extremely hard to beat the Russians."

But a monkey named Ham had made a successful suborbital flight in a craft atop the second Redstone rocket in January, 1961, and on May 5 of that year, Alan B. Shepard, Jr., climbed into the third Mercury-Redstone combination. "I was extremely nervous on this flight," Kraft admits. "It was the first time we had had a man on top of that burning stick. It took a heck of a lot of courage on Shephard's part and on our part, too." When the suborbital flight was successful, Kraft says, "It was one of the most exhilarating experiences I've ever had."

On Nov. 29, 1961, a 37½-pound chimpanzee named Enos made an orbital flight aboard Mercury-Atlas 5. While Enos was in orbit, a mechanical failure caused the spacecraft to begin to burn too much

fuel—fuel needed to maneuver the spacecraft into position for its return to earth. Kraft, directing the flight, was confronted with the decision to end or continue the mission. Russia had already put two men in orbit, and one of them had circled the earth 16 times. The United States was struggling to catch up. Officials worried about how it would look to the world if Enos came down before completing his scheduled three orbits.

For Kraft, there were other considerations. He felt that endangering Enos would be just like jeopardizing the lives of the men who would follow him into space on orbital flights. He asked his engineers and flight controllers for a decision. None came. Finally, he snapped, "You have 12 seconds to decide." "Go ahead," was their vague reply. They were passing the ball to him. "Eleven seconds to retrofire," Kraft said without hesitation, "ten, nine, eight...." Bringing that chimp down was the toughest decision he had ever made, says Kraft.

About three months later, on Feb. 20, 1962, John H. Glenn, Jr., became the first American to orbit earth. During his second revolution, instrument readings indicated that the clamp holding the heat shield onto the Mercury spacecraft had been accidentally released. If the shield did not stay in place during re-entry into the earth's atmosphere, America's first manned orbital space flight would end tragically with the incineration of the astronaut.

Chris and Betty Anne
eat at the country
club, *left,* before he
plays golf. Out on the
course, Kraft carefully
lines up a putt, *above.*
Waiting at the next tee,
right, he thinks about
how he should have
played the last hole.

If the instrument readings were correct, then the shield was held on only by the metal straps of the retrorocket pack—a unit containing rockets used to bring the spacecraft down out of orbit. The pack normally drops off after the rockets fire. Some flight controllers insisted on retaining the pack to keep the heat shield from falling off. Others wanted to get rid of it because they felt the instrument readings were false and leaving it on would do more harm than good. Kraft and the director of flight operations, Walter C. Williams, had to make the decision. "I was gravely concerned that heat building up on that pack might burn a hole in the shield," Kraft says, "but we decided to leave it on."

As Glenn plunged into the earth's atmosphere, the heat on the outside of his spacecraft built up to 3000° F. He saw blazing chunks flying by his window and thought his heat shield was disintegrating. Then the capsule went into the blackout zone, where ionized air prevents radio contact with the ground. "That four minutes in blackout felt like it lasted a lifetime," Kraft remembers. But Glenn made a safe and triumphant landing. Later, the engineers found that a faulty switch had given a false indication that the shield was loose. The blazing chunks Glenn saw were pieces of the retrorocket pack and the normal peeling of heat-shield layers as the craft entered the atmosphere.

As the Mercury program progressed, Kraft and his men accumulated a tremendous amount of practical knowledge about flying in space. At the same time, Kraft developed a highly trained team of flight controllers. Before each space flight, every possible problem was handled in simulations, or rehearsals, of the real thing. The problems were programmed into computers, and no one, not even Kraft, knew what situation the team would face at any particular time in their practice runs. Kraft credits these simulations for the ability he developed to maintain his balance and remain cool-headed no matter what critical situation occurred.

In November, 1963, at the end of the Mercury program, Kraft became director of flight operations. But he still served as flight director on the two-man Gemini missions. On the Gemini 5 mission in August, 1965, L. Gordon Cooper, Jr., and Charles (Pete) Conrad, Jr., had not even completed their first revolution around the earth when the pressure that fed oxygen into their fuel-cell power supply began to drop. The oxygen was needed for a chemical reaction in the fuel cells to produce electrical power. Engineers were not sure that the power system could operate with pressure below 250 pounds per square inch, but it dropped to 186, to 125, to 95, then to 71. Kraft told the astronauts to turn off everything that was not necessary for survival.

In this powered-down situation, the spacecraft made six revolutions, then an important decision had to be made. Beyond this point it would be another 18 hours before the astronauts could come down in an area where recovery ships could easily pick them up. Kraft noted that the pressure had stabilized at 71 pounds per square inch. This provided enough power to keep the spacecraft running, and the risk to the astronauts was slight. Therefore he evoked one of his most famous maxims: "If you don't know what to do, don't do anything." He decided to let the flight continue. It went on for almost eight days, more than twice as long as any previous U.S. mission, and accomplished its primary purpose of proving that man could survive that long in space.

Gemini 8, America's first docking mission, in March, 1966, was the last in which Kraft was directly involved in the hour-by-hour direction of a space flight. Management responsibilities were piling up, and other people were waiting to take his place. "There were a lot of young guys around who were just as good engineers and as good or better at being flight director as I was, so I decided to give them a chance," he says. "It was a tough decision. To this minute, I miss the direct involvement in space flight and the excitement of mission control."

Kraft, however, did not give up making valuable contributions to space missions, particularly the Apollo flights aimed at landing men on the moon. This program had hardly started when it suffered its greatest setback. Virgil (Gus) Grissom, Edward H. White II, and Roger B. Chaffee were killed by a fire in their spacecraft during a launch-pad test on Jan. 27, 1967. For Kraft, this was not only a "horrible experience" professionally but a personal tragedy as well; he and Grissom were extremely close friends. The United States had been rushing so fast to beat Russia and to land men on the moon before the end of the decade that safety and careful workmanship had been sacrificed. "The

Three generations of Krafts–Chris, his mother, and daughter, Kristi-Ann–sit and chat in the spacious backyard of their home in Friendswood, Tex.

Looking to the future, the Manned Spacecraft Center's new director discusses Skylab with crew station branch chief, Louie G. Richard. The space station is scheduled to go into earth orbit in 1973.

accident," says Kraft, "brought about such a change in attitude, atmosphere, and performance, that without it we would never have landed men on the moon by the end of the 1960s."

While the United States labored to pick up the pieces of the Apollo program, the Russians sent unmanned spacecraft looping around the moon. These vehicles were large enough to carry men. "We were concerned that the Russians would make a manned flight around the moon in the spring of 1969," Kraft admits. NASA still was working on earth-orbital flights. Apollo 8, due to blast off in the winter of 1968, was slated to test the lunar module for the first time in earth orbit. But, as time passed, it became apparent that the lunar module would not be ready for the flight. So, instead, NASA officials considered sending Apollo 8 on a loop around the moon.

George M. Low, who then managed the Apollo program, called Kraft and asked him how he felt about this. Kraft thought it over for a few days and came up with a bold idea that guaranteed the United States first place in the moon race. He replied that the three astronauts should go into orbit around the moon, not just loop around it. "I insisted we go into orbit because we needed the information such a mission could provide to do what we had to do on lunar-landing missions.

"At that time, the Apollo spacecraft hadn't flown yet, not even in earth orbit," Kraft points out. Also, getting into lunar orbit required firing an engine on the back side of the moon that had never been tested in space before. The engine would have to be fired again to get the astronauts out of lunar orbit. If it failed either time, the astronauts could be stranded 250,000 miles from earth. But Frank Borman, James A. Lovell, Jr., and William A. Anders agreed to fly the mission.

The new spacecraft was successfully tested in earth orbit with the flight of Apollo 7, and plans to go to the moon on the next mission were "cast in concrete," says Kraft. Many engineers wanted Apollo 8 to orbit 200 miles above the moon, to be safe. They argued that so little was known about navigating to the moon that the spacecraft might go off course and crash on the lunar surface. But Kraft insisted they cut the distance to 60 miles, the actual altitude to be flown on a lunar-landing mission. Kraft had his way and proved he was right.

Apollo 8 was launched on Dec. 21, 1968. The first important engine firing took place on Christmas Eve, behind the moon and out of radio contact with earth. For 30 tense minutes no one knew whether it had been successful or not. Then the spacecraft reappeared, precisely on schedule. "It's hard to put into words the relief and excitement we felt when they came out from behind the moon," Kraft says.

After 10 revolutions, the spacecraft disappeared behind the moon for the last time. The crucial burn that would start Apollo 8 home would be made out of sight and contact with the earth. "I don't believe I've ever spent a more tense time in my life," Kraft recalls. "I can remember like it was yesterday waiting for that spacecraft to come around the back. When it appeared again, right on the money, Jim Lovell's first words were: 'There really is a Santa Claus.' That was the most thrilling

Kraft drops by to check with astronauts and engineers in a mock-up of the orbital workshop for the space station.

moment of my career. The mission really scooped the Russians, and gave us a leg-up on our capability to land on the moon. Apollo 8 was the whole key to the Apollo program—no question about that."

The next two Apollo missions were successful beyond anyone's expectations, and Kraft regards them as milestones in his career. When he finally watched Apollo 11 leave the launch pad on July 16, 1969, for the first moon landing, with Neil A. Armstrong, Edwin E. Aldrin, Jr., and Michael Collins aboard, he regarded this achievement as the "culmination of 10 years' labor."

He was in the control center on July 20 when Eagle, the lunar module, began its final descent to the lunar surface. Everything went smoothly until a red warning light flashed in Eagle, indicating that the computer guiding them to a landing was overloaded—more information was being fed in than it could handle. Eugene Kranz was flight director at the time, and Kraft let him handle the situation. As far as anyone could tell, the computer was correctly steering Eagle to the landing site in spite of the red warning light. So, Kranz applied the Kraft maxim and calmly told the astronauts to proceed. They made a perfect landing on the moon. "It was a fascinating experience," Kraft remembers—with a slight shudder.

Four months after Apollo 11, in November, 1969, Kraft was appointed deputy director of the Manned Spacecraft Center. In January, 1972, he took over as director of this $400-million facility, where manned spacecraft are developed, astronauts trained, and manned space missions planned and controlled. Instead of day-to-day involvement in space flight, Kraft now has to deal with such things as long-range planning, jurisdictional disputes between NASA centers, budget problems, personnel problems, and reductions in force as the space program winds down. "Sometimes I have to let people go who have devoted just as much of their lives to the space program as I have," he says. "When I have to do that, it doesn't make for a very good day.

"However," he says, "people who can do management-type jobs ought to do them. To avoid such responsibility is unfair to both NASA and to the people working for you who are waiting to move up."

Kraft still has unreached goals. "I have always wanted to fly," he says. "I just never got around to taking the time to learn. I hope I have the opportunity to fly in space someday. That may be optimistic, but I don't think it's impossible."

With the switch to a managerial position, Kraft's personal life changed a great deal, too. There is no longer the struggle for enough money to live comfortably as there was in Phoebus, no more of the frantic pace of space flights every few months. He and his wife are settled in a sprawling ranch-style house on a tree-lined street in Friendswood, Tex. He designed the house himself. It has an attached bedroom, sitting room, and kitchen where his mother lives. His son Gordon, 20, is studying law and philosophy at Colorado State University. His daughter Kristi-Anne, 17, attends a local high school. The family room and Kraft's small den are crowded with memorabilia from the space program—and baseball: pictures, plaques, trophies, and models. A cabinet below the crowded bookshelves holds thousands of newspaper clippings about his career that Betty Anne has collected and will "do something with someday."

The Krafts' favorite recreation is going out to dinner together. Chris has gourmet tastes and likes to order exotic dishes and eat in well-known gourmet spots. On vacations, they travel through the West by car. "We never really saw the West until we came to Texas," Kraft says. "And we always end our trips in San Francisco, our favorite city."

Kraft has become a "fanatic golfer." "It's as great as baseball was when I was a kid," he says. "I play it every chance I get. If I were a young man starting out today, I'd consider being a golf pro."

Kraft has been tempted by big money and new challenges in private industry. However, he says, "The things I still get out of my job are more rewarding than what I would get out of industry. I've got the best job in NASA, if I can't be a flight director. I was involved in all the flights. Now I'm involved in everything."

Even as director of the Manned Spacecraft Center, burdened with management and administrative problems, Kraft still gets into the tricky decisions that have to be made when a mission looks like it's about to fall apart. His long experience and quick, logical thinking can turn impending failure into success, as was the case with Apollo 16. When Kraft learned that the faulty backup system, if used, would cause the engine to oscillate, he immediately ordered an identical engine and control system at North American Rockwell's facility in Downey Calif., to simulate conditions aboard the Apollo 16 command module. The test indicated that the oscillations would not throw the spacecraft out of control. Also, someone remembered that this kind of situation had been simulated in a vacuum chamber at a NASA test facility in Tullahoma, Tenn., in August, 1969. Kraft had the data checked. It showed that the oscillating engine had run for five minutes without causing any damage to itself.

Next, he had engineers at the Manned Spacecraft Center calculate how stresses from the oscillations would affect the command module's structure. They found that it could easily withstand the strain. Then, the question of what the oscillations would do to the guidance system was fed into a computer at the Massachusetts Institute of Technology in Cambridge, Mass. The conclusion: They would not harm the system. It was only after analyzing all this data that Kraft decided it was safe to continue with the mission, and Dr. James C. Fletcher, head of NASA, accepted his recommendation to go ahead.

"Solving problems like this gives me a whole lot of satisfaction—satisfaction I would never get at any other job," says Kraft. "Apollo 16 was typical of the decisions that were made on the Mercury, Gemini, and earlier Apollo flights. And I'm sure such things will happen on future flights. It's part of the game."

Awards
And Prizes

A listing and description of major science awards and prizes, the men and women who won them, and their accomplishments

Earth and Physical Sciences

Chemistry. Major awards in the field of chemistry included:

Nobel Prize. Gerhard Herzberg, a German-born Canadian scientist, was awarded the 1971 Nobel prize for chemistry "for his contributions to knowledge of the electronic structure and geometry of molecules, particularly free radicals." Herzberg leads a group of scientists at the National Research Council of Canada. The Swedish Academy of Science, in making the announcement of his prize, said his leadership has made his laboratory the world's foremost center for molecular spectroscopy.

Free radicals, molecular fragments with free electrons, react with other molecules so easily, rapidly, and violently that scientists had difficulty collecting enough for spectroscopic analysis. Herzberg developed a method of collecting them. Another scientific team used this in work for which its members were awarded the Nobel prize in 1967.

Herzberg was born in Hamburg, Germany, on Dec. 25, 1904, and studied there and in England. He left Germany in 1935, after Adolf Hitler came to power.

His wife and co-worker, physicist Luise Herzberg, died in 1970. Colleagues say this caused him to submerge himself still more deeply in his work, putting in 80-hour workweeks.

Herzberg's published works have become basic to spectroscopy. They include *Atomic Spectra and Atomic Structure* (1937), the three-volume *Molecular Spectra and Molecular Structure*, published in 1939, 1945, and 1966; and *The Spectra and Structures of Simple Free Radicals* (2nd Ed. 1944).

Chemical Industry Medal. Jesse Werner, chairman and president of GAF Corporation, was named Chemical Industry medalist for 1972 by the Society of Chemical Industry, American Section. He pioneered in the field of high-pressure acetylene chemistry. Werner began his career with GAF Corporation as a research chemist in 1938.

Priestley Medal. George B. Kistiakowsky, chemist and international authority on explosives, shock waves, thermodynamics, and the speeds of chemical reactions, was awarded the Priestley Medal in recognition of his

Nobel prize winners (from left) Dr. W. Earl Sutherland, Jr.,
chemist Gerhard Herzberg, and physicist Dennis Gabor stand
beneath bust of Alfred Nobel at award ceremonies in Stockholm.

George B. Kistiakowsky

Stafford L. Warren

Shields Warren

distinguished service to chemistry. It is the highest honor given by the American Chemical Society.

Kistiakowsky designed the triggering device that set off the first atom bomb at Alamogordo, N. Mex., on July 16, 1945. After the test, he called the explosion "the nearest thing to Doomsday that one could possibly imagine."

He served as President Dwight D. Eisenhower's special assistant for science and technology and chairman of the President's Science Advisory Committee. He also organized the National Academy of Sciences Committee on Science and Public Policy and was its first chairman.

In disagreement with the Vietnam War, he discontinued all activities relating directly to it in early 1968. He retired from his position as the chairman of the Harvard University department of chemistry in 1971.

Physics. Awards for important work in the field of physics included:

Nobel Prize. Dennis Gabor, Hungarian-born engineer and inventor was awarded the Nobel prize for physics. He developed three-dimensional, or holographic, photography.

Gabor is both professor emeritus at the Imperial College of Science and Technology in London and a staff scientist at CBS Laboratories in Stamford, Conn. He spends about half his time in the United States and the rest in London and a new villa near Rome.

As a 17-year-old student in Budapest, Gabor wondered about the possibility of seeing through frosted glass. From that point, he developed a lifelong interest in the nature of photographic images and invisible light beams. The idea on which holography is based came to him while he was waiting for a tennis court at Rugby, England, in 1947.

At the time, Gabor was seeking a way to improve the resolution of the electron microscope, overcoming the limits imposed by its lens. He said holography "was not pure science. I consider it an invention."

Atomic Energy Commission Enrico Fermi Award. Drs. Shields Warren and Stafford L. Warren, two pioneers in the medical and biological aspects of atomic energy, won the Atomic Energy Commission's 1971 Enrico Fermi Award. It marked the first time the $25,000-award

was given for contributions to medicine and biology. The two physicians, among the world's leading experts on radiation, are unrelated.

Stafford L. Warren served from 1943 to 1946 as director of the health and medical program of the Manhattan Engineer District of the U.S. Army Corps of Engineers, which produced the first atomic bomb. Warren was studying the possible environmental effects of radioactivity from atomic energy plants even before construction of the first one, the Hanford (Wash.) plant on the Columbia River, began. He helped plan the program to determine the effects of discharging cooling water for the plant into the river, one of the world's greatest salmon-fishing areas.

In 1947, Dr. Warren became founding dean of the medical school at the University of California, Los Angeles. There, he organized the environmental study program. He is currently active in smog-control work for Los Angeles County and the state of California.

Shields Warren was the first director of the Atomic Energy Commission division of biology and medicine. He was chief of the U.S. Navy's medical field team that investigated effects of the atomic bombs on Hiroshima and Nagasaki, Japan.

Einstein Award. Eugene P. Wigner won the Albert Einstein Award of the Institute for Advanced Study, Princeton, N.J., for his outstanding contributions to the physical sciences. A former co-worker with Enrico Fermi when Fermi produced the first controlled nuclear chain reaction, Wigner was born in Hungary and became a United States citizen in 1937.

Wigner developed many practical uses for atomic energy. He shared the 1959 Atoms for Peace award and the 1963 Nobel prize in physics, and won the Atomic Energy Commission's Enrico Fermi Award in 1958. The Einstein Award consists of a gold medal and an honorarium of $5,000.

Franklin Medal. Hannes O. G. Alfven of the Department of Applied Physics and Information at the University of California, San Diego, won the 1971 Franklin Medal, the highest award of the Franklin Institute of Philadelphia. Alfven was recognized for his "outstanding pioneer work in establishing the field of magnetohydrodynamics and for his many revolutionary contributions in

Earth and Physical Sciences

Continued

Eugene P. Wigner

that field to plasma physics, space physics, and astrophysics." Alfven shared the 1970 Nobel prize in physics.

Geology. Awards recognizing major work in geology included:

Geological Society of America (GSA) Day Medal. Hans P. Eugster, professor of geology at Johns Hopkins University, Baltimore, was awarded the GSA Arthur L. Day Medal. The medal honors "outstanding contributions through the application of physics and chemistry to geological problems." Eugster is best known for original, innovative contributions to experimental geology and petrology. He developed a way to control the oxygen content in laboratory experiments to influence certain chemical reactions. Eugster was born in Switzerland in 1925.

National Academy of Sciences (NAS) Day Prize. Hatten S. Yoder, Jr., director of the Carnegie Institution's Geophysical Laboratory in Washington, D.C., received the first NAS Arthur L. Day prize and lectureship. The prize is named for the laboratory's first director.

Yoder's early work surrounded building high pressure equipment for "growing" minerals by simulating conditions believed to exist in the earth's crust. Later, he studied both natural and synthetic minerals and rocks at high pressures and temperatures. The new prize includes a $10,000 award and the invitation to deliver a series of lectures at the NAS.

Penrose Medal. Marshall Kay of Columbia University received the 1971 gold Penrose Medal from the Geological Society of America. His recognition and delineation of major basins of North American sedimentary rocks and his reconstructions of continental behavior, the society said, paved the way for the "new, global tectonics." This view of the earth's crust superseded the "land-and-sea paleogeography" of the early 1900s.

Among Kay's early professional interests was volcanic-ash beds. This led to work on geosynclinal belts, now called eugeosynclinal, and their marginal thrust sheets. The similarities of these belts to modern island arc and trench systems were recognized in the 1940s.

Life Sciences

Biology. Important research in biology resulted in the following awards:

Louisa Gross Horowitz Prize. Hugh E. Huxley, a molecular biologist at Cambridge University in England won the $25,000 Louisa Gross Horowitz Prize for 1971. It honored his development of the "sliding filament hypothesis" to explain the chemical mechanism of muscle contraction. The prize citation said the theory "provides for the first time an understanding of the manner in which chemical energy can be transformed into mechanical energy in the living organism."

NAS Microbiology Award. The 1972 NAS microbiology award went to Professor Charles Yanofsky of the Stanford University department of biological sciences. It honored Yanofsky's work in helping to explain how information in genes is translated into chemical action by body cells. He used the bacterium *Escherichia coli* to determine the precise relationship between a gene's chemical sequence and the corresponding sequence among the amino acids that make up a protein.

NAS U.S. Steel Foundation Award. Howard M. Temin, a University of Wisconsin cancer researcher, won the NAS U.S. Steel Foundation Award in molecular biology for 1972. He demonstrated experimentally that some cancer viruses transfer genetic information by a reverse process, contradicting the traditional theory that this transfer is strictly a one-way process.

The sequence in which deoxyribonucleic acid (DNA) provides a pattern for the synthesis of ribonucleic acid (RNA) was believed to be irreversible. But the RNA-to-DNA path now known to take place in some cases reverses the flow of information.

The work has since become the focus of scientific research throughout the world. It is important in cancer research, molecular genetics, molecular biology, and genetic engineering. The award includes an honorarium of $5,000.

NAS Walcott Medal. Elso S. Barghoorn, a Harvard University paleobotanist and curator of Harvard's extensive paleobotanical collections, was awarded the Charles Doolittle Walcott medal.

Howard M. Temin

Life Sciences

Continued

Charles Yanofsky

Hugh E. Huxley

Kimishige Ishizaka

The award, presented no oftener than every five years, honored his work in ascertaining what life on earth was like more than 3.4 billion years ago.

Barghoorn has used modern experimental techniques to study fossil plants and has discovered many indications of life at the dawn of earth's history. His discovery of a South African sediment containing preserved cellular life from more than 3.4 billion years ago is the oldest evidence of life yet found.

Barghoorn was a principal investigator in the National Aeronautics and Space Administration lunar research program. He studied specimens brought back from the moon by Apollo astronauts from 1969 to 1971.

Medicine. In the field of medical science, major awards included:

Nobel Prize. Dr. Earl W. Sutherland, Jr., won the $90,000 1971 Nobel prize in physiology and medicine for his investigations into the mechanisms of hormone activity. Sutherland is a professor of physiology at Vanderbilt University in Nashville, Tenn.

Sutherland discovered cyclic adenosine monophosphate, or cyclic AMP, an important substance in the mechanism by which many hormones control metabolic activity in the human body. Cyclic AMP was later found in bacteria and such primitive life forms as flatworms and amebas. Sutherland noted, "It's obviously something that Mother Nature came up with a long time ago when she was figuring out how to make living cells operate."

Cyclic AMP acts as an intermediary to control many cell functions that were once thought governed by hormones. In cells that make bone, for example, it increases calcium intake. It increases the secretion of digestive juices in the stomach. Changes in cyclic AMP levels are also thought to affect the storage of long-term memory in the brain and the development of allergic reactions.

Cyclic AMP is thought to play a major role in cell differentiation, the general process by which cells become specialized or change their form or function. Cancer researchers are excited because when it is added to tumor cells in test tubes, the cells become normal.

Dr. Sutherland received a rare honor when he was made a career investigator

of the American Heart Association in 1967. The action assured financial support for his research as long as he lives.

Gairdner Awards. Celebrating the 50th anniversary of the discovery of insulin, the Gairdner Foundation of Willowdale, Ontario, Canada, confined all six of its annual $5,000 awards to contributions in the field of insulin.

Dr. Charles Best, director emeritus of the Banting and Best Institute at the University of Toronto, was one of the winners. Working with the late Dr. Frederick Banting, he discovered insulin. He received his award from Banting's widow, Lady Henrietta Banting, and in turn presented awards to the other five recipients.

The others were Drs. Solomon A. Berson and Rosalyn S. Yalow of Mount Sinai Medical Center in New York City; Dr. Rachmiel Levine of the City of Hope Medical Center, Duarte, Calif.; Dr. Frederick Sanger, head of the Protein Chemistry Division at the University of Cambridge in England; and Dr. Donald F. Steiner, professor of biochemistry at the University of Chicago.

Lasker Basic Research Award. Three pioneers in explaining the genetic code on strands of deoxyribonucleic acid (DNA) shared the 1971 Albert Lasker Award in basic medical research. They are Seymour Benzer of the California Institute of Technology, Sydney Brenner of Great Britain's University of Cambridge, and Charles Yanofsky of Stanford University.

Their work also helped to explain the nature of mutations, in which the DNA molecular sequence, or "genetic code," changes. This can result in genetic information being decoded into abnormal proteins. In some cases, it can result in individuals with inherited abnormalities and diseases.

Biologists had believed genes, individual units of heredity, were irreducible entities, but Benzer's research showed genes are composed of subunits. Based on this, Brenner and Yanofsky learned how the "message" of an inherited characteristic is "read" by the cell and how it guides the formation of various protein molecules.

The Lasker Award includes a $10,000 honorarium. Twenty-two Lasker Award winners have won Nobel prizes. Yanofsky also was the 1972 recipient of the

Life Sciences

Continued

Teruko Ishizaka

National Academy of Sciences Award in Microbiology.

Lasker Clinical Research Award. Edward D. Freis, who found that cases of moderately high blood pressure, or moderate hypertension, can be dangerous even though it produces no easily recognized symptoms, won the Albert Lasker Award in clinical medical research. Freis proved that proper treatment in such cases can greatly reduce the high risk of stroke and heart failure.

Such risks have long been associated with high blood pressure, but many physicians believed that moderate hypertension did not require treatment. They felt treatment was necessary only as pressure increased, often years later.

In 1969, Freis completed a five-year study of 380 moderately hypertensive patients. He found that major complications could be reduced by 67 per cent if treatment began while pressure was only moderately high.

In hypertensive patients under the age of 70, adequate drug treatment could prevent seven out of every eight stroke deaths, Freis estimated. Tens of thousands of lives could be saved each year, he said, if more doctors recognized moderate hypertension as dangerous and treated it properly. About 30 per cent of the patients he studied had been under treatment but decided to stop it. Many did this because they had felt no symptoms of their high blood pressure.

Freis is a senior medical investigator at the Veterans Administration Hospital in Washington, D.C. His award includes a $10,000 honorarium.

Passano Award. A husband-and-wife medical research team at the Johns Hopkins University School of Medicine, Baltimore, Md., won the 1972 Passano Foundation Award for American medical research. The award includes an honorarium of $7,500.

The award winners, Kimishige Ishizaka and his wife, Teruko Ishizaka, were honored for their study of allergic reactions and the antibodies that cause them. Their research led to the discovery of a new class of immunoglobulin, which they called IgE, for immunoglobulin E. It is believed responsible for hay fever and other allergic reactions.

Space Sciences

Roman A. Schmitt

Aerospace. Major awards in the aerospace sciences included:

American Institute of Aeronautics and Astronautics (AIAA) Goddard Award. The AIAA presented its 1971 Goddard Award to a team of three engineers for gas turbine research. Two of the team worked at Wright-Patterson Air Force Base in Ohio. They are Squadron Leader Brian Brimelow of the Royal Air Force and Howard E. Schumacher, chief of the performance branch. The third member is Gary A. Plourde, a project engineer for Pratt & Whitney Aircraft Division of United Aircraft Corporation.

The three were honored for their original research leading to the first understanding of stalling in gas turbine compressors caused by turbulent inlet flow conditions. They were presented commemorative gold medals and shared the $10,000 honorarium.

Hill Space Transportation Award. Dr. Hubertus Strughold won the $5,000 Louis W. Hill Space Transportation Award of the American Institute of Aeronautics and Astronautics. Strughold, emeritus professor of medicine at Brooks Air Force Base in Texas, and chief scientist at the Aerospace Medical Division there, pioneered in the development of space medicine. He received the award for the foresight and talent with which "he established the rational biomedical foundations for manned exploration of space."

Strughold was born in Westtuennen, Westphalia, now a part of West Germany. He came to the United States in 1947 and became a naturalized citizen in 1956. Strughold wrote *The Green and Red Planet: A Physiological Study of the Possibility of Life on Mars* (1953). The Aerospace Medical Association's Space Medicine Branch created an award in his name, the annual Hubertus Strughold Award, in 1961.

Astronomy. Awards for important work in astronomy included:

NAS Arctowski Medal. Francis S. Johnson, director of the Center for Advanced Studies at the University of Texas in Dallas, won the National Academy of Sciences Henryk Arctowski Medal

Hubertus Strughold, winner of the Hill Space Transportation Award, sketches a cross section of Mars's crust. He was honored for pioneering efforts in space medicine.

Space Sciences

Continued

for studies of solar activity. Johnson is an expert on high atmosphere, particularly the lunar atmosphere. He designed a special experimental device for use on the moon by the Apollo 14 astronauts in February, 1971. It was designed to respond to changes in gas pressure. This would allow the device to detect the presence of lunar atmosphere.

Johnson has also described how the atmospheric transport processes regulate their responses to changes in solar activity as well as to changes in the seasons.

NAS Merrill Award. The NAS presented the second George P. Merrill Award to Roman A. Schmitt, a professor of chemistry at Oregon State University and an authority on the abundance of rare earth elements in meteorites. The $1,000 Merrill Award is given every two years for studies of meteors, meteorites, and space.

Schmitt's work has provided clues toward the solution to a variety of geochemical and cosmochemical problems. He helped to clarify many discrepancies that existed in the experimental values given to the abundance of rare earth ele-

ments. Scientists seek accurate values for these in order to better determine and understand the abundance of chemical elements in general that exist throughout the cosmos.

Watson Medal. André Deprit, a research associate at the National Aeronautics and Space Administration's Goddard Space Flight Center won the 1971 National Academy of Sciences James Craig Watson Medal for astronomical research. The award was for work that he performed with computers, his "adaptation of modern computing machinery to algebraic rather than arithmetical operations."

The Watson Medal Committee said he had solved the main problem of lunar theory – the motion of the moon around the earth as affected by the gravitational force of the sun.

Born in Belgium, Deprit came to the United States in 1964. In mid-1972, he became a professor of astronomy in the graduate faculty of the college of arts and sciences at the University of Cincinnati, and began research work at the Cincinnati Observatory.

420

General Awards

Edwin H. Land

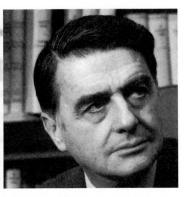

Margaret Mead

Science and Man. Outstanding contributions to science and mankind brought the following awards:

Kalinga Prize. Anthropologist Margaret Mead became the first woman to win the Kalinga Prize awarded by the United Nations Educational, Scientific, and Cultural Organization (UNESCO). She was chosen unanimously by a three-member committee made up of a Dane, an Indian, and a Yugoslav. The prize is awarded for contributions to the popularization of science.

In accepting it, Miss Mead outlined a five-point program to achieve world culture, which she said was essential to human survival. It includes a shared, spoken, natural language that all people would adopt as a second language. By a natural language, she said she meant one already spoken on earth.

New urban systems are needed, she said, that would allow small, face-to-face communities to continue. They would allow everyone access to the countryside and wilderness areas. In industrialized nations, the small family unit does not meet present needs, she said. Miss Mead also called for a single, worldwide system of measurements for time-space measurement of the physical universe, and currency as well as common scientific and engineering terminology.

The Kalinga Prize includes 1,000 British pounds sterling (about $2,500). The prize was founded by an Indian industrialist, Bijoyanand Patnaik, and named for an empire that existed in what is now India more than 2,000 years ago. Kalinga was one of the first empires to pursue a policy of peaceful coexistence.

National Academy of Engineering Founders Medal. Edwin H. Land, developer of cameras, films, and processes for instant photography and inventor of synthetic polarizers for light, received the Founders Medal of the NAE. The medal honors contributions to engineering and to society. Land is a visiting professor at the Massachusetts Institute of Technology, Cambridge, Mass.

Although he has honorary doctorate degrees from Harvard, Yale, and Columbia universities and eight other institutions, Land never earned an undergraduate degree. He attended Harvard twice. After a few weeks as a freshman, he took 18 months off to study the problem of how to make light polarizers. Later, he returned and was given a laboratory in which to continue his investigations, but he left a second time to devote all his time to research.

During the 1930s, Land developed a series of noncrystalline polarizers. He also worked out the theory and methods to produce three-dimensional colored motion pictures, polarized sunglasses and camera filters, and nonglare automobile headlights. In 1947, at a scientific society meeting, he demonstrated the first instant-picture camera.

NAS Applied Mathematics Award. Kurt O. Friedrichs won the first National Academy of Sciences Award in Applied Mathematics and Numerical Analysis. The $5,000 award was established by the IBM Corporation.

As a major research goal, Friedrichs has been seeking some unifying mathematical principle that would characterize all equations that describe natural processes. According to the academy, his most recent work seems to approach that goal closely.

Friedrichs was born in Kiel, Germany. He came to the United States in 1937 and was naturalized in 1944. He joined the faculty of New York University in New York City, where he is now distinguished professor of mathematics at the university's Courant Institute of Mathematical Sciences.

NAS Cotrell Award. Arie J. Haagen-Smit, emeritus professor of bio-organic chemistry at the California Institute of Technology in Pasadena, won the $5,000 National Academy of Sciences Cotrell Award in environmental improvement and pollution control.

Haagen-Smit was a pioneer investigator of atmospheric pollution in the Los Angeles area. He linked both industrial and automobile-exhaust gases to eye irritation, plant damage, and potential health hazards. He has been active in air-pollution research at the local, state, and federal levels.

NAS Hartley Public Welfare Medal. Leonard Carmichael, National Geographic Society vice-president for research and exploration, won the Hartley Public Welfare Medal. The only award of the National Academy of Sciences for other than direct contributions to scientific knowledge, the gold medal is for "eminence in the application of science to the public welfare."   [Ed Nelson]

421

Major Awards and Prizes

*Award winners treated more fully in the first portion of this section are indicated by an asterisk (*)*

Acoustical Society of America Biennial Award:
Robert Finch

Agassiz Medal (oceanography): Seiya Uyeda

American Association for the Advancement of Science
Socio-Psychological Prize: David C. Glass and
Jerome E. Singer

American Chemical Society (ACS) Award for Creative
Invention: H. Tracy Hall

ACS Award for Nuclear Applications in Chemistry:
Anthony Turkevich

ACS Award in Enzyme Chemistry: Ekkehard K. F.
Bautz

ACS Garvan Medal (chemistry): Jean'ne M. Shreeve

*ACS Priestley Medal: George B. Kistiakowsky

American Heart Association Research Achievement
Award: Earl W. Sutherland, Jr.

*American Institute of Aeronautics and Astronautics
Goddard Award: Brian Brimelow, Gary A. Plourde,
and Howard E. Schumacher

American Physical Society High-Polymer Physics
Prize: Anton Peterlin

*Atomic Energy Commission (AEC) Fermi Award
(physics): Dr. Shields Warren and Dr. Stafford L.
Warren

AEC Ernest O. Lawrence Memorial Awards: Charles
C. Cremer, Sidney D. Drell, Marvin Goldman, David
A. Shirley, and Paul F. Zweifel

AEC Bertner Foundation Award (cancer research):
Howard M. Temin

Bonner Prize (nuclear physics): John D. Anderson
and Donald Robson

Buckley Solid State Physics Prize: James C. Phillips

*Chemical Industry Medal: Jesse Werner

Collier Trophy (astronautics): Robert R. Gilruth,
James B. Irwin, David R. Scott, and Alfred M.
Worden

Davisson-Germer Prize (physics): Erwin W. Mueller

*Einstein Award (physics): Eugene P. Wigner

*Franklin Medal (physics): Hannes O. G. Alfven

*Gairdner Awards (medicine): Dr. Solomon A. Berson,
Dr. Charles H. Best, Dr. Rachmiel Levin, Dr. Frederick
Sanger, Dr. Donald F. Steiner, and Dr. Rosalyn Yalow

*Geological Society of America Day Medal: Hans P.
Eugster

Heineman Prize for Mathematical Physics:
J. D. Bjorken

*Hill Space Transportation Award (astronautics):
Dr. Hubertus Strughold

Hoover Medal (engineering): Luis A. Ferre

*Horowitz Prize (biology): Dr. Hugh E. Huxley

Ives Medal (optics): A. Francis Turner

*Kalinga Prize: Margaret Mead

Kepler Gold Medal (astronomy): Hannes O. G. Alfven,
Gerard Kuiper, Boris Y. Levin, E. J. Opik,
Harold C. Urey, and Fred Whipple

Kosekisho Achievement Award (metal physics):
Masahiro Meshii

Langmuir Award in Chemical Physics: Harden M.
McConnell

*Lasker Awards (basic research): Seymour Benzer,
Dr. Sydney Brenner, and Charles Yanofsky; (clinical
research): Dr. Edward D. Freis

Meggers Award (optics): Louise L. Sloan

*Merrill Award (astronomy): Roman A. Schmitt

*National Academy of Engineering Founders Medal:
Edwin H. Land

*National Academy of Sciences (NAS) Applied
Mathematics Award: Kurt O. Friedrichs

*NAS Arctowski Medal (astronomy): Francis S. Johnson

*NAS Cotrell Award (pollution control): Arie J.
Haagen-Smit

*NAS Day Prize (geology): Hatten S. Yoder, Jr.

*NAS Hartley Public Welfare Medal: Leonard Carmichael

*NAS Microbiology Award: Charles Yanofsky

*NAS Walcott Medal: Elso S. Barghoorn

*NAS U.S. Steel Foundation Award (molecular biology):
Howard M. Temin

*Nobel Prize: chemistry, Gerhard Herzberg; physics,
Dennis Gabor; physiology and medicine, Earl W.
Sutherland, Jr.

Oppenheimer Memorial Prize (physics): Robert Serber

*Passano Foundation Award (medicine): Dr. Kimishige
Ishizaka and Dr. Teruko Ishizaka

*Penrose Medal (geology): Marshall Kay

Public Health Service Commendation Medal
(epidemiology): Dr. Frank E. Lundin, Jr.

Royal Astronomical Society Gold Medals: Fritz
Zwicky and H. I. S. Thirlaway

Soviet Academy of Sciences Lomonosov Gold Medals:
Viktor A. Ambartsumyan and Hannes O. G. Alfven

Warner Prize (astronomy): Kenneth Kellerman

*Watson Medal (astronomy): André Deprit

Deaths of Notable Scientists

Notable scientists who died between June 1, 1971, and June 1, 1972, include those listed below. An asterisk (*) indicates that the person has a biography in *The World Book Encyclopedia.*

*Adams, Roger (1889-July 6, 1971), chemist, was awarded the 1965 National Medal of Science for his work in organic chemistry. He headed the University of Illinois department of chemistry and chemical engineering from 1926 to 1954 and chaired the American Chemical Society from 1944 to 1950.

Albright, William F. (1891-Sept. 19, 1971), Biblical archaeologist, directed excavations in the Middle East and identified Tirzah, an ancient capital of Israel. He developed the method of dating archaeological discoveries by using information obtained from potsherds, broken pieces of earthenware.

Andrews, E. Wyllys IV (1916-July 3, 1971), archaeologist, excavated the ancient Mayan city of Dzibilchaltun in Yucatán, Mexico, and was noted for his studies of Mayan culture.

Beal, George D. (1887-Jan. 3, 1972), chemist, helped with revisions of the United States Pharmacopeia and prepared a cumulative index for all revisions dating from 1820. He was director of research at the Mellon Institute in Pittsburgh from 1951 to 1958.

Bernal, John Desmond (1901-Sept. 15, 1971), Irish crystallographer, pioneered in protein crystallography and determined the structure of graphite. He also wrote many popular books, including *The World, the Flesh, and the Devil* (1929), *The Origin of Life* (1967), and *Science in History* (1971).

Berson, Solomon A. (1918-April 11, 1972), physician, helped develop the technique of using radioisotopes to detect and measure hormones in the blood and was noted for applying this method to study insulin.

Blegen, Carl W. (1887-Aug. 24, 1971), archaeologist, directed excavations at the site of ancient Troy and uncovered Linear B tablets, examples of an ancient Greek language, at Pylos.

*Boyd Orr, Lord (1880-June 25, 1971), British physiologist and nutritionist, won the 1949 Nobel Peace prize for his work on combating world hunger. He helped to found the United Nations Food and Agriculture Organization in 1945 and was its first director.

Bragg, Sir William Lawrence (1890-July 1, 1971), British physicist and crystallographer, shared the Nobel prize for physics in 1915 with his father, Sir William Henry Bragg, for determining the structure of crystals through the use of X rays. Along with his father, he developed the X-ray spectrometer.

Castellani, Marquis Aldo (1879-Oct. 3, 1971), Italian physician, was an expert on tropical diseases and proved that a parasite causes sleeping sickness. He established the School of Tropical Medicine at Tulane University in New Orleans in 1926.

Croneis, Carey (1901-Jan. 22, 1972), geologist, designed the geology section of the Museum of Science and Industry in Chicago. He was formerly president of Beloit College and also chancellor of Rice University.

De Vaux, Roland (1903-Sept. 10, 1971), French archaeologist and Roman Catholic priest, headed the team that edited the Dead Sea Scrolls and took part in excavations of the caves in the Qumran Valley in Jordan.

Dreikurs, Rudolf (1897-May 25, 1972), Austrian-born psychiatrist, was an associate of Alfred Adler and an originator of multiple psychotherapy. He was professor emeritus of psychiatry at the Chicago Medical School and director of the Alfred Adler Institute.

Dyer, Rolla E. (1886-June 2, 1971), physician, was an expert on rickettsial diseases such as Rocky Mountain spotted fever and was among the first scientists to link cancer with cigarettes. He was director of the National Institutes of Health from 1942 to 1950.

Forsythe, George E. (1917-April 9, 1972), mathematician and computer expert, was chairman of the Stanford University department of computer science from 1965 to 1971. He also wrote or co-authored several books on higher mathematics.

Germer, Lester H. (1896-Oct. 3, 1971), physicist, with Clinton J. Davisson, discovered the diffraction of electrons by crystals, which led to the development of the electron microscope.

Gilbreth, Lillian M. (1878-Jan. 2, 1972), engineer, was an early researcher, along with her husband Frank, in time-motion studies for industrial management. She also developed rehabilitation devices for disabled persons.

Lord Boyd Orr

Sir William Lawrence Bragg

Bernardo A. Houssay

Deaths of Notable Scientists

Continued

Edward C. Kendall

Maria Goeppert-Mayer

Wendell M. Stanley

Halpern, Julius (1912-May 13, 1972), nuclear physicist, helped establish the Pennsylvania-Princeton particle accelerator and tried to prove the existence of the theoretical particle called the quark. He was professor of physics at the University of Pennsylvania from 1952 to 1972.

Heiser, Victor G. (1873-Feb. 27, 1972), physician, introduced modern preventive-medicine techniques in the Far East and wrote *An American Doctor's Odyssey* (1936).

*Houssay, Bernardo A. (1887-Sept. 21, 1971), Argentinian physiologist, shared the Nobel prize for physiology and medicine in 1947 with Carl F. and Gerty T. Cori for discovering the role of pituitary hormones.

*Karrer, Paul (1889-June 18, 1971), Swiss chemist, shared the 1937 Nobel prize in chemistry with Sir Walter N. Haworth for his research on carotenoids, flavins, and vitamins.

Katchalsky, Aharon Katzir (1914-May 30, 1972), Israeli biophysicist, was renowned for his work on the electrolytic properties of chain molecules and for developing an engine that converts chemical energy into mechanical energy. He also did research on brain functions and the molecular basis of memory.

*Kendall, Edward C. (1886-May 4, 1972), biochemist, shared the 1950 Nobel prize for physiology and medicine with Philip S. Hench and Tadeus Reichstein for his research on the adrenal cortex and his subsequent discovery of cortisone. He studied the effects of cortisone and ACTH on rheumatoid arthritis and rheumatic fever.

Kurchatov, Boris V. (1907-April 13, 1972), Russian nuclear chemist, studied nuclear isomerism, and, with his brother Igor V., discovered isomerism in artificial elements. He became the first director of research in nuclear chemistry at the Russian Atomic Energy Institute in 1943.

Martí-báñez, Felix (1912-May 24, 1972), Spanish-born physician, lectured extensively on psychiatry and the history of medicine. He founded *MD* magazine in 1950.

*Mayer, Maria Goeppert- (1906-Feb. 20, 1972), German-born physicist, shared the 1963 Nobel prize in physics with Eugene P. Wigner and J. Hans Jensen. She and Jensen developed the shell theory to explain the structure of atomic nuclei.

Parin, Vasily V. (1903-June 15, 1971), Russian physiologist, was an expert on how conditions in outer space affect the human body.

Phillips, George E. (1913-Jan. 28, 1972), biochemist, developed drugs for treating and preventing coronary thrombosis and hemorrhages. He was director of the biochemistry department of the Warner-Lambert Research Institute in Morris Plains, N.J., from 1947 to 1972.

Rado, Sandor (1890-May 14, 1972), Hungarian-born psychiatrist, studied under Sigmund Freud and edited early psychoanalytic journals. He established the graduate school of psychoanalysis at Columbia University's college of physicians and surgeons in 1944 and headed it until 1955.

Sarabhai, Vikram A. (1919-Dec. 30, 1971), Indian physicist, headed India's space-research efforts since 1962 and was chairman of the Indian Atomic Energy Commission from 1966 to 1971.

*Stanley, Wendell M. (1904-June 15, 1971), biochemist, was the first scientist to isolate a virus. For this work, he shared the 1946 Nobel prize in chemistry with James B. Sumner and John H. Northrop.

*Tiselius, Arne (1902-Oct. 29, 1971), Swedish physical chemist, developed methods for analyzing proteins and won the 1948 Nobel prize in chemistry for his research on proteins in blood serum.

Whitby, G. Stafford (1887-Jan. 10, 1972), chemist, was noted for his research on rubber and for developing a low-temperature process for making synthetic rubber. He was professor of rubber chemistry at the University of Akron from 1942 to 1954.

Wintersteiner, Oskar P. (1898-Aug. 15, 1971), Austrian chemist, isolated the crystalline forms of penicillin-G and streptomycin. He was director of biological chemistry at the Squibb Institute of Medical Research in New Jersey from 1959 to 1963.

Witschi, Emil (1890-June 9, 1971), Swiss-born biologist, was noted for his research on embryology and reproductive physiology. He was senior scientist on the Population Council of Rockefeller University in New York City from 1967 to 1971.   [Darlene R. Stille]

Index

This easy-to-use index covers all sections of the
1971, 1972, and 1973 editions of *Science Year,*
The World Book Science Annual.

An index entry which is the title of an article
appearing in *Science Year* is printed in boldface italic
letters as: ***Archaeology.*** An Entry which is not an
article title, but a subject discussed in an article
of some other title is printed: **Amino acids.**

The various "See" and "See also" cross references
in the index list are to other entries within the index.
Clue words or phrases are used when the entry needs
further definition and when two or more references to
the same subject appear in *Science Year*. These make
it easy to locate the material on the page, since
they refer to an article title or article subsection
in which the reference appears, as:

Heart surgery: coronary artery by-pass,
73-329, *72*-178, 330, *71*-325;
endarterectomy, *72*-185, 330, *71*-324,
325; revascularization, *72*-182, 330;
saphenous vein by-pass graft, *72*-188,
330; transplantation, *73*-329, *72*-331,
332, *71*-324, *il.*, 324. See also **Heart,
artificial ; Heart, mechanical**

The indication *"il."* means that the reference
is to an illustration only.

Index

A

Abouna, George, 73-328, 329
Abscisic acid (ABA), 71-276
Acanthaster planci, 71-112
Accelerator, particle: Argonne
National Laboratory, 71-218;
European Center for Nuclear Research
(CERN), 73-238, 72-347, 71-30; Los
Alamos, 73-351, *il.*, 350; National
Accelerator Laboratory (NAL),
73-349, *il.*, 348, 72-205–211, 347;
Princeton-Pennsylvania (PPA), 71-215,
343; proton synchrotron, 73-248;
Serpukhov (Russia), 71-343, 344;
Stanford Linear, 73-346, 72-348,
71-343; tandem Van de Graaff, 71-347
Acetylcholine: 72-274; role in
memory, 71-353; sensitivity of
muscle cells to, 73-275
Achondroplastic dwarfism, 71-100
Acne: *Special Report*, 72-152; *Science
Year* Trans-Vision, 72-162; acne
conglobata, 72-159, *ils.*, 173; diet and,
72-159, 174
Acoustical Society of America
Biennial Award, 73-422
Actin: role in muscle building,
73-155, 156, *ils.*, 156, 158; role in
muscle contraction, 73-275, 276
Actinomycin D, interaction with
DNA, 72-289, *il.*, 289
Actomyosin, 73-157, 72-395
Acupuncture: 73-330, *il.*, 328, 72-13,
il., 13; anesthesia, 73-192, 193, *il.*,
194, 330; theory, 73-192
Adams, Roger, 73-423
Adams Award, (organic chemistry),
72-422
Adenine (A): 71-99; 3', 5' adenosine
monophosphate (cyclic AMP), gene
expression, 72-338, 340
Adenosine (A), 72-289
Adenosine diphosphate (ADP), role
in muscle power, 73-157, *il.*, 157
Adenosine triphosphate (ATP),
73-157, 275, 276, *il.*, 157, 72-395,
71-271
Adipose tissue: fat production,
73-122; glucose level and food
intake, 73-122
Aeronautics, *Books of Science*,
71-273
Aerospace industry, unemployment,
73-360, 72-361, 364
Afar Triangle, 73-170, *il.*, 173
Agassiz Medal (oceanography),
73-422, 71-397
Aggression, population and, 72-305
Agranoff, Bernard, 72-145
Agriculture: 73-254, 72-254, 71-252;
archaeological findings, 72-259, 262;
asphalt moisture barrier, 72-256;
Books of Science, 73-277, 71-273;
effect of farming methods on climate,
72-101, 102, *il.*, 103; effect on
population, 71-83, 84; in China,
73-197, 198, *ils.*, 198, 199; in Japan,
71-226; nitrogen supply, 71-285;
plant cell fusion, 72-203, *il.*, 202;
Triticale, synthetic grain species,
72-255. See also Aquaculture; Cell
fusion; *Chemical Technology;
Climate; Pesticides*
Agriculture, Department of, 71-30
Air-circulation patterns, 72-104
Air-cushion vehicle, 73-373, *il.*,
371, 72-375
Air Delivered Anti-Pollution
Transfer System (ADAPTS), 72-344

Air pollution: 72-34, 284, 333, 374,
71-304, 374; and acidic rainfall,
73-297; atmospheric monitoring
stations, 73-331, 332; automobile
exhaust, 73-311, 71-158, 160; dust,
72-99, 100; increased levels of CO
and CO_2, 73-331; jet aircraft exhaust,
71-304; lead poisoning, 72-302, 303,
natural gas, 73-306; ozone absorption
by plants, 72-282; particulate matter
in atmosphere, 73-331, 332; public
health, 73-325; utility plants, 71-304.
See also **Pollution**
Alaska pipeline, 72-34, 372
Alaskan oil fields, environmental
threat, 71-298
Albinism, 71-100, *il.*, 101
Albrecht, Erich, 73-252, *il.*, 250
Albright, William F., 73-423
Alcator, plasma device, 71-307
Alcoa Seaprobe, *il.*, 73-342
Aldrin, Edwin E., Jr., 71-418
Alexander, Richard D., 72-417
Alfven, Hannes O.G., 73-416,
72-416, *il.*, 416
Algae: 73-109, 71-129, *il.*, 128;
source of protein, 72-284
Alikhanov, Abram I., 72-423
Allelopathy, 73-300
Allosteric transition, 71-392
Alpha-bungaro, acetylcholine
receptor isolation, 72-274
Alpha waves, *il.*, 73-338
Altschul, Aaron M., 73-322
Aluminaut submersible, retrieval of
Alvin, 71-338, *il.*, 337
Alvin submersible, 71-338, *il.*, 337
Amalthea, Jupiter satellite, 71-63
Ambartsumian, Viktor A., 71-38
American Association for the
Advancement of Science (AAAS):
73-302; *Special Report*, 72-240
AAAS Socio-Psychological Prize,
73-422
American Association for Thoracic
Surgery, 72-189
American Cancer Society Award,
72-422
American Chemical Society Awards:
Analytical Chemistry, 72-422, 71-420;
Biological Chemistry, 72-422, 71-420;
Creative Invention, 73-422; Enzyme
Chemistry, 73-422; Inorganic
Chemistry, 72-422, 71-420; Nuclear
Applications in Chemistry, 73-422;
Pure Chemistry, 72-422, 71-420
American Heart Association
Research Achievement Award,
73-422, 72-422, 71-420
American Institute of Physics, 71-359
American Institute of Physics-
U.S. Steel Foundation Award
(science writing), 72-422, 71-419
American Medical Association,
drug evaluation, 72-298
American Medical Association
Awards: Distinguished Service,
72-422, 71-420; Scientific Achievement,
72-422, 71-420; Sheen, 72-422
American Physical Society
High-Polymer Physics Prize, 73-422
American Sign Language, 71-66
American Telephone and Telegraph
Company (A.T.&T.), 73-291, 292,
72-293
Amino acids: Al protein, 72-274, 276;
glutamic acid, 71-269; histidine,
71-269; in Murchison stone meteorite,
72-315, 316, *il.*, 316; proline,
71-269; tryptophan, 72-276, 277
Ammonia, atmospheric: 71-59;
absorption by plants, 73-255
Ammonia molecules, interstellar, 72-67

Ammonia synthesis, 71-286, *il.*, 286
Amnesia, experimental, 71-352, 353,
354
Amniotic fluid, prenatal diagnosis,
72-315
Amphetamines, 72-299
Amtrak, 72-374
Anabolic steroids, 73-161
Anand, Bal K., 73-120
Anatomy, medical illustration, 72-11—25
Anders, Edward, 72-420
Anders, William A., 71-418
Andrews, E. Wyllys IV, 73-423
Andromeda Nebula, 71-34, 36, 38
Anesthesia, by acupuncture,
73-192, 193, *il.*, 194
Angina pectoris, *Special Report*,
72-176
Angiography, 72-178, 187
Angiosperms, 72-281
Angiosperms, primitive, 73-281
Anliker, James, 73-118
Anorthosite, in lunar rocks, 73-313, 314
Antarctic current, 71-337
Antarctic Ocean, 72-118
Anthropology: 73-256, 72-256,
71-254; *Books of Science*, 73-277,
72-278, 71-273; early man, *Special
Report*, 73-224. See also *Archaeology*
Antibiotics: acne therapy, 72-172;
actinomycin D, interaction with DNA,
72-289, *il.*, 289; as amnesic agents,
71-353, 354; in combination, 72-298,
71-292; novobiocin, 71-292;
Panalba, 71-292; protein synthesis
inhibition, 73-114, 72-145; resistance
factors, 71-334, 335; tetracycline,
72-172; tetracycline phosphate
complex, 71-292; trobicin, 73-297
Antiferromagnets, 73-353
Antihemophilic factor (AHF), 71-323
Antilymphocytic globulin (ALG),
use in tissue transplants, 71-324
Antimatter, 71-262
Antiparticles, 71-209
Aphasia, 72-151
Apollo program: 73-360, 362, 363,
366, 398, 400, 409—413, 72-366, 367,
71-362—365, 368; Apollo 11, 71-287;
Apollo 12, 71-362, *ils.*, 364, 365;
Apollo 13, 71-362, 363, 365, *il.*, 363;
Apollo 14, 72-43, 45, 47, 366, 367,
ils., 39, 41; Apollo 15, 73-363—366,
72-367; Apollo 16, 73-366; Apollo 17,
73-366; communications systems,
73-292; lunar seismographs, 71-315,
il., 316
Appetite, 73-117, 128, 129, 134
Applications Technology Satellite 3
(ATS-3), 71-265
Applied Mathematics Award, 73-421
Aquaculture: *Special Report*, 72-107;
raft culture, 72-114, 115, *ils.*, 114,
115; zones of upwelling, marine life
concentrations, 72-118
Archaeology: 73-259, 72-259, 71-256;
archaeological looting, *Close-Up*,
73-260; *Books of Science*, 73-277,
72-278, 71-273. See also
*Anthropology; Megalithic
archaeology*
Arches of Science Award (science
writing), 72-422, 71-419
Architectural engineering, *Books of
Science*, 73-277
Arctic Ice Dynamics Joint
Experiment (AIDJEX), 73-339
Arctic ice floes, 73-339, 340
Arctic tundra, Alaska pipeline,
72-34, 71-298
Arctowski Medal, 73-419
Arees, Edward, 73-120, 122
Arginase deficiency, 71-309

Index

Arm movements, in the infant, 72-359
Armitage, Frank, *il.*, 72-24
Armstrong, Neil A., 72-367, 71-418
Arp, Halton C., 71-45
Arp galaxies, 71-46, *il.*, 44
ARPANET, computer network, 73-295
Arroyo Hondo, 73-262
Arsenic: used in insecticides, 72-291; water pollutant, 72-303
Arterial heart defects, 71-325
Arteriosclerosis: 72-181; and diet, 73-322
Arteriovenous fistula, 72-332, 333
Arthritis, effect on hip joint, 71-326
Artificial Heart Program, National Heart Institute, 71-322, 323
Artificial placenta, *il.*, 71-323
Artificial upwelling, 72-343
Artsimovich, Lev, 72-215, 353, 71-306, 348
Ascenzi, Antonio, 73-256
Ascorbic acid: 72-387—389; and iron availability, 73-145
Ashflow deposits, 71-313
Asphalt moisture barrier, for sandy soils, 72-256
Asthenosphere, 72-322
Astrology, 71-26, 32
Astronautics: *Books of Science*, 73-277, 71-273. See also *Space Exploration*
Astronomy: *Books of Science*, 73-277, 72-278, 71-273; similarity theory, 71-62; X-ray, 73-270, 72-269
Astronomy, High Energy: 73-270, 271, 72-269, 270, 71-266—269; neutrinos, 71-268
Astronomy, Planetary: 73-264, 72-263, 71-260; heliocentric system, 73-266. See also *Space Exploration*
Astronomy, Stellar: 73-268, 269, 72-265, 71-262; Applications Technology Satellite 3 (ATS-3), 71-265. See also **Molecules, interstellar; Pulsars**
Asymptopia, particle physics, 72-347
Athletics, *Special Report*, 73-146—161
Atmospheric motion, Venus, 71-261, 262
Atomic absorption spectroscopy, 73-141, 142
Atomic and molecular physics. See *Physics, Atomic and Molecular*
Atomic Energy Commission (AEC), 73-309, 310, 360, 361, 72-32, 205, 352, 400, 71-307, 358
Atomic fission, 72-214
Atomic fusion, 72-214, 215
Atoms, pictures of, 71-351
Atoms for Peace Award (physics), 71-414
Atwood, Susie, *il.*, 73-160
Aurignacian culture, 71-258
Australia antigen, *Close-Up*, 71-320
Australopithecine remains, 72-257, 258, 71-254, *il.*, 255
Australopithecus, 73-228, 229, 231, 72-257, 71-254, *il.*, 255
Australopithecus africanus, 73-229, 231, 236, 237, 257, *ils.*, 224, 235
Australopithecus boisei, 73-233, *il.*, 234
Australopithecus robustus, 73-229, 231, 233, 257
Auto-Train, 73-373, 374
Automobile: emission standards, 73-311, 72-373, 374; impact on society, 72-28, 29; inadequacy as urban transportation, 71-371;

problems caused by, 71-372; steam car, *Special Report*, 71-156; tire safety test, *il.*, 73-374; Wankel engine, *Close-Up*, 73-372. See also **Air pollution; Transportation**
Autonomic learning, *Special Report*, 71-180—193
Aves Swell, submarine ridge, 72-135
Awards and Prizes, *73-415—422, 72-414—422, 71-412—420*
Axelrod, Julius, 72-418, *il.*, 418
Axons, 73-336, 337, *il.*, 337
Aztec period, 72-262

B

Babbles and coos, 72-359
Babies, test-tube, 72-362
Baboon behavior, 71-69—77, *ils.*, 70—75
Bacteria: *il.*, 73-295; genes incorporated into other cells, 73-333, 334; growth studies, 71-381, 390, 392; used to control oil pollution, 72-344
Bacterial protein, food source, 72-340
Bacteriology. See *Microbiology*
Bacteriophage, 71-388, 389
Bagasse, source of protein, 72-284
Bagnold, Ralph A., 72-417
Baltimore, David, 73-94, 72-417
Bang, O., 73-92
Barabashov, Nikolai P., 72-423
Barbour, John, 73-190, 72-82, 71-397
Bardach, John E., 72-108
Barghoorn, Elso S., 73-417, 71-276
Barnacle glue, use in dentistry, 72-325
Barth, Charles A., 71-72
Barton, Derek H. R., 71-413, *il.*, 414
Bastnasite, 73-349
Bathythermograph, 71-402
Bats: fog avoidance, 72-380; hibernation, 73-376
Batty, John, 72-351
Bay Area Rapid Transit (BART), *il.*, 71-372
Bayard, Donn, 72-259
Baynard, Ralph, 73-259, 261
Baynard's Castle, 73-259, 261
Beal, George D., 73-423
Bean, Alan L., 71-362
Bear seamount, 73-339
Bears, grizzly, *Special Report*, 73-35—49
Beata Ridge, submarine ridge, 72-135
Beberman, Max, 72-423, *il.*, 423
Beck, Claude S., 72-182
Beckwith, Jonathan, 71-27, 308, 415, *il.*, 415
Beetle, bombardier, *il.*, 71-378
Beetle larva, memory studies, 71-375
Beetles, heat experiments, 72-300
Behaviorism, 73-356, 357, 72-150
Behnke, Frances, 73-301
Bell, Charles, 72-12
Bell, Joseph N., 72-400
Bell Telephone Laboratories, 71-301
Bemis, Curt, 72-351
Ben Franklin, submersible, 71-338
Benzer, Seymour, 73-418, 422
Berger, Rainer, 72-257
Bernal, John Desmond, 73-423
Bernstein, Gerald, 73-337
Berry, Charles A., 71-418, *il.*, 418
Berson, Solomon A., 73-418, 423
Bertner Foundation Award (cancer research), 73-422
Besler steam engine, 71-165, *il.*, 166
Best, Charles, 73-418
Bezymianny volcano, 72-318
Bhabha, Homi J., 72-221
Bietak, Manfred, 72-260
Big bang theory, 73-272
Bile, formation of gallstones, 73-327

Billroth, Theodore, 72-182
Binary stars: radio and X-ray radiation from, 73-268, 269; WZ Sagittae, 71-265
Bing, R. H., 72-245
Bioavailability: of drugs, 73-296, 297, 71-178
Biochemistry: 73-274, 72-273, 71-269; *Books of Science*, 73-277; neural, 72-137; trace elements, 73-144, *il.*, 135; trace elements in humans, 73-133—145. See also **Amino acids; Chemistry, Synthesis; DNA; Enzymes; Hormones; RNA**
Biological magnification, pollutant concentration in food chain, 73-52
Biology, *Books of Science*, 73-278. See also *Biochemistry; Botany; Ecology; Genetics; Microbiology; Zoology*
Biomes, grassland, 72-81—93
Biophysics, communication, 72-137
Birch, Francis, 73-163, 172, 71-415
Birds: endangered species, 73-50—59; grassland study, 72-85, 87, *ils.*, 86
Birge, Edward A., 71-127
Birth control, 71-83, 90, 256, 294
Birth defects: 71-95—105; prenatal diagnosis, 72-315; related to thalidomide use, 71-173
Birth rates, underdeveloped countries, 71-82
Bjorken, James, 73-346
Björklund, Anders, 73-336, 338
Black holes: *Special Report*, 73-77—89, 71-47, 263, 342, *il.*, 47; possible discovery by X-ray astronomy, 73-271
Black lung disease, 73-324
Blackheads, 72-158
Bladderworts, *il.*, 73-13
Blakemore, Colin, 72-139
Blance, Beatrice, 72-229
Blastoids, 73-96, 97
Blegen, Carl W., 72-423
Blight watch, 72-254
Bloch, Bruno, 72-159
Blood: antihemophilic factor (AHF), 71-323; artificial, 71-279; clotting (artificial heart interface), 71-327, (oral contraceptives), 71-294, 356; erythropoietin, 72-330; phonoangiography, 72-329, 330, *il.*, 326; red blood cells, 73-155, 378, 380, *il.*, 72-327; red blood cell stimulant, 72-330; sickle cell anemia, 73-323, 72-328, 329, *il.*, 323; T-globulin, cancer test, 72-326; transfusions, hepatitis resulting from, 71-320; 2,3-DPG, 73-155
Blood sugar, insulin deficiency, 73-140
Blue-green algae, 73-109, 110
Blumberg, Baruch S., 71-320
Bodmer, Walter F., 72-202
Body tissue, radiocarbon analysis, 71-259
Boersma, Larry, 72-256
Boffey, Philip M., 71-220
Bohr, Aage N., 71-414
Bolli, Hans, 72-123, *il.*, 133
Bolli, Hans, 72-123, *il.*, 133
Bones: electric fracture treatment, 73-330; repair with ceramics, 73-284, 285
Bonner Prize (nuclear physics), 73-422, 72-422, 71-420
Books of Science, 73-277, 72-278, 71-273
Bootstrap hypothesis, 73-348
Boretsky, Michael, 71-223
Borlaug, Norman E., 72-254, 255, 421
Borman, Frank, 71-418, 420
Bórmida, Marcelo, 72-262
Born, Max, 71-421, *il.*, 421
Boss, Benjamin, 72-423
Bost, A. Crawford, 71-420

427

Index

Botany: 73-280, 281, 72-281, 282, 71-276, 277; *Books of Science,* 73-277, 72-278, 71-273; endangered plants, *Essay,* 73-11—23; nectar guides as aid to pollination, 71-276, 277, *il.,* 277; plant cell fusion, 72-203, *il.,* 202; plant growth regulators, 73-255
Bowie Medal (geophysics), 72-422
Boyd Orr, Lord, 73-423
Bradbury, Norris E., 72-416
Bragg, Sir William Lawrence, 73-423
Brain, Charles K., 71-255
Brain, *Special Report,* 72-137—151, 71-194—205. See also *Neurology; Psychology*
Brain surgery, acupuncture anesthesia, 73-192
Brassins, plant hormones, 71-276
Brauer, Richard D., 72-421
Brazil, communications satellite systems, 73-292
Breccias, Apollo 14 rocks, 72-43
Brenner, Sydney, 73-418, 422
Breslow, Ronald C. D., 72-292
Brett, Robin, 72-43
Brimelow, Brian, 73-419, 422
British soldier lichen, *il.,* 73-12
Brödel, Elizabeth, 72-21
Brödel, Max, 72-13, 18, 19
Bronfman Prize (public health), 72-422, 71-420
Bronowski, Jacob, *il.,* 71-358
Bronze casting, 72-259, 260
Brown, Herbert C., 71-420
Brownouts: 72-213; peak demand, 73-308
Bruce Medai (astronomy), 72-422, 71-419
Brucellosis, 73-256
Bruch, Hilde, 73-129
Bryson, Reid, 72-96
Bubble chamber, 73-209, 71-344, 345
Bucha, Vaclav, 72-231, 71-257
Buckley Solid State Physics Prize, 73-422, 72-422, 71-414
Bumblebees, 73-375
Burbank, Robinson, 72-290
Burkitt's lymphoma, 71-319
Burmester, Ben R., 72-255
Buses, 73-374
Bushman society, 71-78, 83

C

Calcium metabolism, 72-276
Calcitonin: 72-276; salmon, 71-269, 271, *il.,* 271
Calculators, electronic, 72-308
Calculators, portable, 73-293, 294, 72-303
Caldera, 73-62, 65, *il.,* 63
Callisto, Jupiter satellite, 71-62
Calories, utilization of by exercise, 73-125, 126, 127
Camus, 73-118
Canada, communications satellite systems, 73-292
Canada geese, 72-210
Canals, on Mars, 73-62, 63, 69
Cancer: abnormal properties of cells, 73-25; betel nuts, mouth cancer, 72-258; blood test, 72-326; breast, 72-326; Burkitt's lymphoma, 71-319; carcinogens, 73-25, 26, 71-96; caused by virus, *Special Report,* 73-89—103, 26, 27, 72-299, 326, 327, 71-319; cell fusion, 72-198; chemotherapy and cancer research, 73-27; chicken vaccine, 72-255;

clusters of incidence, 73-95, 96, 97, 98; cyclamates, 71-294; detection, *il.,* 72-328; diagnostic test, 73-327; immunization, 72-327, 328; immunological response, 73-26; kaunaoa worm, anticancer ingredient in, 72-377; L-asparaginase, 72-298; National Cancer Act, 73-27, 29; National Cancer Bureau, 73-359; National Cancer Institute (NCI), 73-28, 29, 30; National Cancer Plan, 73-30; nuclear magnetic resonance (NMR), 72-327; occupational, 73-324; oral contraceptives, 71-294; osteosarcoma, 71-319; research, *Essay,* 73-24—32; research, monetary considerations, 73-29, 30, 72-363; RNA-directed DNA polymerase, 72-275; SV-40, cancer-producing virus, 72-201; vaccine, 71-319, 321; vital statistics, 71-319
Cannon, Walter B., 73-118
Capillaries, increase by exercise, 73-155
Carbohydrates, and glycogen content in muscles, 73-158, 159
Carbon, activated, used in waste control, 73-283, 284
Carbon compounds, traces in lunar samples, 71-271
Carbon dioxide (CO_2): increased level in atmosphere, 73-331, 72-98, 99; use in heart surgery, 72-189, 330
Carbon dioxide gas laser, 71-339
Carbon-14 dating, 72-230, 71-257
Carbon monoxide (CO): increased level in atmosphere, 73-331; radio waves from, 71-263, 264
Carcinogens, 73-25, 26, 71-96
Carcinoembryonic antigen, 73-327
Caribbean Sea floor, 72-123, 125, 128, 131, 134, 135
Carlisle, Edith M., 73-144
Carlson, Anton J., 73-118
Carmichael, Leonard, 73-421
Carnap, Rudolf P., 72-423
Carp, 72-108
Carruthers, George, 72-70
Carson, Rachel, 71-107—109, *il.,* 106
Carter, Brandon, 73-88
Cartilage, regeneration, 72-378
Carty Medal, 72-417, 422
Casman, Ezra P., 72-423
Caspersson, Tobjörn O., 72-314
Castellani, Marquis Aldo, 73-423
Cat, visual cortex experiments, 72-139, *ils.,* 142, 143
Cates, Rex G., 73-300
Catfish, 72-109. *il.,* 110
Catheterization, 72-331
Cattle, crossbreeding of, 72-255, 71-252
Cedarwood extract, insect repellent, 73-280, 281
Cell-bound antibodies, and cancer immunity, 73-26
Cell division: 73-332, 333; effects of malnutrition, 71-199, 200, 201, *il.,* 201
Cell evolution: 73-105—115, 71-332; capture theory, 73-333
Cell fusion, *Special Report,* 72-191—203. See also *Genetics; Mutagenics*
Cell-mediated immunological response, and cancer, 73-26
Cells: cancer (detection), *il.,* 72-328, (abnormal properties), 73-25; hybrid cell formation, 72-193, *il.,* 195; multidisciplinary study, 73-25
Cellulose, and protein-forming microbes, 72-284, 286
Centaurus X-3, 73-270
Centaurus XR-4, 71-262
Central Arid Zone Research Institution, 72-101

Central nervous system, 73-336
Centromeres, 72-274
Ceramic, used in bone repair, 73-284, 285
CERN (European Organization for Nuclear Research): *Special Report,* 73-238—252; proposed accelerator, 72-350; proton collisions, 72-347
Cernan, Eugene A., 73-366
Cesium 137, 72-300
Chandrasekhar, Subrahmanyan, 72-420
Chaney, Ralph W., 72-423
Chao, Edward, 72-402
Chao, Li-Pen, 72-276
Chapman, Sydney, 72-423
Charge-coupled device (CCD), 73-304, 305, 71-302
Charney, Jesse, 73-103
Charpak, Georges, 73-239, 240, 241, 242, *il.,* 242
Chemical and biological warfare (CBW), 72-247
Chemical disease, 73-215
Chemical Industry Medal, 73-414
Chemical reaction, angular distribution of product molecules, 73-286, 287
Chemical Technology, 73-282, 72-283, 71-278
Chemistry: Books of Science, 72-278; hyperthermal collisions, 71-283
Chemistry, Dynamics, 73-285, 72-287, 71-281. See also *Biochemistry*
Chemistry, Structural, 73-287, 288, 72-289, 290, 71-283, 284
Chemistry, Synthesis, 73-289, 72-291, 71-285. See also *Biochemistry*
Chenodeoxycholic acid, 73-327
Chesher, Richard H., 71-110
Chew, Geoffrey, 73-348
Chicken, cancer vaccine, 72-255
Chicken leukosis 72-255, 71-319
Child development: 72-357—359; effects of malnutrition, 71-195—205
Childe, V. Gordon, 72-227
Chimpanzee, *il.,* 71-78; linguistic ability, 71-66
China: social structure and scientific progress, *Special Report,* 73-188—203; space program, 72-371, 71-368, 369
Chisholm, G. Brock, 72-423
Chlorinated hydrocarbon insecticides, 72-291
Chloroplasts, 73-106, 109, 110, *il.,* 111
Cholesterol gallstones, 73-327
Christofilos, Nicholas C., 72-354
Chromium, human requirement, 73-139, 140
Chromosome stain technique, 73-311, 312, 72-314, 315, *il.,* 314
Chromosomes, 73-192. See also *Genetics*
Circulatory system, 73-155
Classical conditioning, 71-182, *il.,* 185
Classroom, open, 72-307
Clay analysis technique, 71-256, 257
Claytonia virginica (spring beauty), chromosomal drift, 72-281, 282
Clean Air Act, 72-374
Cleveland Clinic, 72-182, 185, 187
Cleveland Prize (general), 72-422
Climate: animal adaptation, 73-375, 376; Caucasus Mountains, dust-snow studies, 72-95, 99; dust-climate relationship, 72-96; effect of cities on, 73-332; effects of atmospheric pollution, 73-331; farming methods, effect on climate, 72-101, 102, *il.,* 103; India, dust concentrations, 72-99, 100, 101; postglacial vegetation and, 73-298, 299; *Special Report,* 72-95—105; sun, 72-96. See also *Meteorology*

Index

Cloud seeding, *73*-333, *72*-334, 335, *ils.*, 334, 335, *71*-328
Coal : *72*-30, 214, 215; gasification of, *73*-283, 307, *72*-31; natural gas substitute, *73*-306
Coaxial cable system, *72*-293
Cobb seamount, *73*-339
Cocking, Edward C., *72*-203
Coesite, mineral, *72*-402
Cohen, I. Bernard, *71*-24, *il.*, 25
Cold-water eddies, *73*-340
Coliform bacteria, indicator of water quality, *71*-148, 154
Colleges and universities, discriminatory admissions practices, *73*-302, 303
Collier Trophy (astronautics), *73*-422, *72*-422, *71*-418
Collins, Michael, *71*-418
Combustion, automobile, *71*-159, 160, *il.*, 160
Comet Medal (astronomy), *71*-419
Commoner, Barry, *72*-249, *71*-106
Communication, human-chimpanzee, *71*-65, 66, 79
Communications: 73-291, *72*-293, *71*-287; Advanced Research Projects Agency (ARPA), *71*-291; *Books of Science, 72*-278; computer networks, *71*-290, 291; extraterrestrial civilizations, *73*-364, 365; satellites, *72*-293, *71*-287, 288, 289; undersea cables, *71*-288. See also *Computers; Electronics*
Communications Satellite Corporation (Comsat), *73*-291, 292, *72*-293, *71*-289
Community ecology, *71*-296
Computers: 73-293, *72*-295, *71*-289; CCD shift registers, *73*-304; ferrite core memory, *72*-297; grassland study, *72*-82, 93; networks, *71*-290; particle detection, *71*-344; semiconductor memory, *72*-297; simulation, *71*-289, 290, *il.*, 291; simulation of galactic evolution, *il.*, *71*-267; use in supertankers, *73*-216
Concrete mixes, *71*-281
Condensation nuclei, used in fog seeding, *71*-330, 331
Conditioning, classical, *71*-182, *il.*, 185
Condors, *il.*, *73*-375
Conrad, Charles, Jr., *73*-366, *71*-362, *il.*, 365
Conservation: *Books of Science, 72*-278, *71*-273; legislation, *73*-13; new strategy to save birds, *73*-50—59
Conservation of parity, *71*-209
Consumer safety, radiation from television receivers, *71*-303
Continental drift theory: *73*-163, 320, *72*-123, 134, 322, 323, 342, *71*-314, 316; primitive mammals, *73*-318
Contraceptives, oral, *71*-294, 356
Cook, John P., *72*-261
Cook, Thomas B., Jr., *72*-416
Cooper, G. F., *72*-139
Cooper, William E., *72*-303
Coppens, Yves, *73*-235, *71*-254
Copernicus, Nicolaus, *Close-Up*, *73*-266
Coral, photographic essay, *71*-10—23
Coral head, *71*-113, 115, 123, *il.*, 114
Coral reefs, *Special Report*, *71*-110—124
Coriolis force, *71*-345
Corn : blight-resistant seed, *73*-254; experimental hybrid, *73*-254
Corn leaf blight, *72*-254, *il.*, 254
Coronary arteries, *72*-176, 330

Coronary arteriography, *72*-178, 331
Coronary artery by-pass, *72*-330
Corpus callosum, *72*-150
Correal, Gonzalo, *72*-262
Corynebacteria acnes, *72*-158, 172
Cosmetic manufacturing, *71*-177
Cosmic rays, *73*-271, *71*-267, 268, 345
Cosmodom (home in space) project, *72*-365
Cosmology, 73-272, 273, *72*-271—273, *71*-268, 269
Cosmonauts, deaths, *72*-365
Cotrell Award, *73*-421
Cottonseed, *71*-254
Cottonseed protein, *72*-284
Cotzias, George C., *71*-293, 417, *il.*, 417
Council on Environmental Quality, *72*-361, 373
Coupling constants, *73*-347
Cowen, Robert C. *72*-228
Cox, Allan V., *72*-417, *71*-316
CPT Theorem, *71*-209
Crab Nebula : *71*-267, 268; pulsar, *71*-264; X-ray emission, *73*-271
Craighead, Frank C., Jr., *73*-36
Craighead, John J., *73*-36, *il.*, 41
Cravioto, Joaquin, *71*-203
Cretaceous Period, *72*-281
Crick, Francis H.C., *73*-311
Cromie, William J., *73*-62, 400, *72*-38
Croneis, Carey, *73*-423
Cronin, John F., *72*-318, 319
Cross-Florida Barge Canal, *72*-361, 373
Crow, James F., *71*-95
Crown gall, plant tumor, *73*-334, 335
Crown of thorns (*Acanthaster planci*), *Special Report*, *71*-110—124, 335
Crystals, ion penetration, *72*-357
Cucumber, *73*-254
Cultural eutrophication, *71*-126—145
Cultural Revolution (China), effect on science, *73*-191, 201
Currents, measurement of, *72*-343
Cyanoacetylene, *72*-73
Cycladic Islands, archaeological investigations, *72*-229
Cyclamates, *71*-95, 105, 174, 292, 294
Cyclic AMP, *72*-338, 339, 340
Cyclohexaamylose molecule, key to enzyme synthesis, *72*-292
Cyclohexylamine, *71*-96
Cyclones, *72*-333
Cygnus A, *71*-42
Cygnus X-1, *73*-271
Cytidine (C), *72*-289
Cytosine (C), *71*-99

D

Da Costa, Cyro P., *73*-280
Daddario, Emilio Q., *71*-26, 358
Dairy farming, *71*-252, *il.*, 252
Damadian, Raymond, *72*-327
Damon, Albert, *72*-256
Daniel, Glyn E., *72*-238
Danziger, Rudy, *73*-327
Data processing. See *Computers*
Data transmission services, *72*-293
Daugherty, Richard A., *72*-262
David, Edward E., Jr., *72*-362, 363, 364
Davidson, Allan, *72*-204
Davies, Jack, *73*-94, 95
Davis, Roger, *72*-145, *71*-352
Davisson-Germer Prize (physics), *73*-422, *71*-420
Davitaia, Fyodor F., *72*-95
Day, Michael H., *73*-258
Day, William H., *72*-256
Day Medal (geology), *73*-417, *72*-422, *71*-420
DDE, breakdown product of DDT, *73*-53

DDT (dichloro-diphenyl-trichlorethane) : *73*-53, 215, 310, 341, 342, *72*-291, *71*-107—109, 253, 297; concentration in food chain, *73*-342; governmental restriction, *71*-109
Dean, Jim, Jr., *72*-127
Dean, Jim, Sr., *72*-127
Death rate : underdeveloped countries, *71*-82; United States, *73*-322
Deaths of Notable Scientists, 73-423, 424, *72*-423, 424, *71*-421, 422
DeBakey, Michael E., *71*-325
Debye Award (physical chemistry), *72*-422, *71*-420
Decathlon, *73*-148
Decker, Robert, *73*-322
Deep Sea Drilling Project, *72*-122, 323, 342, *71*-336
Deep Submergence Rescue Vehicle (DRV), *71*-338
Deer, *73*-380
Defense, U.S. Department of (DOD), *71*-357, 358, 360
Deffeyes, Kenneth S., *73*-164, *il.*, 172
Dehmelt, Hans G., *73*-343; *71*-420
Deimos, *73*-65, 66, *il.*, 66
DeKruif, Paul, *72*-423; *il.*, 423
Delbruck, Max, *71*-388, 417, *il.*, 413
Del Moral, Roger, *73*-300
De Lumley, Henry, *73*-256
De Lumley, Marie-Antoinette, *73*-256
Dendrochronology, *72*-231
Density : of ice under pressure, *71*-283, 284, *il.*, 284; of neutron star matter, *71*-263
Dental glue, *72*-325
Dental impressions, *73*-39, 40, *il.*, 38
Dentistry: 73-321, *72*-325, *71*-318; adhesive paints and sprays, *71*-318, *il.*, 318; implants, *73*-321; plaque inhibition, *73*-321; preventive, *72*-325, *71*-318; use of lasers to prevent tooth decay, *73*-305, 306
Deoxyguanosine groups, *72*-289
Deoxyribonucleic acid. See DNA
Deprit, André, *73*-420
De Remer, E. Dale, *72*-255
Dermis, *72*-154, *il.*, 155
Descartes Highlands, *73*-366
De Serres, Fred J., *71*-101, 103, *il.*, 103
Detergents. See **Phosphate detergents**
Deuterium : *72*-218; plasma, *71*-349
Deuterium-tritium reaction, *72*-219
Deuteron, *73*-349
De Vaux, Roland, *73*-423
De Vore, Irven, *71*-64; 83, *il.*, 64
Dewey, C. Forbes, Jr., *72*-329
Diabase, *72*-134
Diabetes mellitus, *73*-140, *72*-298
Dial-a-bus, *71*-371, 372
Dialysis, *72*-332, 333
Diamond knife, *73*-382, 384, 385
Diauxia, *71*-389, 390
Dicke, Robert H., *73*-88, *72*-63, 421
Dickerson, Richard E., *72*-290, *71*-283
Diethylstilbestrol, *73*-255, 256
Differential solubility, *73*-212
Diffusionism, *72*-227
Digoxin, *73*-296, 297
Dimers, *71*-271, *il.*, 272
Dimond, Grey, *73*-190, *il.*, 202
Discrete system, hi-fi channels, *73*-306
Disease : animal, *73*-256; chemical, *73*-215; genetic, *72*-315. See also *Medicine; Public Health*; also specific diseases
Dispersants, *72*-344
Dissection, human, *72*-11, 12, 13
Distillation, *71*-140
Dixon, Frank J., Jr., *71*-416
Dixon, R. A., *73*-336
Dmochowski, Leon, *73*-90, 92, *72*-326

Index

DNA (deoxyribonucleic acid):
72-273, 274, 289, 290, 337—340,
71-99—101, 199, 272, 308, 333, *il.*,
98, 309; bacterial, 73-333, 334, 335;
cell evolution, 73-110; chemical
structure, 73-287; *Close-Up,* 72-275;
denaturation of in stain technique,
73-311; drug resistance factor, 71-335;
gene synthesis, 73-313; procaryotic
and eucaryotic cells, 73-106; satellite
DNA, 72-273, 274; synthesis of,
71-333; viral, 73-94, 96, 98, 99, 100.
See also *Genetics;* Mutagenics
Doan, Herbert, 71-158
Dobrovolsky, Georgi, 72-365, *il.*, 366
Doell, Richard R., 72-417, 71-316
Dole, Vincent P., 72-418
Doll, W. Richard S., 72-418
Dolmatoff, G. Reichel, 72-262
Dominant mutation, 71-100
Donaldson, Lauren, 72-107
Donnelly, Thomas W., 72-123
Dopamine, 71-293
Doppler effect, 73-343, 345
Double-clutching, 73-56
Douglas fir, *il.*, 73-14
Draper, Charles S., 71-419, *il.*, 419
Draper Medal, 72-420
Dreikurs, Rudolf, 73-423
Drell, Sidney D., 72-348
Drought, 72-334
Drug-abuse center, *il.*, 73-296
Drug control legislation, 72-299,
71-294
Drug resistance, 71-334, 335
Drugs: 73-295, 72-298, 71-292;
bioavailability, 73-296, 297; *Books of
Science,* 73-277, 72-279, 71-274;
fixed-ratio combinations, 72-298;
over-the-counter, 73-295, 296;
pharmacogenetics, 73-325, 326; role
in athletics, 73-160, 161; testing for
safety, 71-172—179, *il.*, 172. See also
Antibiotics; Cancer; *Medicine*
Dual-mode transit, 71-372
Dubilier, William, 71-421
DuBridge, Lee A., 71-26, 359
Dudrick, Stanley, 72-333, 71-420
Duffield, Wendell A., 73-316, 317
Duke, Charles M., Jr., 73-366, *il.*,
369, 71-362
Dulbecco, Renato, 73-94, *il.*, 93
Dust storms, on Mars, 73-65, 67, 264
Dust studies, 72-95, 99, 100
Dutch elm disease, 73-255
Dwarf coral, *il.*, 71-22
Dwarf galaxies, 72-267, 268
Dwarf grains, 73-254
Dyer, Rolla E., 73-423
Dynamic positioning, 72-122
Dysentery, 71-334
Dyslexia, *il.*, 72-140
Dyson, Freeman, 73-87

E

Earth: age of crust, 72-122; magnetic
field, 72-231; origin of life, 71-56, 57
Earth Day, 72-308, 71-298
Earth sciences, *Books of Science,*
73-278, 72-279, 71-274. See also
*Geochemistry; Geology;
Geophysics; Meteorology;
Oceanography*
Earth Week (1971), 72-308
Earthquakes: building codes, 72-321;
man-made, 73-320, 72-323, 324;
ocean ridge system, 72-322; polar
wobble, *il.*, 72-320; scientific

monitoring, 72-321, 322; studies,
71-317; surface faulting, 72-321;
zones, 73-169
East Pakistan, 72-333, *il.*, 336
Echo sounder, 72-124
Eclipse, solar, 71-265, *il.*, 264, 265
Ecological Society of America, 72-300
Ecology: 73-297, 72-300, 71-295;
Books of Science, 73-278, 72-279,
71-274; coral destruction by starfish,
71-123, 335; cultivation of technology,
72-29, 32, 34; decay of lakes, 71-126;
endangered birds, 73-50; endangered
plants, *Essay,* 73-11; environmental
education, 71-298; geoscience projects,
72-321; grizzlies in Yellowstone, 73-35;
marine environment, pollution of,
71-335; Pawnee Grasslands, 72-81;
pesticides, 71-107, 108, 109;
technological solutions to pollution
problems, 72-283, 284, 285, 286; water
pollution and aquaculture, 72-119. See
also *Education; Environment;
Grassland Biome Program;
Ecology;* Metal pollution;
Phosphate detergents
Economic Opportunity, United
States Office of, 71-354
Economic reciprocity, role in origin
of human family, 71-78, 79
Ecosystems: 72-300, 303, 71-296,
297; fresh-water study, 71-146;
variables in, 72-91; Yellowstone, 73-35
Eddington Medal (astronomy), 71-419
Edgar, N. Terence, 72-122
Edholm, Otto G., 73-152
Edmondson, Walles T., 72-285
Edson, Lee, 73-92, 72-138, 71-206
Education: 73-301, 72-305, 71-298;
behaviorist techniques, 73-356, 357;
China, 73-199, 200; decreasing
enrollment in graduate programs,
73-358; environmental teach-ins,
71-298, *il.*, 301; medical illustration,
72-11; population growth, 71-88, 89;
precision teaching approach, 73-357;
programmed instruction, 73-356;
relevance, 71-299, 300; science
curriculums, 72-305; women in
science, *Close-Up,* 73-302, 303. See
also *Ecology;* Science, rejection of
Edwards, Charles C., 72-298, 71-173,
174, 293
Edwards, Robert G., 72-379
Effler, Donald B., 72-187
Ehrlich, Paul, 71-298
Eightfold way, 71-414, 415
Eimas, Peter D., 72-359
Einstein, Albert, 73-80, 72-52
Einstein, Elizabeth, 72-276
Einstein Award (physics), 73-416,
71-144
Eisinger, Magdalena, 73-95
Electric dipole moment, 71-219
Electrical force, 72-207
Electrocardiography, 71-322, *il.*, 322
Electroconvulsive shock (ECS),
experimental amnesia, 71-352
Electrodialysis, 71-140, *il.*, 140
Electromagnetic waves, 71-342
Electron beam: 72-354; fusion
generator, 73-353
Electron clouds, 71-41, 42
Electron microscope: 73-382, 387,
388, 389, 390; Japanese advances,
71-225; lunar rock investigation, 71-366,
ils., 366, 367; photography, *il.*, 71-334
Electronic medicine, 73-305
Electronic timepieces, 72-309
Electronic Video Recording, 71-303
Electronics: 73-304, 72-308, 71-301;
charge-coupled device (CCD), 73-304,
305, 71-302; computers, 71-289;

desk-top calculators, 71-292;
holographic recording technique,
71-303; large-scale integration (LSI),
72-308; memory technology, 71-291,
292; mini-computers, 71-291, 292;
Selecta Vision, 71-303. See also
Communications; Computers
Electrons: interaction with light,
72-355; magnetism, 73-353;
recombination with ions, 73-343
Electrophysiology, 72-138
Elementary particles, properties,
72-206, 207. See also *Physics,
Elementary Particles*
Ellermann, V., 73-92
Elliot Medal, 72-417
Emergency Core Cooling System
(ECCS), 73-309
Emiliani, Cesari, 72-98
Emission-control devices, 71-159, 160
Emission spectrometry, 73-140, 141
Employment, population growth and,
71-85
Endangered plants, *Essay,* 73-11—23
Endangered species: grizzly bears,
73-35—40; osprey, 73-50
Endarterectomy, 72-185, 330
Endosseous implants, 73-321
Energy: 73-306; *Close-Up,* 71-304;
consumption, 72-30; new scale of,
Close-Up, 73-347; Plowshare project,
71-307; *Special Report,* 72-213, 311
Energy exchange: 72-300; between
animals and environment, 71-295
Engineering, *Books of Science,*
72-279
Engineering medicine, 71-279
Ensign, Jerald C., 71-106
Environment: 73-308; effect on child
development, 72-359; long-range
effect of dust on, 72-105
Environmental education: 72-305,
71-298; Environmental Studies (ES),
73-303; interdisciplinary courses,
71-298, 299, 300
Environmental geoscience, 72-321
Environmental Protection Agency
(EPA), 73-310, 311, 72-34, 364
Environmental Quality, Council on,
72-34
Environmental Science Services
Administration (ESSA), 71-328
Environmental teach-ins, 71-298,
il., 301
Enzymatic adaptation, 71-382, 389,
390, 395
Enzymes: α-chymotrypsin, 73-288;
arginase, 71-308, 309;
beta-galactosidase, 72-339, 340;
inducible, 72-339; L-asparaginase,
cancer therapy, 72-298; lactate
dehydrogenase, 72-202, 71-283; liver
alcohol dehydrogenase, 71-283;
malate dehydrogenase, 71-283;
nuclease, 71-308; peptidase B, 72-202;
reverse transcriptase, 73-94, 98, 99,
313; RNA-directed DNA polymerase,
72-273, 275; RNA polymerase,
72-337, 338; role in DNA duplication,
71-333, 334; sebaceous glands and,
72-158, 159; synthetic, 71-292;
thymidine kinase, 72-202; trace
element functions, 73-134, 135, *il.*,
138; trypsin, structure of, 72-290;
trypsin-trypsin inhibitor, 73-288
Ephrussi, Boris, 72-196, 71-385
Epidermis, 72-154, *il.*, 155
Epilimnion, 71-127
Epstein-Barr virus (EBV), 71-319
Equatorial bulge, on Mars, 73-65, 66
Erythropoietin, 73-330
Escherichia coli, lactose metabolism,
71-308

Index

Eskimos: cheekbones, *73*-258; skeletons, *72*-258
Estrogens, in animal feed, *73*-255, 256
Ethology, *Books of Science,* *71*-274
Eucaryotic cells, *73*-106—115, *il.* 107, 108, *71*-332
Eugster, Hans P., *73*-417
Euler, Leonhard, *73*-170
Europa, Jupiter satellite, *71*-62
European Center for Nuclear Research. See **CERN**
Eutrophication, *72*-285, *71*-126
Evans, Herbert M., *72*-423
Evans, Ronald E., *73*-366
Evaporites, *72*-323
Evapotranspiration, *73*-298
Everett, A. Gordon, *72*-321
Evolution: development of cells, *73*-105; human, *il.,* *73*-236; knuckle-walking, *73*-258, *il.,* 379; mathematical model, *71*-256
Exercise, effects on body, *73*-146, 148, 152, 153, 155
Experimental allergic encephalomyelitis (EAE), *72*-274, 276
Experimental psychology, *71*-352
Exploration, *Books of Science,* *71*-274
Explorer 42, X-ray satellite, *72*-269
Extracorporeal liver perfusion, liver regeneration technique, *73*-328, 329
Extraterrestrial civilization, communication with, *Close-Up,* *73*-364
Eye: hologram, *il.,* *73*-345; visual cortex experiments, *72*-139, *ils.,* 142, 143
Eylar, Edwin H., *72*-274

F

Faber, Herman, *72*-12, *il.,* 17
Fabing, Howard D., *72*-423
Fabry-Perot interferometer, *71*-341
Faget configuration, *71*-247, *il.,* 248
Fairbank, William M., *72*-348
Fairchild, Sherman M., *72*-423
Faissner, H., *72*-348
Fallout pollutants, *73*-209, 213
Family: origin of human, *71*-78, 79; population growth and, *71*-86, 88
Fantz, Robert L., *72*-357, 358
Farming methods, effect on climate, *72*-101, 102, *il.,* 103
Farnsworth, Philo T., *72*-423, *il.,* 423
Farrar, Clarence B., *72*-423
Fat, as energy source, *73*-158
Fatty acids, *72*-172
Fault zone, Santa Susana-Sierra Madre, *72*-319
Favaloro, René G., *72*-186
FDA. See **Food and Drug Administration**
Federal Communications Commission (FCC), *73*-291, 292, 293, *72*-293
Federal Food and Drugs Act (1906), *71*-171
Federal Radiation Council, *72*-224
Federal Water Pollution Control Administration, *71*-138
Federal Water Quality Administration, *72*-306
Feeding ecology, *73*-299, 300
Fergusson, Charles W., *72*-231
Fermi Award (atomic energy), *73*-416, *72*-416, *71*-414
Fernandez-Moran, Humberto, biography (*A Man of Science*), *73*-382—397
Ferree, Gertrude Rand, *72*-423
Ferromagnets, *73*-353

Fertility, social mechanism for controlling, *71*-84, 85
Fertility Testor, *il.,* *73*-322
Fertilization, test-tube, *72*-379, *il.,* 376
Fertilizers, sewage sludge, *71*-141
Feynman, Richard P., *73*-346, *72*-348, *71*-208
Fibers, shrinkproof, *71*-280
Film editing, *72*-296
Finch, Robert H., *71*-95, 292
Fingerprint identification, *72*-295
Fire coral, *il.,* *71*-13
Firn (granular snow), *72*-95
Fish: carp, *72*-108; catfish, *72*-109, *ils.,* 110; flatfish, *72*-109; salt-water, *72*-109; selective breeding, *72*-107; supertrout, *72*-107, *il.,* 108; yellowtail, *72*-111
Fish farming, Japan, *71*-231, 233, *ils.,* 226, 227
Fission-track analysis, anthropological dating method, *73*-229, 230
Flatfish, *72*-109
Flatworms, memory transfer, *72*-145, 148
Fleischer, Robert L., *72*-416
Fleming Award (geophysics), *72*-422, *71*-420
Flerov, Georgii, *71*-347
Fletcher, James C., *73*-360
Flexner, Josepha B., *71*-353
Flexner, Louis, *72*-145, *71*-353
Floating bladderwort, *il.,* *73*-13
Flowers, as seen by insects, *71*-276, *il.,* 277
Flu vaccine, *71*-321, 356
Fluorescent dye, used in chromosome stain technique, *73*-311, *il.,* 312, *72*-314, *il.,* 314
Fluorocarbon, in artificial blood, *71*-279, 280
Fly ash, solid waste, new uses, *71*-279
Fog seeding, *71*-330, 331
Folkman, M. Judah, *73*-328
Food: additives, *73*-274, *71*-95, 171, *il.,* 171, 355; additives, Generally Recognized as Safe (GRAS) list, *73*-295; as energy source, *73*-157; from electricity and gas, *72*-340; processed foods and nutrition, *73*-134, 145; protein-producing hydrocarbons, *72*-284
Food and Agriculture Organization, *72*-119
Food and Drug Administration (FDA): *73*-295, 296, *72*-298, 302; restrictions on feed additives, *73*-255, 256; *Special Report,* *71*-168, 292—294
Food, Drug, and Cosmetic Act (1938), *71*-172
Food chain: concentration of pollutants, *73*-52, 212, 342; consumption and elimination studies, *73*-300
Foraminifera, microfossils, *72*-97, 98, *il.,* 99
Ford Foundation *72*-306
Forecasting, weather, computer-produced, *72*-329, 330
Forest regeneration, *72*-304
Forestry, *Books of Science,* *72*-279
Formaldehyde molecules, in the Milky Way, *72*-67, 68
Fornix bundle, axons, *73*-338
Forsythe, George E., *73*-423
Fossils: animal, *73*-230, 233; human, *73*-236, 237, *ils.,* 228, 229, 232—235, 237; ocean-floor sediments, *72*-127
Founders Medal (engineering), *73*-421, *72*-419, *71*-419
Fox, John W. C., *73*-330
Foxes, foot temperature regulation, *73*-375, 376

Fra Mauro highlands, *72*-366
Fracture, bone, treated with electricity, *73*-330
France, space program, *72*-371
Frank, Marilyn L., *72*-300
Franklin Medal, *73*-416, *72*-416, *71*-420
Freidberg, Charles K., *72*-188
Freis, Edward D., *73*-419
Frequency measurement, *71*-340
Friedrichs, Kurt O., *73*-421
Friends of the Earth, *72*-308
Froshe, Franz, *72*-12, *il.,* 17
Fruit fly (Drosophila melanogaster), *71*-99, 103
Fuel, from wastes, *73*-283, *72*-284, 286
Fuester, Robert W., *72*-256
Fuglister, Frederick, biography (*A Man of Science*), *71*-396—411
Fungus, blue-green algae, *73*-115
Fusion, cell, *72*-191
Fusion, nuclear: energy, *72*-214, 215, *71*-305, 306, 348; quarks, *72*-273

G

Gabor, Dennis, *73*-416, *il.,* 415
Gaillard, Mary, *73*-244, *il.,* 243
Gairdner Awards, *73*-418, *72*-418, *71*-416
Galactic evolution, computer simulation, *il.,* *71*-267
Galaxies: *71*-34; dwarf, *72*-267, 268; expanding universe, *il.,* *71*-34, 266; nucleus, *Special Report,* *71*-34, 266. See also *Astronomy, High Energy; Molecules, interstellar*
Galaxy NGC-4151, *71*-266
Galilean Satellites, *71*-62, 63
Galileo, *72*-38, *71*-28
Gall, Joseph G., *72*-274
Gallstones, *73*-327
Galston, Arthur W., *73*-188, 190, 191, 192, 197, 199, 200, 201, 202, 203
Gamma rays, *71*-267
Gamow, R. Igor, *73*-376
Ganglionectomy, cervical, *72*-182
Ganymede, Jupiter satellite, *71*-62
Garbage, as fuel source, *73*-284, 286
Gardner, Murray, *73*-90, 92, 100, 102, *il.,* 100
Garlic, as pesticide, *73*-280
Garvan Medal (chemistry), *73*-422
Gas, made from coal, *73*-307
Gas, noble, compound synthesized, *73*-290, 291
Gas endarterectomy, *72*-330, *71*-324, 325
Gashwiler, Jay C., *72*-304
Gazzaniga, Michael, *72*-150, 151
Geballe, Theodore H., *71*-414
Gell-Mann, Murray, *72*-348, 360, *71*-345, 414, 415, *il.,* 413
Gene mapping, *72*-192, 202
Genes: *72*-191; drug resistance factor, *71*-335; isolation of, *71*-308; synthesis of, *71*-308; transfer of, *73*-335, 336. See also *Genetics*
Genes, structural: base sequence determined, *73*-311, 312; synthesized, *73*-312, 313
Genetic counseling, *73*-323, *72*-315
Genetic engineering, *72*-202, 203, *71*-309
Genetics: *73*-311, *72*-314, 315, *71*-308; *Books of Science,* *73*-278; cell evolution, *73*-105; cell fusion, *Special Report,* *72*-191; DNA-duplication studies, *71*-333, 334; immunological response to disease, *71*-320; inherited drug reactions, *73*-325, 326; mRNA, *72*-337—340; nucleotides, *71*-272; polypeptides,

Index

71-272; RNA-directed DNA
polymerase, 72-275; sickle cell
anemia, 73-323; *Special Report*,
71-95; synthesis of virus DNA,
71-333; trace element concentration
in nucleic acids, 73-136, *il.*, 139;
Y chromosome identification, 71-309.
See also *Microbiology; Mutagenics*
Gentry, Jerauld R., 71-418
Geochemistry, 73-313, 72-315, 71-310
Geodynamics Project, 72-324, 71-317
Geological Society of America,
Penrose Conferences, 72-320
Geological Survey, U.S., 71-312
Geology: 73-316, 72-318, 71-312;
activity on Mars, 73-61; ashflow deposits,
71-313; *Books of Science*, 71-274;
earth movements, mathematical model,
73-171, 172; plume theory, 73-164,
ils., 165, 167, 168, 170—175;
shell deposits as clue to marine life,
72-342. See also *Oceanography*
Geophysics: 73-319, 320, 72-321,
71-315; *Books of Science*, 71-274;
earth movements, mathematical model,
73-171, 172; gravitational variations,
mapping of, 73-164; plume theory,
73-164, *ils.*, 165, 167, 168, 170—175.
See also *Oceanography*
Gerasimov, Mikhail M., 72-423
German measles (rubella), 71-356
Germer, Lester H., 73-423
Gershon-Cohen, Jacob, 72-423
Gey, George O., 72-423
Geyser, *il.*, 73-174
Ghiorso, Albert, 71-347
Gibberellin, and plant growth, 73-197;
synthesized, 73-289, *il.*, 289
Gibbons, 71-77
Gibbs, J. Willard, 71-27
Gibbs Brothers Medal, 72-422
Giemsa stain, 73-311
Gilbreth, Lillian M., 73-423
Glass, H. Bentley, 72-248
Glass fiber, light guides, 72-295
Glass technology, 72-286
Glenn, John H., 73-406, 408, *il.*, 401
Glial scar, 73-336, 337, *il.*, 337
Global Atmospheric Research
Program (GARP), 72-335, 71-329, 331
Global horizontal sounding
technique (GHOST), 71-331
Glomar Challenger, 72-121—135, *ils.*,
120, 124, 125, 126, 127, 128, 131
Glucose: hunger regulator, 73-121;
metabolism, 73-140; repression of
enzyme synthesis, 72-339, 340
Glueck, Nelson, 72-424
Glycogen, metabolism, 73-158, 159
Gneisses, metamorphic rocks, 72-320
Goddard Award (aeronautics),
73-419, 71-418
Gofman, John W., 72-224, 71-305
Goiter, 73-137
Gold, Phillip, 73-327
Goldberg, Lawrence C., 72-172
Goldberger Award (nutrition),
72-422, 71-420
Goldfish, 72-146, 148, *il.*, 146
Goldman, Henry M., 72-258
Goldstack interferometer, 73-273
Goldwasser, Eugene, 72-330
Golitsyn, G. S., 71-61
Gomez, Edward C., 72-172
Gondwanaland, 71-314
Good, Robert A., 72-419, 71-420
Gordian Knot, Martian caldera, 73-62
Gordon, Richard F., Jr., 71-362
Gorman, Chester F., 72-259

Goss, Richard J., 72-378
Gould, Shane Elizabeth, 73-146
Gould Prize, 72-420
Grad, Harold, 72-222
Graham, Ian, 73-260
Grain, synthetic, 72-255
Grand Tour, space flight, 73-370, 71-63
Grapefruit, Star Ruby, 71-254
Graphite fiber, used in reinforced
plastics, 73-282
Grassland Biome Program, 72-81—93
Gravitational collapse, 73-77—89
Gravitational-radiation experiments,
Special Report, 72-52—65
Gravitational waves, 72-52—65,
71-39, 342
Gravity: as energy source, 71-52;
irregularities, 73-164; irregularities on
Mars, 73-66, 67. See also **Black holes**
Grazing, 72-83, 85
Great Barrier Reef, 71-113, 115
Great Lakes, 73-332
Great Red Spot, Jupiter, 71-51, 61,
ils., 57, 60
Great Rift Valley, anthropological
studies, 73-224, 226, 231, *il.*, 226
Greenland, temperature effect on
glaciers, 72-102
Greenwald, Peter, 73-94, 95
Greer, J., 73-288
Gregory, William K., 72-424
Gretz, Darrell, 72-63
Grimes, L. Nichols, 72-378
Grizzly bears, *Special Report*, 73-35
Gross, Ludwik, 72-327, 328
Gross national product (GNP),
and pollution potential, 73-208
Growth rate, marine plants, 73-214
Guanine (G), 71-99
Guanosine (G), 72-289
Guggenheim Award, 72-419, 71-418
Gulf Stream, 71-397, 406, 411, *il.*, 403;
cold-water eddies, 73-340
Gulf Stream Drift Mission, 71-338
Gum Nebula, 72-267, *il.*, 268
Gunn, James, 72-272, 273
Gussinger, Jordi, 72-262

H

Haagen-Smit, Arie J., 73-421
Haber-Bosch synthesis, 71-285
Habitat II, underwater laboratory, 71-339
Hadley, Neil F., 72-300
Hadley-Apennine region, 73-363,
366, 72-367
Hadley Rille, 73-363, 366
Hair follicles, 72-156, 157, *ils.*, 156, 157
Haise, Fred W., Jr., 71-362, 365
Hajek, K., 72-12
Haley Astronautics Award, 72-422
Hall, Asaph, 73-65
Hall, Donald J., 72-303
Haller, Albrecht von, 73-118
Halpern, Julius, 73-424
Hambidge, K. Michael, 73-133, 134
Hammen, Thomas van der, 72-262
Hamster, hairless, *il.*, 73-378
Handedness, transfer of, 72-144
Handler, Philip, 71-360
Hanel, Rudolph A., 73-71
Hanley, Daniel F., 73-148
Harappan civilization, 72-101
Hardin, Clifford, 71-95, 109, 253
Harmon Trophies (aerospace), 71-418
Harris, Henry, 72-192
Harris, James, 72-303
Harris, John F., 73-376
Hartley Public Welfare Medal, 73-421
Harvey, William, 71-28
Hasler, Arthur D., 71-138
Hassel, Odd, 71-413, *il.*, 414
Hatem, George (Ma Hai-teh), 73-196

Havasupai Indians, 71-256
Hawaiian Volcano Observatory,
71-312
Hawk, red-tailed, artificially
inseminated, *il.*, 72-377
Hawkins, Gerald S., 72-236
Hay, William, 72-123
Hazen, Wayne E., 72-348
Head-turning, infant, 72-358, 359
Health centers, neighborhood, 71-354
Health, Education, and Welfare,
Department of (HEW): 73-302, 303,
71-170
Health insurance, 71-354
Health Maintenance Organization,
71-354
Health-testing center, 72-295
Heart, artificial: 73-329, *il.*, 330,
72-331, 332, 71-327; nuclear-powered,
73-329, 330, *il.*, 330
Heart, mechanical, heart-lung
machine, 72-179
Heart, physiology of: 72-181;
cardiac muscle, 73-153
Heart-assist system, 71-322
Heart diseases: 72-181, 71-325;
angina pectoris, 72-176; arteriosclerosis,
72-181, 330; coronary thrombosis,
72-181; diet, 73-322;
myocardial infarct, 72-181; vital
statistics, 72-182, 300
Heart surgery, coronary artery by-pass,
73-329, 72-178, 330, 71-325;
endarterectomy, 72-185, 330, 71-324,
325; revascularization, 72-182, 330;
saphenous vein by-pass graft, 72-188,
330; transplantation, 73-329, 72-331,
332, 71-324, *il.*, 324. See also **Heart,
artificial; Heart, mechanical**
Heat, plume theory, 73-164—176, *ils.*,
165, 167, 168, 170, 172, 173, 174, 175
Hebard, Arthur, 72-348
Hegeman, George, 73-334, 335
Heineman Prize for Mathematical
Physics, 73-422, 72-422, 71-420
Heinrich, Bernd, 73-375
Heiser, Victor G., 73-424
Helium, interstellar clouds, 72-69
Helium 3, 73-354
Hellas, 73-68
Hellens, Robert L., 72-416
Hemoglobin: effect of iron on, 73-136,
il., 136; gene synthesis, 73-313;
molecule, 72-290; structural
abnormalities in mutants, 73-288
Hemophilia, 71-323
Henkin, Robert I., 73-139
Hepatitis, Australia antigen, 71-320
Herbicides, in Vietnam, 73-308, 72-247
H. erectus species, 73-256, 257
Heredity, child development, 72-359
Hermit crab, 72-376, 377
Héroux, Bruno, 72-12, *il.*, 17
Herpes viruses, 73-95, 96, 97, *il.*, 96
Herrera, Felipe, 71-420
Hershey, Alfred D., 71-388, 417,
420, *il.*, 417
Herzberg, Gerhard, 73-414, *il.*, 415
Hesperia, 73-67
Hexachlorophene, 73-295
Hey, Dick, 73-170, 171, *il.*, 172
Hi-fi channels, 73-306
Hibernation: bat gestation, 73-376;
grizzly bears, 73-40
High Polymer Physics Prize, 72-422,
71-420
Hill Space Transportation Award,
73-419, 72-419, 71-418
Hip joint replacement surgery:
Close-Up, 71-326, *il.*, 326
Hippocampus, protein production and
learning, 72-145
Hirsch, Helmut, 72-139

Index

Hirsch, Jules, 73-124
H.M.S. *Challenger,* 72-122
Hoabinhian people, 72-259
Hodgkin, Dorothy, 71-271
Hodgkin's disease: 72-298, 71-319;
  possibility of contagion, 73-95, 96, 97
Hofheimer Prize, 72-422, 71-420
Hog cholera, 73-256
Hogweed, Giant, skin disorders and,
  72-282, *il.,* 281
Holography: *il.,* 73-345; use in
  videotape production, 71-303
Holt, Sidney, 72-119
Hominid species: 73-226, 229, 230,
  231, 233, 235, 236, 237, *il.,* 237;
  fossil finds, 73-256, 257, 258,
  71-254
Homo-Cooperativity Phenomena,
  73-393
Homo erectus, 73-258
Homo habilis, 73-231, *il.,* 229
Hong Kong flu, 71-356
Hoover Medal (engineering), 73-422
Hopewell culture, American Indians,
  73-262, 263
Hopkins, Sir Frederick Gowland,
  72-387
Hora, Heinrich, 72-355
Hormones: abscisic acid (ABA),
  71-276; acne therapy, 72-161;
  androgens, 72-156; brassins, 71-276;
  calcitonin, 72-276; di-dehydroepi-
  testosterone, 72-172; erythropoietin,
  72-330; human growth hormone
  (HGH), 72-277, *il.,* 277, 71-269;
  human placental lactogen hormone
  (HPL), 72-277; insulin, molecular structure,
  71-271, *il.,* 272; ovine prolactin, 72-277;
  parathyroid, 72-276; plant, 71-276;
  prostaglandins, 71-285; sheep
  lactogenic hormone (OLH), 71-269,
  il., 270; testosterone, 72-172;
  thyrotropin-stimulating, 71-269
Horne, Walter H., 72-255
Horned bladderwort, *il.,* 73-13
Hornig, Donald F., 71-358
Hornykiewicz, Oleh D., 71-293
Horowitz Prize (biochemistry), 73-417,
  72-422; (biology), 71-420
Horticulture, *Books of Science,*
  72-279
Hot accretion theory, 72-50
Housing modules, 71-280, *il.,* 278
Houssay, Bernardo A., 73-424
Howell, F. Clark, 73-226, 71-254
Hoyle, Fred, 71-419
Hsia, Sung Lan, 72-172
Huart, Raoul, 73-333, 334
Hubbard, Robert P., 72-258
Hubble, Edwin P., 73-272, 71-268
Hubble constant, 73-272, 71-268, 269
Hubel, David, 72-137
Huebner, Robert J., 73-102, 103,
  71-420
Hui, Ferdinand, 72-377
Human behavior, relation to baboon
  behavior, 71-77, 78
Human growth hormone (HGH),
  72-277, *il.,* 277, 71-269
Human placental lactogen
  hormone (HPL), 72-277
Hunger: body tissue functions, 73-122;
  brain mechanism, 73-117, 118,
  120—122; effects of malnutrition,
  71-195—205; glucostatic mechanism
  in brain, 73-121
Hunting camps, Peruvian, 72-262
Hurricanes: cloud seeding, 73-333,

72-335; Debbie, 71-328, *il.,* 331;
  Ginger, 73-333
Hurt, Wesley R., 72-262, 71-259
Huxley, Hugh E., 73-417, *il.,* 418
Hyde, James F., 72-414
Hydén, Holgar, 72-144, 145
Hydrocarbons, automobile emission,
  71-160
Hydrodynamic instabilities, 72-222
Hydrogen: interstellar clouds, 72-69,
  70; Lamb shift, 73-343; molecular,
  72-70; planetary composition, 72-263,
  71-55, 56, 57, 61
Hydrogen cyanide, interstellar
  molecule, 72-265
Hylander, William, 73-258
Hyperalimentation, intravenous,
  72-333
Hypolimnion, 71-127
Hypophysation, 72-111, 113
Hypothalamus: lateral areas, feeding
  centers, 73-120; ventromedial areas,
  satiety centers, 73-118, 120, 121,
  122, *il.,* 120

I

Ice, high-pressure, 71-283, *il.,* 284
Ice floes, 73-339, 340
Icecap, Mars, 71-260, *ils.,* 260, 261
Iceland, drops in temperature, 72-104,
  il., 104
Image enhancement, hurricane study
  technique, 72-336
Image sensor, function of CCD, 73-304
Imbrie, John, 72-98
Inactivated Sendai virus (ISV), 72-192
Incubator, monitoring, *il.,* 71-321
India, dust concentrations, 72-99,
  100, 101
Indian, American, 73-36
Indian dwellings, Washington coast
  excavations, 72-262
Indium antimonide (InSb), Raman
  scattered radiation, 72-347
Infectious mononucleosis, 71-319
Influenza: Hong Kong flu, 71-356,
  vaccine, 71-321, 356
Infrared interferometer
  spectrometer (IRIS), 71-329
Infrared radiation, 71-266
Infrared spectroscopy, used to study
  Martian dust, 73-11
Inhibitor I, plant protein, 73-281
Inland Sea, 72-115
Innis, George, 72-93
Insecticides, 72-291, 292. See also
  Pesticides
Insects, reaction to chemicals in
  plants, 73-280
Institute of Muscle Research, 72-393
Institute of Nutrition of Central
  America and Panama, 71-205
Insulin: effect of chromium, 73-139,
  140; molecular structure, 71-271, 283,
  il., 272; self-powered infusion pump,
  il., 72-333
Integrated circuits, surplus, 72-308
Intelligence, protein-calorie
  malnutrition and, 71-196, 197, 201—205
Intelsat, communications satellite,
  73-291, 72-293, 71-287, 288, 289
Inter-American Institute of
  Ecology, 72-300
Inter-Society Commission for
  Heart Disease Resources, 72-182
Intermediate boson, 73-347
Internal-combustion engine (ICE),
  71-157—161, 163—165, *il.,* 160
International Association for
  Dental Research, 72-325
International Atomic Energy
  Agency, 72-214

International Biological Program
  (IBP), 72-81, 300, 71-296, 297, 298
International Council of Scientific
  Unions, 71-316
International Field Year for the
  Great Lakes (IFYGL), 73-332
International Maize and Wheat
  Improvement Center, 72-255
International Symposium on
  Comparative Leukemia Research,
  71-319
International Telecommunications
  Satellite. See Intelsat
International Transportation
  Exposition, 73-370
International Union of Geodesy
  and Geophysics, 71-317
International Union of Geological
  Sciences, 71-317
Interrogation recording and
  location system (IRLS), 71-329
Intrauterine device (IUD), *il.,* 72-330
Intravenous hyperalimentation,
  72-333
Io, Jupiter satellite, 72-264, 71-62
Iodine, human requirement, 73-137, 138
Ioffe, M. S., 72-222
Ion-exchange, water treatment, 71-140
Ion implantation, 72-357, *il.,* 71-229
Ions: crystals, 72-355; storage at
  constant energy, 73-343
Ionicity, measure of, 72-356
Iraqi paintings, 72-261
Iron, human requirement, 73-136,
  137, *il.,* 137
Iron Age chieftain, burial site
  uncovered, 73-259, *il.,* 261
Irrigation: prehistoric, 73-263,
  trickle, 72-255, 256
Irtrons, 71-262
Irwin, James B., 73-363, 366, 72-367
Isaac, Glynn, 73-231, 233, 261, *il.,* 230
Iselin, Columbus O'Donnell,
  72-424, 71-402, 403, 407
Ishizaka, Kimishige and Teruko,
  73-419, *il.,* 418, 419
Isobutyl cyanoacrylate, dental
  adhesive spray, 71-318
Isocyanic acid, 72-73
Isothermal plateau, solar layer, 72-265
Isotopes, 71-347
Isotopic power sources, 72-312
Ivanhoe, Francis, 72-256
Ives Medal (optics), 73-422, 72-422,
  71-420

J

Jacob, François, 71-382, 392, 393
Jacobi Award (pediatrics), 72-422,
  71-320
Jacobs, Nichel, 73-333, 334
Jacobson, Willard J., 72-305
Janssen Research Foundation, 72-185
Japan, space program, 72-371,
  71-369, 370
Japanese National Railways, 72-374
Japanese technology, *Special
  Report,* 72-220—235
Java Man, 71-255, 256
Jeffries Award (aerospace), 71-418
Jentschke, Willibald K., 73-251
Jerne, Niels K., 72-418
Jet Propulsion Laboratory (JPL),
  73-62
John, E. Roy, 72-141
Johnson, Clarence L., 72-419
Johnson, Francis S., 73-419
Johnson, W. Dudley, 72-187
Joint Oceanographic Institutions
  for Deep Earth Sampling (JOIDES),
  73-319, 72-122, 71-317
Jones, Charles M., 73-280

Index

Jones, K. W., 72-274
Jones, Tom, 72-13, il., 20
Jordan, Pascual, 72-63
Josephson, Brian D., 71-420
Jovian planets, 71-51
Juan Comas Award, 71-256
Jupiter: 72-263, 264; giant satellites, 73-264; occultation of Beta Scorpii, 73-265, il., 265; space probe, 73-370; Special Report, 71-50
Jupiter V, 71-62, il., 59
Jusczyk, Peter, 72-359
Justice, U. S. Department of, 72-299
Juvenile hormones, pesticides, 73-255, 72-292

K

Kalinga Prize (science writing), 73-421, 72-422, 71-419
Kao, Fa-Ten, 72-196
Karrer, Paul, 73-424
Kashiwagi, Midori, 72-377
Katchalsky, Aharon Katzir, 73-424
Katz, Sir Bernard, 72-418
Kaunaoa worm, anticancer ingredient, 72-377
Kawakami, Thomas, 73-100
Kay, Marshall, 73-417
Kefauver-Harris Drug Amendments, 71-173, 177
Keller, Fred S., 73-357
Kellum, Robert E., 72-158
Kelsey, Frances O., 71-172, 173, il., 170
Kelvin, Lord, 72-249
Kendall, Edward C., 73-424, il., 424
Kennedy, Edward M., 72-363
Kennedy, John F., il., 73-401
Kennedy, Robert S., 73-55, 56
Kenward, Rory, 73-50, ils., 149, 150, 152—154
Kepler, Johannes, 71-28
Kepler Gold Medal (astronomy), 73-422
Kerwin, Joseph P., 73-367
Khorana, Har Gobind, 71-308
Kidney, artificial, 72-332
Kidney machines, 72-332, 333
Kilauea volcano, 73-316, 317, 71-312
Killian, James R., Jr., 71-358, 359
Kinetic energy, 72-207
Kistiakowsky, George B., 73-414, il., 416, 71-358
Kligman, Albert M., 72-158
Kluyskens, Paul, 71-327
K-meson, 71-211
Koch, Lauge, 72-104
Kornberg, Arthur, 71-333
Kosekisho Achievement Award (metal physics), 73-422
Kovalenko Gold Medal (medical science), 71-420
Kraft, Christopher, biography (A Man of Science), 73-398, 72-419
Krakatoa volcano: 72-318, effect on atmosphere, 72-99
Krebs, Hans A., 72-387
Kung, Charles, 72-330
Kurchatov, Boris V., 73-424
Kutner, Marc, 72-75
Kwashiorkor, 71-195

L

L-asparaginase (L-asparaginase aminohydrolase), cancer therapy, 72-298
Lactate dehydrogenase (LDH), 71-283

Lafferty, Robert, 72-340
LaFond, Eugene C., 71-405
Laing, Phillip A., 73-65, 66
Lake: oligotrophic, 71-128, 129; seasonal stratification, 71-127, 128
Lake decay: 71-126—145; Science Year Trans-Vision, 71-129
Lake Rudolf, anthropological studies, 73-231, 233, 235
Lake Tahoe, 71-138, ils., 142, 143
Lamb shift, 73-343
Lamont-Doherty Geological Observatory, 72-122
Land, Edwin H., 73-421, il., 421
Landau, Lev Davidovich, 71-211
Lange, Paul, 72-145
Langmuir Award in Chemical Physics, 73-422, 72-422
Language barrier, Japan, 71-234
Langur monkey, il., 71-77
Laplace, Pierre Simon, 73-80
Lapp, Ralph E., 72-26, il., 27
Large-scale integration (LSI), 73-293, 72-308
Lark bunting, 72-85, 87, il., 86
Laser radiation frequency, 71-341
Laser-saturation spectroscopy, resolution of hydrogen spectrum, 73-345
Lasers: Books of Science, 71-275; development of, 71-339; excitation of molecules in chemical reaction, 73-287; fusion reactor, 73-351; gas molecules, excitation of, 72-287; improvements, 72-311; in the far-ultraviolet range, 72-345; methane stabilized, 71-340; pulsed ruby, 71-339; molecular fluorescence, laser-induced, 72-287, 288; picosecond pulses, 72-288; plasma production, 71-348; satellite tracking stations, 73-319, 320; solid state, 72-311; solid state, optical signal generators, 72-295, il., 293; used to prevent tooth decay, 73-305
Lasker Award: basic research, 73-418, 72-418, 71-416; clinical research, 73-419, 72-419, 71-417
Lassa fever, 71-321, 322
Latham, Gary, 72-45, 47, 49
Lava, 71-312, 313, 73-168, 169
Lavine, Leroy S., 73-330
Lawick-Goodall, Jane van, 71-69
Lawrence, Ernest O., 72-207
Lawrence Memorial Award, 73-422, 72-416, 71-420
Lawson, J.D., 72-221
Lawson criterion, 72-221
LD 50, pollutant concentration, 73-214
L-dopa, 71-293, 294
Lead poisoning, 72-302, 303
Lead pollution, in Greenland, 73-209, il., 213
Leakey, Louis S.B., 73-228, 230, 231, 257
Leakey, Mary D., 73-226, 227, 228, 230, 231, 234, 71-254
Leakey, Richard E., 73-233, 72-257, 71-254
Lear, William P., 71-165
Learning, 72-141, 142, 144; autonomic, 71-180—193; instrumental, 71-182, il., 185; visceral, 71-180—193
Leather coral, il., 71-23
Lederberg, Joshua, 72-192
Ledoux, Lucien, 73-333, 334
Lees, Robert S., 72-329
Leloir, Luis F., 72-414, il., 416
Length, standard of, 71-340, 341
Lennette, Edwin H., 71-420
Lepier, Erich, il., 72-22
Leptospira, 73-256
Leukemia: caused by virus, 71-319; myeloid, chromosomal identification,

72-314, 315; tuberculosis vaccine (BCG), 72-328
Levine, Rachmiel, 73-418
Lewis, Oscar, 72-424
Lewis, Walter H., 72-281
Ley, Willy, 71-421, il., 421
Li, Choh Hao, 72-277, 71-269, 420, il., 270
Libby, Willard F., 72-229
Life expectancy, in America, 73-322
Light: animal cycles and, 71-375, 376; galaxies, 71-36, 38, 39, 41—45; interaction with electrons, 72-355; velocity of, 71-341
Light-emitting diodes, 72-309, 310
Lilbourn Village, 73-264, il., 263
Lilly Award (microbiology), 72-417, 71-415
Limnology, 71-127
Linear induction motor, 73-373, 374
Lipsitt, Lewis P., 72-358
Liquid-crystal devices, 72-310, 311
Liquid thread, 72-286
Lister, Joseph, 72-12
Lithium, 72-221
Lithium carbonate, treatment of manic-depressive psychosis, 71-294
Lithosphere, 72-322
Liver alcohol dehydrogenase, 71-283
Liver damage, 73-328, 329
Lobsters, 72-342
Loch Ness monster, 71-378
Lodgepole pine, il., 73-15
Long, C.N. Hugh, 72-424
Long Island Sound, dwindling osprey population, 73-50—59
Lorell, Jack, 73-66
Los Angeles, air pollution, 71-158
Los Angeles Man, 72-257
Lothagam Hill, 73-224, 236
Lothagam Mandible, 72-257, il., 258
Lovell, James A., Jr., 73-362, 363, 365, 418
Low, Frank J., 71-39, 52, 262, 266
Low, George M., 71-418
Lowell, Percival, 73-63
Lowry, John, 73-306
Lunar basalt, and age determinations, 73-314, 315, 71-310, il., 311
Lunar geophysics, 73-320
Lunar mapping, 72-403
Lunar rocks: age determinations, 71-310—312; chemical composition, 72-316, 317, 71-310; Close-Up, 71-366, ils., 366, 367; geochemical research, 73-313—315, il., 314, 315; Rock 12013, 73-316, 317, il., 317; Special Report, 72-36—51
Lunar Rover, 73-363, 366, 72-367, il., 367
Lunar seismic velocities, 73-320
Lunar seismometer, 73-366, 71-315, 362, 363
Lunar soil: age data, 71-311; chemical composition, 72-317; plant studies, 72-255
Lunokhod, 72-312, 369
Luria, Salvador E., 71-388, 417, il., 416
Lwoff, André, 71-381
Lyden-Bell, D., 71-48
Lynch, Thomas, 71-262
Lyot, Bernard F., 72-264
Lysenko, Trofim, 71-387

M

Macaques, 71-76, 77, ils., 68, 69
Macelwane Award (geophysics), 72-422, 71-420
MacNeish, Richard S., 71-259
Macrocosms, 72-303
Magasanik, Boris, 71-392
Magnesium, deep-sea nodules, 71-347

Index

Magnetic bottle, 72-222
Magnetic bubbles, 71-301, 302, il., 302
Magnetic field: carbon-14 and, 71-257; in galaxy M-87, 71-39, 41
Magnetic field irregularities, continental drift caused by, 71-316, 317
Magnetism: and superconduction, 71-349; nucleus, 73-353
Magnetohydrodynamic (MHD) generators, 72-313
Magnetometer, superconducting, 71-351
Maize agriculture, 72-262
Malaria, 71-109
Malate dehydrogenase, 71-283
Malnutrition: among the affluent, 73-322; effect on brain, 71-196—205; overpopulation and, 71-86; protein-calorie, 71-195—205
Malthus, Thomas R., 71-80
Malthusian Doctrine, 71-80
Manic-depressive psychosis, 71-294
Mansfield, Mike, 71-357
Marasmus, 71-195
March, Robert H., 73-240
Marchantia polymorpha (liverwort), 71-276, il., 276
Marek's disease, 72-255, 71-319
Marine plants, 73-214
Marine worms, 71-377
Mariner series, 73-61—75, il., 61, 72-265, 370, 71-260, 287, 290, 374, ils., 260, 261, 290
Marpessa, tanker, 73-223
Mars: *Special Report,* 73-61—75, 72-265, 370, 71-260, 261, 287, 290, 374, ils., 260, 261; andesitic material on, 73-315; chaotic terrain, 73-64, 71-261; computer clarification of photographs, 71-290, il., 290; cratered terrain, 71-261; featureless terrain, 71-261; Mariner 8, 72-370; Mariner 9, 73-264, 315, 368, 72-371; seasonal changes, 73-68, 69, 71; volcanic activity, 73-62
Martí-báñez, Felix, 73-424
Maslow, Abraham H., 72-424
Mass spectroscopy, 71-281, 282
Masters, Arthur M., 72-188
Masursky, Harold, 73-62, 67, 74, 75
Mathematics, *Books of Science,* 71-274
Matrixed system, hi-fi channels, 73-306
Matthews, Drummond, 71-316
Matthias, Berndt T., 73-355, 71-414
Mattingly, Thomas K., 73-366, 398, 71-362
Matulane (procarbazine-hydrochloride), Hodgkin's disease treatment, 72-298
Maurasse, Florence, 72-123
Mayer, Jean, 73-118
Mayer, Maria Goeppert, 73-424, il., 424
Mayer-Oakes, William J., 72-262
Mayr, Ernst, 71-420
McAllister, Robert, 73-90, 92, 100, 102
McClintock, Barbara, 72-421, il., 421
McConnell, James V., 72-145, 148
McCormick, Robert, 72-99
McCown's longspur, 72-87
McCulloch, Warren S., 71-422
McCurdy, Paul R., 72-329
McCusker, Charles B.A., 72-348
McDivitt, James A., 72-366
McGowan, Alan, 71-146
McGowan, John, 73-206

McKenzie, Norrey, 73-251, il., 250
McKhann, Guy, 71-200
Mead, Margaret, 73-421, il., 421, il., 72-246
Meadows, Dennis L., 73-362
Measles, in cattle, 73-256
Median forebrain bundle, 73-338
Medicaid, 71-354
Medical art, *Essay,* 72-11—25
Medical genetics, 72-315
Medicare, 71-354
Medicine: acupuncture anesthesia, 73-192; automated physical examination, 72-295; *Books of Science,* 73-278, 72-279, 71-274; cancer research, *Essay,* 73-24—32; cancer research, *Special Report,* 73-89—103; dentistry, 73-321, 72-325, 71-318; home-testing kits, il., 73-322; in China, 73-191, 192, 193, 196; internal, 73-322—326, 72-326—330, 71-319—323; medical illustration, 72-11—25; neurology, 73-336—338; prenatal diagnosis, 72-315; surgery, 73-327—330, 72-330—333, 71-324—328; use of paraprofessionals in China, 73-191
Medulla, 72-144
Megalithic archaeology, *Special Report,* 72-228
Meggers Award (optics), 73-422
M-82 galaxy, 71-41, il., 35
M-87 galaxy, 71-39, 45, 266, ils., 40, 45
Melatonin, role in animal cycles, 71-376
Melloni, Biagio John, 72-11, il., 24
Memory: 72-141, 144, 145, 148, 150, 151, il., 146, 71-352, 374, 375; fragile period, 71-352, 353
Memory technology: 72-297, 71-291, 292, 301, 302
Mental health, 71-355
Mental retardation, related to arginase deficiency, 71-309
Mercury, superconductivity in, 71-349
Mercury poisoning, 72-302
Mercury pollution, 73-377, 72-249, 302
Merrifield, R. Bruce, 72-418, 71-416, 417
Merrill Award (aerospace), 73-420, 71-418
Mertz, Walter, 73-134
Meselson, Matthew S., 72-247
Mesons, 72-348, 71-211—213
Metabolism, rate of, and fetal development in bats, 73-376
Metal-oxide semiconductor (MOS) technology, 73-293, 294, 72-297, 308
Metal pollution: *Close-Up,* 72-302
Metal reclamation, 72-284
Metallic compounds, 73-288
Meteorite amino acids, 72-315, 316
Meteorology: 73-331, 72-333, 71-328; Global Atmospheric Research Program (GARP), 71-329, 331; global horizontal sounding technique (GHOST), 71-331; Improved Tiros Operational Satellite, 72-336, 71-328; Nimbus 4, 72-335; storm mapping, il., 71-330; Stormfury Project, 73-333, 72-335, 71-328. See also **Climate**
Methane stabilized laser, 71-340
Metropolitan Meteorological Experiment (Metromex), 73-332
Meyerhoff, Arthur, 72-323
Mice, "double" litter, 72-378, 379
Michelson-Morley Award (physics), 72-422
Microbiology: 73-333, 72-337, 71-332. See also **Biochemistry; Cell fusion;** *Genetics*
Microcosms, 72-300
Microscope technology, 73-396, 397

Microsurgery, 73-397
Microwave ovens, 71-303
Microwave signals, interstellar molecules, *Special Report,* 72-67—79
Mid-Ocean Dynamics Experiment (MODE), 73-341
Migration, and ocean pollution, 73-213
Mikoyan, Artem I., 72-424
Mileti, R., 72-274
Milking, automatic, 71-252, il., 252
Milky Way: 72-67, 68, 75, 79, ils., 69, 71, 71-34, 36, 39; from moon, il., 73-270
Miller, Neal E., 73-118, 71-180, ils., 182, 188, 192
Millon, René F., 72-262
Mind, *Special Report,* 72-138—151
Mineralogy, *Books of Science,* 73-278
Mini-computers, 72-297, 71-291, 292
Mini school, 73-301, 303
Minimata Bay (Japan), mercury poisoning, 72-302
Miocene epoch, 72-131
Mitchell, Ben F., 72-189
Mitchell, Edgar D., il., 73-403, 72-366
Mites, 72-83, 85, 256, il., 91
Mitochondria: 73-106, 109, 110, il., 111; and ATP production, 73-157, il., 156; energy production, 73-274
Mitosis, 72-193
Molecular beam chemistry, 73-285, 286
Molecular fluorescence, 72-287, 288
Molecules, cancer detection, 72-327; collision experiments, 71-281—283
Molecules, interstellar: *Special Report,* 72-67—79, 265, 267
Molinoff, P., 72-274
Molnar, Peter, 72-322
Molotov, Vyacheslav M., 72-391
Momentum order, 71-355
Mongolism (Down's syndrome), 72-314, 315, il., 314, 71-99
Monocab, 73-301
Monod, Jacques, biography (*A Man of Science*), 71-380—395, il., 358
Monosodium glutamate (MSG), 71-355
Montalto, Joseph, 72-176, 177
Montelius, Oskar, 72-227
Moon. See headings beginning **Lunar . . .**
Moonquake studies, 72-324
Moore, Dan H., 73-97
Moore, Raymond C., 71-415
Moore, Richard E., 72-376
Moore, Robert, 73-338
Morgan, Jason, 73-163—176
Morgan, Thomas Hunt, 71-27, 385
Morphactin, 72-282
Morrison, Philip, 71-34, ils., 46, 47, 48
Morrison, Seoras D., 73-120
Moss campion, il., 73-23
Motion picture, used in obesity studies, 73-129
Mottelson, Ben, 71-414
Mount Asama, ils., 71-230, 231
Moxibustion, 73-191
M-S plasma stabilization, 71-348
M-31 galaxy, 71-36
Mu-mesons, 72-348
Mueller, George, 72-421, 71-248, 250
Mummies, X-ray analysis, il., 73-258, 72-260, 261
Murayama, Makio, 72-329
Murchison stone meteorite, extraterrestrial amino acids, 72-315, 316, il., 316
Murphy, James L., 72-262
Murray, Bruce, 73-73, 74, 75
Muscle, 73-152, 153, 155, 158, ils., 156, 157, 158
Muscle, cardiac, 73-153
Muscle, skeletal, 73-152, 153

Index

Muscle, smooth, 73-152
Muscle, striated, 73-152, 153
Muscle cells, contact with nerve cells, 73-275
Muscle contraction, 73-275, 276, 72-393
Museums, *Books of Science,* 71-274
Mushroom coral, *il.,* 71-18
Musk oxen, *il.,* 73-380
Mussels, 72-113, 114, 115, *ils.,* 114
Mutagenics, *Special Report,* 71-95
Mutation, 71-96—105
Mycenaean civilization, 72-232
Myelin membrane proteins, 72-274
Myocardial infarction, 72-181
Myofibril, 73-155, *ils.,* 156, 158
Myosin: muscle building, 73-155, 156, *ils.,* 156, 158; muscle contraction, 73-275, 276

N

Nader, Ralph, 71-360
Najafi, Hassan, 72-178
Nalbandian, Robert M., 73-323, 72-329
Nambu, Yoichiro, 71-420
Narcan (naloxone hydrochloride), 73-297
Narcotics, 72-299
NASCOM (National Aeronautics and Space Administration communications network), 71-287
National Academy of Engineering, 71-361
National Academy of Sciences (NAS), 73-211, 71-361
National Academy of Sciences Awards: Day Prize, 73-417; Microbiology Award, 73-417, 71-415; Walcott Medal, 73-417; NAS-U. S. Steel Foundation Award, 73-417
National Accelerator Laboratory (NAL), *Special Report,* 72-205—211, 71-343
National Acne Association, 72-175
National Advisory Committee for Aeronautics, 71-401, 402, 404
National Aeronautics and Space Administration (NASA), 73-358, 359, 360, 405, 410, 412, 413, 72-255, 335, 367, 404, 71-237, 238, 247, 248, 328, 357, 358, 362
National Air Pollution Control Administration, 72-99
National Association of Independent Schools, 72-306
National Cancer Bureau, 73-359
National Cancer Act, 73-27, 29
National Cancer Plan, 73-30
National Cancer Institute (NCI), 73-28, 29, 30
National Drug Analysis Center, 71-177
National Environmental Policy Act, 73-360, 72-34, 71-304
National Heart Institute (NHI), 71-322, 323
National Institutes of Health: 71-355, 358; cancer research, 72-363
National Medal of Science, 72-421, 71-420
National Meteorological Center (NMC), 73-333
National Oceanic and Atmospheric Administration (NOAA), 73-331, 333, 339, 340, 72-334, 335 ,336, 341, 71-339

National Parks and Conservation Association, 73-23
National Research Council, 71-360
National Science Board, 71-359
National Science Foundation, 71-358; educational grants, 73-301, 303; Manpower Survey, 73-302; Very Large Array project, 73-269
National Science Teachers Association (NSTA), 72-305
Natural gas, 72-30, 31; supply, 73-306
Natural resources, fossil fuels, 72-30
Naval Experimental Manned Observatory, 71-339, *il.,* 338
Navigational aids, 72-122
Neanderthal Man, 73-256, 72-256
Nectar guides, 71-276, 277, *il.,* 277
Ne'eman, Yuval, 71-414
Néel, Louis E. F., 72-416
Negev Desert, 72-255
Neighborhood Health Center Plan, 71-354
Neptune, giant satellites, 73-264
Nerve cells, acetylcholine receptors, 72-274; contact with muscle cells, 73-275; myelin damage, 71-200
Nerve fibers, severance and connection of, 72-150
Nervous system: autonomic, 71-182—193, *il.,* 185; basic operation of, 72-144; somatic, 71-182, *il.,* 185
Netter, Frank H., 72-13, *il.,* 23
Neumann, Gerhard, 71-418
Neuquense Chopper complex, 72-262
Neuroanatomy, 72-138, 148, 150, 151
Neurology: 73-336
Neurons, 73-336, 337, *il.,* 337
Neutrinos, 73-268, 72-348, 71-268
Neutron activation analysis, 71-256, 257
Neutron star, *Close-Up,* 71-263
Nevis, S., 72-358
New York City Rand Institute, 71-290
New York State Identification and Intelligence System, 72-295
Newton, Sir Isaac, 71-28, 29
Niall, Hugh, 72-277
Nikolayev, Andrian, 72-419, 71-370
Nisseki Maru, 73-216, 219, 220, *ils.,* 220, 223
Nitrates, and water pollution, 73-310
Nitrilo triacetic acid (NTA), phosphate replacement, 72-285
Nitrogenase genes, 73-335, 336
Nitrogen fixation, 71-252, 253, 285, *il.,* 286
Nitrogen oxides, emission standards, 73-311; nitrogen dioxide, 71-158
Nitroglycerin, angina pectoris therapy, 72-176
Nix Olympica, 73-62, *il.,* 64, 71-260, *il.,* 260
Nixon, Richard M., 73-358, 361, 72-345, 361, 373, 71-158, 357, 365, 368
Nobel Prizes: chemistry, 73-414, 72-414, 71-413; peace, 72-254, 255, 421; physics, 73-416, 72-416, 71-414; physiology and medicine, 73-418, 72-418, 71-417
Nomura, Masayasu, 72-418
Nondisjunction, 71-99, *il.,* 97
Noradrenalin, 73-338
Nordihydroguairetic acid, 73-274
Nordman, Christer, 72-289
North American Osprey Research Conference, 73-52
Norton, Ted. R., 72-377
Norton culture complex, 71-259
Nuclear core, 71-346
Nuclear engineering, *Books of Science,* 71-274

Nuclear explosives, 72-313; Natural gas production, 73-307
Nuclear magnetic resonance (NMR): 73-353, 354; cancer detection, 72-327
Nuclear plant discharges, 72-224
Nuclear power: as energy source, 72-31, 32; atomic fusion, 73-308
Nuclear power plants: arguments against, 72-32; energy crisis, 73-307; nuclear research, 72-352; safety programs, 73-309, 310
Nuclear-powered pacemaker, 73-305, 71-307, 328
Nuclear reactors, fast breeder, 73-307
Nuclear structure, early investigations, 72-207
Nuclear thermionic system, energy conversion, 72-311, 312
Nuclei, heavy, 71-345; magnetism, 73-353
Nucleons, 71-345, 346
Nucleotides, chemical structure, 73-287
Nucleus: high energy proton probes, 73-349; reaction to added energy, 73-350
Nutrients, effect on lakes, 71-126, 127, 139—145
Nutrition: chromium deficiency, 73-140; dietary supplementation of trace elements, 73-145; identifying trace elements, 73-140, 141, 142; iodine deficiency, 73-137, 138; iron deficiency, 73-136, 137, *il.,* 136; silicon deficiency, 73-144; trace elements, *Special Report,* 73-133—145; zinc deficiency, 73-133, 134, 138, 139

O

Oats, multiline, 71-254
Obesity: *Special Report,* 73-117—131; calorie utilization, 73-125, 126, 127; childhood, 73-129, 130, 131, 322, *il.,* 128; "epidemic," 73-322; genetic, 73-123, 124, 129, *il.,* 124; metabolic, 73-123; regulatory, 73-123; role of exercise in weight control, 73-124—131
Obsidian: 71-260; stone tools, 72-262
Occultation, 73-265
Ocean currents, mathematical model, 73-341
Ocean-floor drilling, *Special Report,* 72-121—135. See also *Oceanography*
Ocean pollution: *Special Report,* 73-205, 341, 342, 72-343—345; mercury in fish, 73-377; plant life monitoring device, 73-342
Ocean ridge systems, 71-316, 317
Oceanographer, research ship, 73-339
Oceanography: 73-339, 72-341, 71-335; acoustic signals, 71-337; Alexander Agassiz Medal, 71-397; *Books of Science,* 73-278, 72-279, 71-274; Gulf Stream Drift Mission, 71-338. See also *Coral reefs; Ecology; Geology; Ocean pollution*
Oil consumption, 72-30, 31
Oil pollution: 73-219, 220; attempts at prevention, 71-304; *Close-Up,* 72-344; spills controlled with rice hulls, 73-284
Oil transport, supertankers, 73-216, 223
Okazaki, William, 72-255
Olduvai Gorge, anthropological studies, 73-227, 229, 230, 231, 233, 234, *ils.,* 228, 229, 71-254, 255
Olmec culture, 71-260
Omentum, 72-184
Omo project, 73-257, 71-254
Omo Research Expedition, 73-234, 235, 236
Omo River basin, anthropological

Index

studies, 73-233, 234, 235, 236, ils., 232, 233
Open classroom, 72-307
Oppenheimer, Robert, 73-80
Oppenheimer Memorial Prize (physics), 73-422
Optical activity, in amino acids, 72-315
Optical frequencies, measurement of, 71-339, 340, 341
Optical microscope, limitations, 73-387
Optical resonator, 71-341
Optical signals, 72-70
Optical transmission, in communications industry, 72-295, 311
Oral implantology, 73-321
Orang-utans, ils., 73- 378, 71-76
Orbiting Solar Observatory (OSO), 73-268, 71-265
Organ transplants. See Transplantation
Organic fiber, 73-282
Organometallic superconductors, 71-350, il., 350
Orion Nebula, 72-77, il., 78
Ormak, plasma device, 71-307
Orthopedic surgery, 73-330, 71-327
Osborn, Fairfield, 71-422
Osprey: Special Report, 73-50—59
Ovine prolactin, 72-277
Oxidation, 72-387
Oysters, 72-113, 115, ils., 115
Ozone, air pollutant, 71-158
Ozone, atmospheric, absorption by plants, 73-255, 72-282
Ozone layer, (stratosphere), reduction by SST, 72-333

P

Pacemaker, heart: 73-305, 71-307, 328; nuclear-powered, 73-305
Pacific Plate, 73-169, 339, il., 166
Page, Irvine H., 72-419
Paine, Thomas O., 72-367, 71-363
Paint, polymer-based, 73-282, 283
Painted shrimp, 71-122, il., 336
Paleobotanical investigations, 72-281
Paleolithic man, soil imprint, 71-258, il., 258
Paleontology, Books of Science, 72-279, 71-275
Palmoxylon, new genus, 72-281
Palytoxin, 72-376
Pandas, il., 73-380
Pangaea, supercontinent, 73-318
Panofsky, Wolfgang K. H., 72-416, 71-420
Paper, plasticized, 72-286, 71-280, il., 278
Paraprofessionals, in China, 73-191
Parasites, used to control wasps, 73-197
Parathyroid hormone, 72-276
Parin, Vasily V., 73-424
Parity, 71-209
Parker, Hamilton, 72-259
Parkinson's disease, L-dopa treatment, 71-293, 294
Parthenocarpy, 72-282
Particle behavior, 72-209, 210
Partons, 72-348; and quarks, 73-346, 348
Passano Foundation Award (medical science), 73-419, 72-422, 71-420
Patsayev, Viktor I., 72-365, il., 366
Patten, Bernard, 72-91, 93
Patterson, Bryan, 72-257
Pauli, Wolfgang, 71-210
Pauling, Linus, ionicity of solids, 72-356

Pawnee National Grassland, 72-81—93
PCBs (polychlorinated biphenyls), 73-53, 208, 209, 215, 341, 342
Peat bogs, 73-297, 298
Peck, Paul, il., 72-22
Pendray Award (aerospace), 71-419
Penguins, il., 73-377
Penrose Medal (geology), 73-417, 72-417, 71-415
Penzias, Arno, 72-73, 71-264
Perception, in infants, 72-357, 358, 359
Periodic table, elements essential for nutrition, 73-144, il., 135
Perkin Medal, (chemistry), 72-414, 71-413
Personal Rapid Transit (PRT) system, 73-370, il., 371
Perutz, Max, 72-290
Pest control, integrated, 72-256
Pesticides, 71-107—109, 177, 178, 253, 297, 298; bacillus thuringiensis, bacterial pesticide, 73-255; cedarwood extract, 73-280, 281; effect on bird population, 73-53, 54, 71-335; garlic, 73-280; necessity for, 73-214; pheromones, 73-255
Peterson, Donald W., 71-313
Peterson, Peter G., 73-361
Peterson, Roger Tory, 73-52
Petroleum, search for new sources, 72-321, 71-314
Petrov, Boris N., 72-368
Pfeiffer, Egbert W., 73-308
Pharaohs Seti I and Thutmose III, 72-261
Pharmacogenetics, 73-325, 326
Pheromones, 73-255
Philippine Sea Plate, 73-339
Phillips, George E., 73-424
Phillips, James C., 72-355, 356, 357
Phobos, Martian moon, 73-65, 66, il., 66, 72-265
Phonoangiography, 72-329, 330, il., 326
Phosphate detergents and eutrophication, 73-310, 311, 71-140, 141; Close-Up, 72-285
Photofragment spectroscopy, 71-282, 283
Photography, detecting supertransuranic elements, 71-347
Photons, 73-355, 72-345, 346
Photosynthesis, 71-295, 296; cell evolution, 73-108
Physics: Books of Science, 73-278, 72-279, 71-275; in China, 73-199
Physics, Atomic and Molecular: 73-343, 72-345, 71-339; energy scale, 73-347. See also Lasers
Physics, Elementary Particles: 73-346, 72-347, 71-343; CERN, 73-244—249; energy scale, 73-347. See also National Accelerator Laboratory
Physics, Nuclear: 73-349, 72-351, 71-345; energy scale, 73-347
Physics, Plasma: 73-351, 72-353, 71-348; power from atomic fusion, 71-306. See also Plasma
Physics, Solid State: 73-353, 72-355, 71-349
Piaget, Jean, 72-357
Piccard, Jacques, 71-338
Pickering, Sir George W., 72-419
Picosecond pulses, 72-288
Picturephone, 72-293, 71-288
Piel, Gerard, 71-419
Piezoelectricity, use in healing fractured bones, 73-330
Pigmentation genes, 72-196
Pimples, 72-158
Pineal gland, 71-376

Pink lady's-slipper, il., 73-20
Pioneer 10, 73-292, 293, 364, 365, 370, il., 363
Pioneers F and G, 71-63
Pions, 73-350
Pithecanthropus VIII, 73-257
Plagioclase, lunar soil, 72-42
Planetary systems, probing by occultation, 73-265, il., 265; theories of origin, 71-52. See also Astronomy, Planetary
Plankton, 72-113
Plant cell fusion, 72-203, il., 202
Plant chemicals, and insects, 73-280
Plant hormones, 71-276
Plant tumor, crown gall, 73-334, 335
Plants, primitive, 73-281
Plaque, inhibition of, 73-321
Plasma, 72-353; generation by laser, 73-351; magnetic confinement, 73-351. See also Physics, Plasma
Plastic sealant, tooth decay prevention, 72-325
Plasticized paper, 72-286, 71-280, il., 278
Plastics, disposal of, 72-283; reinforced, polymer, 71-281
Plastics Science Award, 71-419
Plate tectonics: 73-319, 339, 72-322, 323, 324; and plume theory, 73-167—176; and volcanoes, 73-169, 316, 317; origin of, 73-169; Pacific Plate, 73-169, il., 166
Plourde, Gary A., 73-419
Plowshare project, 71-307
Pluto, 71-51
Plutonium: naturally occurring, 73-349; nuclear power plants, 73-307
Plutonium 239, 72-218
Pochi, Peter E., 72-158, 172
Poincaré, Henri, 71-32
Polar bear, il., 71-376
Polarized light, from galaxy M-87, 71-39
Pollen, in primitive angiosperms, 73-281
Pollination, nectar guides, 71-276, 277, il., 277
Pollution: atmospheric fallout, 73-209, 211, il., 211; danger to plants, 73-11, 12; determining safe usage levels for pollutants, 73-214; diffused, 73-205, 206, 211; LD 50, 73-214; point-source, 73-205, 207, 208, 209, il., 207. See also Air pollution; Ecology; Ocean pollution
Pollution control: liquid wastes, 71-278, 279; solid wastes, 71-279
Pollution potential index, related to GNP, 73-208
Polychlorinated biphenyls (PCBs), 73-53, 208, 209, 215, 341, 342
Polyester fibers, 71-280
Polymer plastics, 71-281
Polypropylene cans, 71-281
Polyps: Calliactis, 72-356; coral, 72-112, 113, 115, 120, 122, ils., 116, 117; limu-make-o-Hana, 72-376
Pomeranchuk, Isaak Y., 73-354
Ponderosa pine, il., 73-15
Population: Special Report, 71-80—93
Pories, Walter J., 73-138, 139
Postgate, John, 73-336
Postglacial vegetation, 73-298, 299
Potassium-argon dating method, 73-229, 230
Potato, frost-resistant, 71-254
Potter, L. T., 72-274
Pottery, neutron activation analysis, 71-256, 257
Powdermaker, Hortense, 72-424
Powell, Cecil F., 71-422
Powell, Wilson M., 72-348

Index

Precambrian rocks, 72-320
Precision teaching approach, 73-357
Pregnancy, T-globulin in blood, 72-326
Premack, David, 72-151, 71-66
Premoli-Silva, Isabella, 72-123, il., 133
Prenatal nutrition, 71-204
Preplanetary clouds, 71-52, 54, 55
Preventive dentistry, 72-325, 71-318
Price, P. Buford, Jr., 72-416
Priestley Medal, 73-414, 72-414, 71-413
Primate behavior, *Special Report,* 71-64–79
Primordial organic soup, 73-106
Prince, Alfred, 71-320
Priori, Elizabeth S., 73-90, 92, 72-326
Pro Sil, feed supplement, 71-252
Proctor, Richard J., 72-319
Programmed instruction, 73-356
Promethium, stellar radioactivity, 72-265
Prostaglandins, 71-285
Protein: as energy source, 73-158; chemical structure, 73-287, 288; fish protein concentrate (FPC), 73-283; from cattle manure, 73-283
Protein-calorie malnutrition, related to chromium deficiency, 73-140
Protein synthesis, 73-114
Proton beam, 72-208, ils., 209
Procaryotic cells, 73-106–115, il., 107, 108, 71-332
Proton synchrotron, 73-248, 72-208
Protons, head-on collision, 73-248, 249, 72-347; nuclear research, 73-349
Psychiatry, amphetamine therapy, 72-299
Psychobiology, molecular, 72-137
Psychology: 73-356, 72-357, 71-352; *Books of Science,* 73-279, 72-280; building self-esteem in students, 73-303, 304; population and aggression, 72-305
Public Health: 71-354; chemical mutagens, 71-95–105; Food and Drug Administration, *Special Report,* 71-168–179; in China, 73-196; in the United States, 73-322, 324, 325, 326; Neighborhood Health Center Plan, 71-354. See also Cancer; *Medicine*
Public Health Service Commendation Medal (epidemiology), 73-422
Puck, Theodore T., 72-196
Pulsars: 71-49; and X-ray astronomy, 73-270; discontinuity in rate, 71-264; neutron star, 71-263. See also *Astronomy, High Energy*
Pulsed ruby laser, 71-339
Purchase, H. Graham, 72-255
Purdy, J. M., 72-379
Puromycin, protein synthesis inhibition, 72-145
Purser, Douglas B., 71-252
Pye, J. D., 72-380
Pyridine, 71-350
Pyridoxine, food additive, 71-355

Q

Q-beta virus, 71-271, 272
Quantum mechanics, 71-209
Quantum theory, resolution of contradictory theories, 73-349
Quarks: 72-348, 71-345; and partons, 73-346, 348; fusion reactions, 72-273
Quasars: 72-268, 271, 272, 273, 71-42–45; association with galaxies, 73-273

Queen Notmit, 72-261, il., 261
Quinacrine dihydrochloride, chromosome-stain technique, 71-309
Quinacrine hydrochloride, chromosome-stain technique, 72-314, 315, il., 314
Quinacrine mustard, chromosome-stain technique, 71-309

R

Racey, P. A., 73-376
Radcliffe Institute, 73-303
Radiation: cosmic, 71-267, 268; far-infrared, 71-262; hazards of, 72-32, 71-305, 307; infrared, 71-266; Raman scattered, 72-347; vitamin D-producing, 72-256. See also Molecules, interstellar
Radiation belts, 71-59, il., 59
Radiation Control for Health and Safety Act, 71-303
Radio astronomy, 73-269, 71-42–45; double stars, 73-271
Radio emission, from galactic nuclei, 71-39; from quasars, 71-42
Radio interferometers, 72-77
Radiocarbon dating, 72-97, 98, 228, 71-257
Radioisotopic thermoelectric generators (RTG), 72-312
Rado, Sandor, 73-424
Radot, Louis Pasteur-Vallery, 72-424
Railroads: 72-374, 71-373, 374; in Europe, 73-374; auto-train, 73-373, 374
Rainfall, acidic, 73-297; measurement of, 72-88
Raisman, Geoffrey, 73-338
Raman scattered radiation, 72-347
Raman, Sir Chandrasekhara Venkata, 72-424, il., 424
Randal, Judith, 72-384, 71-195
Read, Kenneth R. H., 71-115
Recessive mutation, 71-100
Recycling, 71-145, il., 73-309
Red blood cells, regeneration of, 73-378, 380. See also Blood
Red mangrove, il., 73-18
Red Paint Indians, 72-261
Red shift: 73-272; galaxy, 72-268, 273; quasar, 72-268, 271, 272
Red-shift distance relation, 71-268
Reed, Michael, 71-11
Regeneration, 72-377, 378; of damaged axons, 73-336, 337, 338
Reinforcement, extrinsic rewards, 73-356; intrinsic rewards, 73-356
Reisner, Ronald M., 72-172
Renfrew, Colin, 72-228, il., 238
Rennie, Compton A., 71-414
Reproduction, studies of deer cycles, 73-380
Research, applied, 73-361
Research, basic: 73-361; lack of, in China, 73-198, 199
Research and development (R & D): federal funding, 73-359, 72-363, il., 364, 71-30, 357, 358, 359, il., 359
Respiration, cell evolution, 73-109
Respiratory system, effects of exercise on, 73-153, 155
Retinoblastoma, 71-100
Retinoic acid, 72-172
Revascularization, 72-182, 330. See also Heart surgery
Revelle, Roger, 71-80, 405, 410
Reverse osmosis, water treatment, 71-139, il., 141
Reverse transcriptase, enzyme, 73-94, 98, 99, 102
Rhizotron, 71-253, il., 253
Rho factor, mRNA synthesis termination, 72-338

Ribonucleic acid. See RNA
Ribosomes: model, 73-274, 275, il., 274; cell identification, 73-110, 114; protein synthesis, 73-275
Rice, Stuart Arthur, 71-422
Rice: hulls used to control oil spills, 73-284; IR-8, 71-254; IR-20, 71-254
Rice-stem borer, 71-228
Rich, Saul, 72-282
Rickets, in Neanderthal Man, 72-256
Ricketts Award (medical science), 72-422, 71-420
Riedel, William, 72-123, il., 133
Riemsdyk, Jan Van, il., 72-16
Rifampin (rifampicin), 72-299
Ring, Peter A., 71-327
RNA (ribonucleic acid): 72-144, 145, 273, 289, 290, 337–340, 71-272, 308; alanine-transfer gene, 71-308; *Close-Up,* 72-275; gene synthesis, 73-313; messenger RNA (mRNA), 73-275, 72-337–340
RNA, viral, 73-94, 96, 98, 99, 100
RNA-DNA hybridization, 73-335
Robbins, Louise M., 71-259
Roberts, Verne L., 72-258
Robinson, Arthur B., 72-274
Robinson, George, 71-325
Rock magnetism, 72-323
Rock shelters, 72-262
Rockefeller Public Service Award, 72-422, 71-420
Roebling Medal (mineralogy), 72-422
Roelofs, Wendell, 73-255
Roels, Oswald, 72-118
Roemer, Elizabeth, 72-420
Roentgen, Wilhelm K., 71-31
Rogers, Stanfield, 71-308
Roosa, Stuart A., 72-366
Rosen, Samuel, 73-190, 192, il., 202
Rosenstiel Institute of Marine and Atmospheric Sciences, 72-122
Roseroot, il., 73-23
Ross, D. M., 72-376
Ross, Donald, 72-188
Ross, John, 72-96
Rossini, Frederick D., 72-414, il., 416
Rous, Francis Peyton, 73-92, 93, 71-422, il., 422
Roussy, 73-118
Rovee, Carolyn K., 72-359
Rovee, David T., 72-359
Royal Aeronautical Society Gold Medal, 72-422, 71-420
Royal Astronomical Society Gold Medals, 73-422, 72-422, 71-420
Rubbia, Carlo, 72-347
Rubella (German measles), 71-356
Rubidium-strontium method, lunar rock dating, 72-41, 317
Ruckelshaus, William D., 73-310
Ruddle, F. H., 72-202
Runcorn, S. Keith, 72-417
Russell, Bertrand Arthur William, 71-422, il., 422
Russia: communications satellite systems, 73-291, 292; plasma device, *Close-Up,* 71-306; space program, 73-75, 367, 368, 72-365, 368–372

S

Sabin, Albert B., 72-421
Saccharin, 73-295
Sachs, Robert G., 72-209, 210
Safar, Fuad, 72-261
Sagan, Carl, 71-51
Salmon, memory studies, 71-374, 375
Salt, iodization, 71-138
Salt-water fish, problems of domestication, 72-109
Salter, R. B., 71-416
Salyut space station, 72-365

Index

Sampson, Milo B., *72*-424
San Andreas fault, *73*-319, 320, *ils.,* 166, 319
San Fernando earthquake, *72*-319, 321
Sandage, Allan R., *73*-272, *72*-421, *il.,* 421, *71*-269
Sanders, F. Kingsley, *73*-95, *ils.,* 96, 97
Sanger, David, *72*-261
Sanger, Frederick, *73*-418
Sansone, Gail, *72*-172
Santa Barbara oil spill, *71*-314
Sarabhai, Vikram A., *73*-424
Sarcoma, *71*-319, *il.,* 319
Sarepta, *il.,* *73*-259
Satellite, Communications, *72*-293
Satellite, Meteorological, *72*-335, 336, *71*-328, 329, 331
Satellite infrared spectrometer (SIRS), *71*-329
Satellite, transponder, *72*-293
Satellites: systems, *71*-62; giant, of outer planets, *73*-264
Satiety, brain functions, *73*-118, 120, 121, 122
Saturn: *71*-51; giant satellites, *73*-264
Scale invariance, *73*-346
Schaper, Wolfgang, *72*-185
Scheuer, Paul J., *72*-376
Schiaparelli, Giovanni, *73*-62
Schistosomiasis, *73*-196
Schlegel, Hans, *72*-340
Schlesinger, James R., Jr., *73*-360, 361
Schlesinger, William, *72*-306
Schmidt, Maarten, *72*-271, 272
Schmitt, Harrison H., *73*-366, *72*-409
Schmitt, Roman A., *73*-420, *il.,* 419
Schneidermann, Nahum, *72*-123
Schumacher, Howard E., *73*-419
Schwartz, Melvin, *72*-348
Schwarz, Helmut, *72*-355
Schwarzschild radius, *73*-80, 81, 83, 84
Schwarzkopf, Paul, *72*-424
Science, General, *Books of Science,* *73*-279, *72*-280, *71*-275
Science, rejection of: *72*-28, 33, 360, 361, 362; *Essay,* *71*-24, 357—361
Science and ethics, *73*-361, 362
Science and religion, *Books of Science,* *72*-280
Science and technology, reordered federal priorities, *73*-358, 359
Science education, *73*-301—304
Science, Research, and Development, House Subcommittee on, *71*-357, 358
Science Support: *73*-358, *72*-360, *71*-357. See also *Ecology; Education; National Accelerator Laboratory; Space Exploration; Transportation*
Scientism, *71*-31
Scorpion, heat experiments, *72*-300, *il.,* 300
Scotophobin, peptide, *72*-148
Scott, David R., *73*-363, 366, *72*-367
Scribner, Belding H., *71*-416
Scripps Institution of Oceanography, *72*-122
Scyllac, *71*-348
Sea-floor nodules, source of ores, *72*-342, 343
Sea floor spreading: *73*-163, *72*-322, 323, *il.,* 322; and primitive mammals, *73*-318. See also **Continental drift theory**
Sea snakes, salt glands, *73*-377

Sea whips, *il.,* *71*-16
Seabed treaty, *72*-345
Seaborg, Glenn T., *73*-360, *72*-248
Sears, William H., *72*-262
Sebaceous glands, *72*-154, *ils.,* 155, 157
Sebum, *72*-154
Sediments, ocean-floor, *72*-122
Seed bank, *73*-13
Seedless fruit, cucumber, *73*-254; cultivated by parthenocarpy, *72*-282
Segre, Aldo G., *73*-256
Seismic profiler, *72*-123, 124
Seki, Tsutomu, *71*-419
SelectaVision, *71*-303
Selye, Hans, *73*-391, 394, 397
Semiconductors, *71*-281
Sensory processing, *72*-139
Sevastyanov, Vitali I., *72*-419, *71*-370
Sewage fish culture, *72*-108, 109
Sewage sludge, used as fertilizer, *71*-141
Sewell, William H., *72*-184
Seyfert, Carl, *71*-36
Seyfert galaxies, *71*-38, *il.,* 38
Shadowgrams, chromosome examination, *72*-314
Shalita, Alan R., *72*-172
Shapiro, Arthur, *71*-422
Shapiro, James, *71*-27, 308, 357
Sheep lactogenic hormone (OLH), *71*-269, *il.,* 270
Shelbourne, James E., *72*-109
Shemyakin, Mikhail M., *72*-424
Shepard, Alan B., Jr., *72*-366
Shia-yu, Hsu, *73*-191, 192
Shigella dysenteriae, *71*-334
Shirey, Earl, *72*-185
Shoemaker, Eugene: biography (*A Man of Science*), *72*-398—413; 45
Shope papilloma virus, *71*-308, 309
Sickle cell anemia, *Close-Up,* *73*-323, *72*-328, 329, *il.,* 327
Sidel, Victor, *73*-188, 191, 193, 200, *il.,* 202
Sigma factor, mRNA synthesis, *72*-337, 338
Signer, Ethan, *73*-188, 197, 199, 202
Silberman, Charles E., *72*-307
Silent Spring, *71*-107—109
Simmons, Henry E., *72*-298
Singstad, Ole, *71*-422
Sioux village, *73*-263
Siqueland, Einar R., *72*-358, 359
Skin, structure and function, *72*-154, *il.,* 155
Skin peeling, acne therapy, *72*-161, 172
Skylab, *73*-366, 367, *72*-365, 366, *71*-368
Slater, John C., *72*-421
Slayton, Donald K., *73*-363, *il.,* 403
Smallpox, *il.,* *71*-356
Smell, identification of mother, *72*-379, 380
Smith, Alan, *72*-352
Smith, Alfred, *72*-377
Smith, Bradford, *73*-62, 64, 67, 68, 72, 73, 74, 75
Smith Medal (geology), *72*-420, *71*-415
Snakes, ability to detect warm-blooded prey, *73*-376, 377; salt glands, *73*-377
Snider, Arthur J., *72*-176
Snowmobiler's back, *72*-258
Snyder, Hartland, *73*-80
Sobell, Henry, *72*-289
Sodium fluoride, bone growth and, *71*-256
Soil, lunar. See **Lunar soil**
Solar clouds, *73*-268
Solar eclipse, *71*-265, *il.,* 264, 265
Solar heating, related to dust storms on Mars, *73*-67
Solar observatory: orbiting, *73*-268,

71-265; underground, *73*-268
Solar polar caps, *73*-268
Solheim, Wilhelm G., *72*-259
Solid state diode laser, *72*-346
Solomon, Philip M., *72*-75
Sones, F. Mason, *72*-185, *71*-416
Sorghums, *71*-252
Soriano, M., *71*-255, 256
South Spot, Martian crater, *73*-62
Southern, E.M., *72*-273
Soviet Academy of Sciences Lomonosov Gold Medals, *73*-422
Soybeans, male-sterile, *73*-254
Soyuz series, *72*-365, *71*-370
Space, geometry of, *72*-55, *il.,* 55
Space Exploration: *73*-362, *72*-365, *71*-362; Apollo program, *73*-360, 362, 363, 366, 398, 400, 409—413, *72*-366, 367, *71*-362—365; 368; *Books of Science,* *71*-273; British spacecraft, *72*-371; Chicom I, *71*-368, 369; galaxy nucleus, *71*-34—49; Gemini 5, *73*-408, 409; gravity-assisted missions, *71*-63; Luna 16, *72*-368; Luna 17, *72*-369; Luna 18, *73*-368; Luna 19, *73*-368; Luna 20, *73*-368; Lunar Rover, *73*-363, 366, *72*-367, *il.,* 367; Lunokhod, *72*-369, *il.,* 369; Mariner series, *73*-61—75, 315, 368, *72*-265, 370, *71*-260, 287, 290, 374, *ils.,* 260, 261, 290; pictures from space, *73*-306; Pioneer 10, *73*-292, 293, 364, 365, 370, *il.,* 363; Pioneers F and G, *71*-63; Project Mercury, *73*-405, 406, 408; Ranger Project, *72*-39, 40; Russian space program, *73*-367, 368, *72*-365, 368—372, *71*-370; Skylab, *73*-366, 367, *72*-365, 366, *71*-368; spectrographic mapping, *71*-261; Stratoscope II, *72*-263, 264, *il.,* 263; U.S.-Russian manned mission, *73*-360, 362, 363; Venera 7, *72*-264, 370, *il.,* 264; Venera 8, *73*-370; Viking series, *73*-74, 75; Zond 7, *71*-370; Zond 8, *72*-369. See also *Apollo program; Astronomy, Planetary; Science Support*
Space shuttle: *73*-360, 367; *Special Report,* *71*-237—250
Space Station, orbiting platform, *72*-366
Spartina, grass, *72*-300
Spectrographic mapping, *71*-261
Spectroscopy, photofragment, *71*-282, 283
Spencer, Frank C., *72*-189
Sperry, Roger, *72*-148, 150, 151
Spherator, *71*-348
Sphygmometrograph, *il.,* *73*-322
Spiegelman, Sol, *73*-98, *il.,* 98
Spilhaus, Athelstan F., *71*-402
Spin-flip laser, *72*-347
Spinar, *71*-48, *il.,* 48
Spinelli, D. Nico, *72*-139
Spitzer, Lyman, *71*-405
Spitzer, Paul, *73*-50—59, *ils.,* 55, 56, 58
Split-brain technique, *72*-151
Squirrels, feeding ecology, *73*-299, 300
Stadtman, Earl R., *71*-415
Stafford, Thomas P., *73*-363
Stainless steel, *71*-281
Stanley, Wendell M., *73*-424, *il.,* 424
Starfish, coral-devouring, *71*-110—124, 335
Starquakes, *71*-264
Stars: *71*-34—49; binary, *73*-268, 269, *71*-265; changes in mass, *73*-268, 269; communication with, *Close-Up,* *73*-364; HR 465, *72*-265; infrared, *73*-269; neutron, *71*-263; X-ray, *72*-268, *71*-262
Steam bath, sperm-killing effect, *71*-256
Steam-powered buses, *73*-374
Steel, stainless, *71*-281
Steinbacher, Robert H., *73*-62, 73

Index

Steiner, Donald F., *73*-418
Steinfeld, Jeffrey I., *72*-288
Steptoe, P.C., *72*-379
Stern, Otto, *71*-422
Stewart, Sarah E., *73*-90, 92
Stock market quotations, computerized, *72*-296
Stomata: and water loss in plants, *71*-296; potassium content of guard cells, *73*-281
Stonehenge, *72*-232, *il.*, 234
Storey, Arthur, *72*-260
Stouffer Prize (medical research), *72*-419, *71*-417
Strain, William H., *73*-138, 139
Stratoscope II, *72*-263, 264, *il.*, 263
Stratosphere, volcanic discharges into, *72*-318, 319
Strauss, John S., *72*-158, 172
Streuver, Stuart, *72*-261
Stroun, M., *73*-335
Strughold, Hubertus, *73*-419, *il.*, 420
Stuiver, Minze, *71*-257
Subperiosteal implants, *73*-321, *il.*, 321
Sucking responses, *72*-359
Suess, Hans E., *72*-232, *71*-257
Sulfur, *73*-285, 306
Sulfur dioxide, *73*-307, *71*-304
Sulzberger, Marion B., *72*-154
Sun: climatic effects, *72*-96; isothermal plateau, *72*-265. See also headings beginning Solar . . .
Sunflowers, hybrid, *73*-255
Superaccelerators, *72*-205, 210
Superconducting electron microscope, *73*-382, 387—390
Superconductivity, *71*-349
Superconductors, *73*-355
Supersonic transport (SST) project, *72*-333, 360, 361, *71*-360, 374
Supertankers, *Special Report*, *73*-216—223
Supertransuranic elements, *71*-347
Surfactants, detergent ingredients, *72*-285
Suri, Kailash G., *72*-258
Surveyor, research ship, *73*-339
Sutherland, Earl W., Jr., *73*-418, *il.*, 415, *72*-418, *71*-416
Svedberg, Theodor H.E., *72*-424
Swartkrans, anthropological studies, *71*-254, 255, *il.*, 255
Sweat glands, *72*-154
Sweet, Haven C., *72*-255
Swigert, John L., *73*-363, *71*-362, 365
Swordfish, mercury content, *73*-378, *72*-302
Sykes, Lynn, *72*-322, *71*-317, 420
Symmetry, *71*-209
Synapse, *73*-336, 337, *il.*, 337
Synchrotron, *71*-215
Synthetic suture, *il.*, *72*-332
Syphilis, in Neanderthal Man, *72*-256
Szent-Györgyi, Albert, biography (*A Man of Science*), *73*-382—397
Szent-Györgyi, Martha, *73*-390

T

Tabrah, Frank L., *72*-377
Tal, Chloe, *72*-326
Tamm, Igor Y., *72*-424, *il.*, 424
Tamplin, Arthur, *72*-224, *71*-305
Tantalum disulfide, *71*-350
Taste, sense of, *73*-133, 134, 139
Tatum, Lloyd Allen, *72*-254
Tautavel Man, *73*-256, *il.*, 257
Taylor, Gordon Rattray, *71*-381

Teacher education, *72*-305, 307, *71*-300
Technology: *Books of Science*, *71*-275; *Essay*, *72*-26—34; foreign industrial competitors, *72*-365; Japanese advances; *71*-220—235; nonmilitary uses for nuclear explosives, *72*-313; rejection of, *72*-360
Tectonic plate. See **Plate tectonics**
Tektite project, *72*-341, 342, *71*-338
Telephones, *73*-293, *71*-288, 289
Television, pictures from space, *73*-306
Teller, Edward, *72*-242
Temin, Howard M., *73*-94. 417, *il.*, 417, *72*-273
Teminism, *72*-273
Temperature inversion, *73*-172
Teotihuacán (Mex.), *72*-262
Test-tube embryos, *72*-379, *il.*, 376
Thaddeus, Patrick, *72*-68, *il.*, 75
Thalidomide, *71*-172
Thar Desert, *72*-99, 100, 101
Theorell, Hugo, *72*-390
Thermal pollution, *72*-32, 256, *71*-304
Thermionic converter, *72*-311, 312
Thermocline, *71*-127
Thermodynamics, (Second Law of), *71*-208
Thermonuclear fusion, *73*-351, *71*-348
Thom, Alexander, *72*-228, *il.*, 238
Thorium 232, *72*-218
Thorne, Kip S., *72*-245, *71*-419
3C-273 quasar, *72*-270, 271, *71*-43, 266
3C-279 quasar, double quasar, *72*-271, *il.*, 272
Thrombosis, coronary, *72*-181
Thymidine (T), *72*-289
Thymine (T), *71*-99
Thyroidectomy, *72*-182
Thyrotropin-stimulating hormone, *71*-269
Till, J. E., *71*-416
Tilleux, Wade, *73*-216
Tilton program, *72*-305, 306
Time, *Books of Science*, *71*-275; standard of, *71*-340, 341
Time-reversal invariance, *Special Report*, *71*-206—219
Time symmetry, *71*-209
Tiselius, Arne, *73*-424
Tishler, Max, *71*-413, *il.*, 414
Tissue culture, *72*-192
Titanium, lunar rocks, *72*-40
Tobolsky, Andrew V., *71*-419
Todaro, George J., *73*-102, 103
Tokamak: *73*-351, *72*-353, 354, *71*-348; *Close-Up*, *71*-306, *il.*, 306; *Special Report*, *72*-215, 223
Tolbutamide, *72*-298
Tomlinson, Harley, *72*-282
Tomonaga, Sin-itiro, *71*-221, *il.*, 235
Toomey, Bill, *73*-148, 152
Tooth impressions, used to determine age of bears, *73*-39, 40, *il.*, 38
Topography, Mars, *73*-267; Venus, *73*-267
Torrey Canyon, *73*-220
Toxicology, drug testing, *71*-176
Trace elements: *Special Report*, *73*-133—145; dietary supplements, *73*-145; experiments in producing deficiency, *73*-143, 144, *il.*, 141
Tracked air cushion vehicles (TACV), *73*-373
Transatlantic cable, *73*-291
Transfer ribonucleic acid (tRNA), *72*-290
Translocation, genetics, *71*-97, *il.*, 97
Transplantation: heart, *73*-329, *72*-331, 332, *71*-324, *il.*, 324; larynx, *71*-327, 328; lung, *71*-328; tissue rejection, *73*-378, *71*-324
Transpo 72, *73*-370

Transportation: *73*-370, *72*-372, *71*-371; mass transit, *71*-371; rapid transit systems, *73*-370, *il.*, 371
Transportation, U.S. Department of, *72*-375, *71*-371; Urban Mass Transportation Administration (UMTA), *73*-373
Transuranic elements, *72*-351
Trapezium, *72*-77
Trauma, *73*-326
Tree growth, affected by motion, *73*-281, *il.*, 280; monkeys and, *71*-277
Trisomy, *71*-99
Triticale, synthetic grain species, *72*-255
Tritium, *72*-218
Triton (*Charonia tritonus*), *71*-121, 122, 335, *il.*, 122
Trobicin, antibiotic, *73*-297
Trout, *72*-107
TSH-releasing factor, *71*-269
Tuberculosis: BCG vaccine, protection against leukemia, *72*-328; drug therapy, *72*-299
Tuck, James, *72*-223
Tumor angiogenesis factor (TAF), *73*-328
Tumors, *73*-328
Tuna, mercury content, *73*-377, 378
Tunable laser, *73*-345, *72*-345
Turkevich, Anthony L., *72*-36, *71*-414
Turner, Barry E., *72*-73
Turner, Earl, *72*-303
Turner, Grenville, *71*-311
2, 4, 5,-T, weed and brush killer, *71*-253
2, 3-DPG (*2, 3-diphosphoglycerate*), *73*-155

U

Uhuru satellite, *73*-270, 273, *72*-269, 270
Ultraviolet lasers, *72*-345
Ultraviolet light: patterns in plants, *71*-276, 277; study of Jupiter, *71*-58; used to harden dental sealant, *72*-325
Underdeveloped countries, overpopulation, *71*-82, 85, 86, 88, 89, 90
Underground solar observatory, *73*-268
Unemployment: among scientists and engineers, *73*-358, *72*-350, 364, *71*-359, 360; population growth and, *71*-85
Ungar, Georges, *72*-148, 247, *il.*, 147
Unger, H., *72*-12
United Nations, seabed treaty, *72*-345
United Nations Conference on the Human Environment, *73*-309
United Nations Educational, Scientific, and Cultural Organization (UNESCO), *73*-260
United States Steel Foundation Award, *72*-418, *71*-415
Universe, expansion, *73*-272
Universities Research Association, *72*-205
University Group Diabetes Program (UGDP), *72*-298
Upper Mantle Project (UMP), *71*-315, 316, 317
Upwelling, artificial, *72*-343
Uranium-235, *72*-214
Uranus, *72*-263, *71*-51
Urea, *73*-323, *72*-329
Urey, Harold C., *71*-271, 420
Urschel, Harold C., *72*-189, *71*-325
Uterus, tissue transplants, *73*-378
Uvarov, Sir Boris P., *71*-422

V

Vaccines: for animal diseases, *73*-256; low U.S. vaccination rate, *73*-324

Index

Vagus nerve, 73-121
Van Allen, James A., 72-369
Van de Graaff accelerator, tandem, 71-347
Van Dyne, George M., 72-82
Van Slyke, Donald D., 72-424
Van Tamelen, Eugene E., 71-285, 286
Vegetation, postglacial, 73-298, 299
Vela X pulsar, 71-264
Venezuelan equine encephalomyelitis (VEE), 73-256
Venezuelan Institute of Neurological and Brain Research, 73-385, 387
Venketswaren, Subramaniaan, 72-255
Venus, space probe, 73-370, 72-264, 265, 370, 71-261, 262
Vesalius, Andreas, 72-11, 12, il., 15
Vetlesen Prize, 72-417
Vianna, Nicholas, 73-94, 95
Video picture storage, 72-295
Videotape playback units, 71-303
Vidicon, CCD vidicon, 73-304, 305
Vietnam, environment, 73-308
Vigorito, James, 72-359
Villegas, Evangelina, 72-255
Vinci, Leonardo da, 72-11, ils., 10, 14
Vineberg, Arthur M., 72-184
Viral vaccines: 71-319, 321; cancer, 72-327; chicken (cancer), 72-255; German measles (rubella), 71-356; influenza, 71-356
Virchow, Rudolf L.K., 72-256
Viruses: Australia antigen, 71-320; B-type, 73-97, 98, 99, 100; breast cancer, 73-326; C-type, 73-100, 102, 103, 72-326, 71-319; cancer-producing, Special Report, 73-89—103, 26, 27, 72-299; Epstein-Barr virus (EBV), 71-319; herpes virus, 73-95, 96, 97, il., 96; Hong Kong flu, 71-321, 356; inactivated Sendai virus (ISV), 72-192; lambda H80, 72-340; Lassa fever, 71-321, 322; Q-beta, 71-271, 272; R17, 71-271, 272; RNA-directed DNA polymerase content, 72-275; rubella, 71-356; Shope papilloma, 71-308, 309; SV-40, 72-201. See also Cancer; Leukemia; Microbiology
Visceral learning, 71-180—193
Visual perception, in infants, 72-357, 358, 359
Vitamin A acid, 72-172
Vitamin C: 72-382, 387—389; and iron availability, 73-145
Vitamin E, 73-297
Volatile pollutants, 73-209
Volcanoes: 72-318, 71-312, 313; Kilauea, 73-316, 317; located over plumes, 73-168, 169, ils., 168, 170, 175
Volkov, Vladislav N, 72-365, il., 366
Von Braun, Wernher, 73-362
Von Euler, Ulf S., 72-418
Voris, Frank B., 71-418
Voroshilov, Kliment, 72-391

W

Walker, Robert M., 72-416
Walkinshaw, Charles H., 72-255
Wankel, Felix, 73-372
Wankel engine, Close-Up, 73-372

Warburg, Otto H., 72-424
Warner Prize (astronomy), 73-422, 72-422
Warren, Shields, 73-416, il., 416
Warren, Stafford, 73-416, il., 416
Washburn, Sherwood L., 71-69
Wasserburg, Gerald, 72-38, 71-311, 312
Waste, human, 73-197, 198
Waste disposal: computerized, 72-295, 296; sewage-disposal system, 73-207; of radioactive materials, 73-308, 318
Water, on Mars, 73-71, 72, 74
Water pollution: Special Report, 73-205—215, 72-119, 284, 285, 302, 306, 344, 345, 71-125—155, 304; animal wastes, 71-252; liquid wastes, 71-278, 279; phosphates, 73-310, 311, 72-285; Tilton program, 72-305, 306. See also Ocean pollution
Water pollution control, 71-278, 279
Water Quality Improvement Act of 1970, 71-304
Water reclamation, 72-286
Water testing, Special Report, 71-146—155
Watkins, J.F., 72-194
Watson, James D., 72-362, 417, il., 417
Watson-Farrar, John, 71-327
Watson Medal, 73-420
Watt, William, 71-420
Weather balloons, 71-331
Weather modification, 73-333, 72-334, 335, 71-328, 330
Weather prediction: computerized, 71-329, 330; long-range, 72-295; numerical, 73-333, 72-336
Weber, Joseph, 72-54, ils., 53, 58, 62, 71-39, 342
Weeks, Kent, 72-260
Weinberg, Alvin M., 72-214
Weitz, Paul J., 73-367
Wells, Herbert, 72-258
Werner, Jesse, 73-414
Westing, Arthur H., 73-308
Wheat, Shabarti Sonora, 71-254
Wheeler, John A., 72-421, 71-420
Whitby, G. Stafford, 73-424
White, Paul Dudley, 73-188, 192, 193, 202, 203, il., 202
White, Robert M., 72-341
White House Conference on Food, Nutrition, and Health, 71-205
Whiteheads, 72-158
Whooping cranes, 71-377
Wiesel, Thorsten, 72-138, 139
Wiesner, Jerome B., 71-358
Wigner, Eugene, 73-416, il., 417, 71-209
Wilderness Society, 72-308
Wildlife telemetry system, 73-37, 38, il., 39
Wiley, Harvey W., 71-171, ils., 171
Wilkes, Charles, 71-30
Williams, Carroll M., 72-245
Willows, Dennis, 72-144
Wilson, Robert Rathbun, 72-211
Wilson, Robert W., 72-73, 71-264, 343
Wind: effect on tree growth, 73-281; on Mars, 73-67
Winick, Myron, 71-198, 200
Winstein, Saul, 72-421
Wintersteiner, Oskar P., 73-424
Wire plane, particle detection, 71-344
Wissler, Robert, 73-322

Witkamp, Martin, 72-300
Witschi, Emil, 73-424
Witten, Victor H., 72-161
Witter, Richard L., 72-255
Wolves, 73-375, 376
Wood, John A., 72-36
Wood, W. Barry, Jr., 72-424
Woods Hole Oceanographic Institution, 72-122, 71-397, 401
Worden, Alfred M., 73-366, 72-367
World Meteorological Organization (WMO), 72-334
Wren, Sir Christopher, 72-12
Wright, D. J. M., 72-256
Wrist watch, electronic, 72-309

X

X-Ogen, radio signal of, 72-73
X-ray astronomy, 73-270, 72-269, 270
X rays: acne therapy, 72-161; analysis of mummies, 72-260, 261; angiography, 72-178; cosmic, 72-269; Cygnus X-I, 72-270; diffraction methods, 72-289, 290; federal limits on television emission, 71-303; galactic emissions, 72-270, 71-41, 46; identifying atoms, 72-351; insulin examination, 71-283; radiation from binary stars, 73-268, 269; stellar emissions, 71-262

Y

Yellowstone National Park, 73-35, 36, il., 41
Y chromosome identification, dye technique, 71-309
Yajko, David, 73-334, 335
Yalow, Rosalyn, 73-418
Yamashiro, Donald, 72-277
Yang, Chen Ning, 73-188, 190, 198, 201, 203, 71-210, 213
Yanofsky, Charles, 73-417, 418, il., 418
Yellowtail fish, 72-111
Yoder, Hatten S., Jr., 73-417
Yoga, visceral control, 71-193
Young, John W., 73-366, il., 369, 71-362
Yukawa, Hideki, 71-221, il., 234

Z

Zenkevich, Lev A., 72-424
Zillig, Wolfram, 72-337, 338
Zimmerman, David R., 73-52
Zinc, function in humans, 73-133, 134, 138, 139
Zinc, superplastic, 71-280, 281
Zinjanthropus specimen, 71-254
Zinn, Walter H., 71-414, il., 415
Zond (probe) series, 72-369, 370
Zones of upwelling, marine life concentrations, 72-118, il., 118
Zoology: 73-375, 72-376, 71-374; Books of Science, 73-279, 72-280, 71-275; dominance hierarchy in baboons, 71-74, 75, ils., 72, 73; Gnathostomulida, new phylum, 71-377; grizzly bears, 73-35—49; Mariner 6, 71-374; mating behavior of baboons, 71-76; new ways to save endangered birds, 73-50—59; pineal gland, 71-376; primate behavior, 71-64—79; social structure among bears, 73-41, 42
Zooplankton, 72-303

Acknowledgments

The publishers of *Science Year* gratefully acknowledge the courtesy of the following artists, photographers, publishers, institutions, agencies, and corporations for the illustrations in this volume. Credits should be read from left to right, top to bottom, on their respective pages. All entries marked with an asterisk (·) denote illustrations created exclusively for *Science Year*. All maps were created by the *World Book* Cartographic Staff.

Cover
Frank C. Craighead, Jr.

Essays
10 Boyd Norton; Ernest Baxter, Black Star; Kitty Kohout; Fred Leavitt; Doug Wilson, Black Star
12 Les Blacklock; John Mathisen
13 Lincoln Nutting, NAS; Torkel Korling; H. L. Parent, NAS
14-15 David Muench
16 Torkel Korling; Durward L. Allen, Dept. Forestry and Conservation, Purdue University; Torkel Korling; Thase Daniel, Bruce Coleman, Inc.
17 Grant Heilman
18-19 Grant Heilman; Walter Dawn; Grant Heilman
20 Les Blacklock
21 U.S. Forest Service; John Kohout
22 David Muench
23 J. Foott, Bruce Coleman, Inc.; Harold R. Hungerford; E. R. Degginger
24-31 Albert Fenn*

Special Reports
34 John J. Craighead
37 Lance Craighead
38-42 Frank C. Craighead, Jr.
43 Frank C. Craighead, Jr.; A. Stephen Johnson; Frank C. Craighead, Jr.
44-45 Frank C. Craighead, Jr.
46 A. Stephen Johnson
47 Frank C. Craighead, Jr.
50-59 Gilbert L. Meyers*
60-61 R. B. Leighton, California Institute of Technology; Jet Propulsion Laboratory
63-66 Jet Propulsion Laboratory
68-69 Donald G. Mieghan*
70-74 Jet Propulsion Laboratory
76-77 Byron Davies*
78 Princeton University
79-88 Byron Davies*
90-91 Albert Fenn*
92-93 Robert Keys*; Darrel K. Miller, The Salk Institute
96 Albert Fenn*; Bernard Roizman, University of Chicago; Dr. K. Nazerian, USDA, East Lansing, Mich.
97 Robert Keys*; Albert Fenn*
98 Donald Getsug; Nurul H. Sarkar, Institute for Medical Research; Nurul H. Sarkar, Institute for Medical Research
99 Robert Keys*
100 Jack Hamilton, Globe*; Dr. Robert M. McAllister, Children's Hospital of Los Angeles; Dr. Robert M. McAllister, Children's Hospital of Los Angeles
101 Robert Keys*
104-107 Jean Helmer*
108 Norma J. Lang, University of California at Davis; Carolina Biological Supply Company
111-113 Jean Helmer*
114 Margit Nass, University of Pennsylvania; William T. Hall, Electro-Nucleonics Laboratories, Inc.
116 Gilbert L. Meyers*

118 Ivan Massar, Black Star*
119 Jean Helmer
120 Barry Sandrew, Dept. of Nutrition, Harvard School of Public Health; Barry Sandrew, Dept. of Nutrition, Harvard School of Public Health; Ivan Massar, Black Star*
122 Dr. Jules Hirsch, Rockefeller University
123 Jean Helmer*
124-127 Ivan Massar, Black Star*
128 Steve Eagle from Nancy Palmer
130 Sonja and Angelo Lomeo, Rapho Guillumette
132 Gilbert L. Meyers*
135 Mas Nakagawa*
136 Eric V. Gravé
137 Reproduced by permission from *The Structure and Action of Proteins* by Richard E. Dickerson and Irving Geis
138-139 Mas Nakagawa*
141-142 Gilbert L. Meyers*
143 Joseph Erhardt*
144 Gilbert L. Meyers*
146-154 Barry King*
156 Lou Barlow*
157-158 Mas Nakagawa*
159 Roger K. McKay, University of Nebraska
160 Albert Schoenfield*
162 Werner Wulf, Black Star*
165 Herb Herrick*
166 NASA
168 Herb Herrick*
170 Carl Wilmington
172 Joseph Erhardt*; Werner Wulf, Black Star*
173 Herb Herrick*; NASA
174 Richard Kehrwald, Black Star
175 NASA; Bob Thuener, Black Star; G. R. Roberts from Carl Östman
177 Richard S. Fiske, U.S. Geological Survey
179-185 Herb Herrick*
187 NASA
189 Brian Brake, Rapho Guillumette
190 Paolo Koch, Rapho Guillumette
193 Hugues Vassal, Photoreporters
194-195 Victor W. Sidel, M.D.
196 *The Globe and Mail*, Toronto; Ilsa Sharp, Woodfin Camp, Inc.
198 R. C. Hunt
199 Audrey Topping, Rapho Guillumette
200 Brian Brake, Rapho Guillumette
201 R. C. Hunt
202 Courtesy Victor W. Sidel, M.D.
204-205 Dennis Brack, Black Star; Bob Marin, Tom Stack & Assoc.
206-207 Dave Davis, Tom Stack & Assoc.; Weston Kemp*; G. William Shallop*; J. R. Eyerman
208-209 G. William Shallop*
210 Serge DeSazo, Rapho Guillumette; Gilbert L. Meyers*; Tom Myers, Van Cleve Photography; G. William Shallop*; Dennis Brack, Black Star; Gilbert L. Meyers*
212 Southern California Coastal Water Research Project

213 Robert Duce, University of Rhode Island; Clair
C. Patterson, California Institute of Technology
217 Orion Press from SCALA
218 Kyodo News Service; Ishikawajima-Harima
Heavy Industries Co., Ltd.
219 Kyodo News Service; Orion Press from SCALA
220 Kyodo News Service
221 Tokyo Tanker Co., Ltd.
222-223 Standard Oil Company (New Jersey)
224-225 Jay H. Matternes
226 F. Clark Howell
228 F. H. Brown; Donald G. Mieghan*
229 F. Clark Howell; H. R. Haldemann;
Donald G. Mieghan*
230 Glynn Isaac
232-233 F. Clark Howell
234 Richard Leakey; F. Clark Howell
235 Donald G. Mieghan*
237 V. Magglio; F. Clark Howell
238-239 CERN
240-243 Adam Woolfitt*
244 Mas Nakagawa*
245-250 Adam Woolfitt*

Science File

254 United Press Int.
257 United Press Int.; Drawing by W. Miller
© 1972 The New Yorker Magazine, Inc.
258 University of Michigan Medical Center
259 University of Pennsylvania Museum
261 Wide World
263 University of Michigan
266 Muzeum Okregowe, Toruń, Poland
269 Herbert Gotsch Studio*
270 NASA
272 Reprinted by permission of the Bulletin of
the Atomic Scientists, 1971
274 Yoshiaki Nonomura, Gunter Blobel, and
David Sabatini, Rockefeller University;
Herbert Gotsch Studio*
276 George W. Stroke, State University of New
York at Stony Brook and Howard University
280 Richard W. Harris, University of California, Davis
282 Hercules, Inc.
283 A. I. McMullen, Microbiological Research
Establishment, Porton Down, England,
Crown Copyright
284 AB Akerlund & Rausing, Sweden
285 The American Museum of Natural History
286 Dawes in The Christian Science Monitor
© TCSPS
288-290 Mas Nakagawa*
292 Bell Telephone Laboratories
294 Hewlett-Packard Company
295 Dr. Albert S. Klainer, Ohio State
University College of Medicine
299 Samuel M. McGinnis, California
State College, Hayward
301 Robert H. Johns, The Physics Teacher
305 General Electric Company
307 Argonne National Laboratory
309 Cartoon by Al Kaufman © 1972 by The
New York Times Company
310 Calvin A. DeViney
312 Dr. Herbert A. Lubs, University of
Colorado Medical School
314-315 NASA
317 Colorado Department of Highways
319 Charles G. Bufe
320 U.S. Atomic Energy Commission
321 Dr. A. Norman Cranin, The Brookdale
Hospital Center, Brooklyn
322 Weston Laboratories; Sears, Roebuck and Co.
323 Patricia N. Farnsworth; C. E. Bremer, IBM
324 R. M. Albrect, P. L. Sandok, R. D. Hinsdill,
A. P. MacKenzie, and I. B. Sachs, Experimental
Cell Research
325 Dr. Irving Redler and Marilyn L. Zimny
Orthopedic Research Laboratory, New Orleans
326 Dr. J. U. Schlegel, Tulane University
School of Medicine
327 Sydney J. Harris © 1972 Saturday Review, Inc.
328 Anne Keatley; Sally Reston, The New York
Times
329 Pennsylvania State University
330 National Institutes of Health
331 Mas Nakagawa*
332 Drawing by Chon Day © 1972 The New
Yorker Magazine, Inc.
334 Charles Dulong De Rosnay and Michael
Mercier, University of Bordeaux, France
337 Mas Nakagawa*
338 Gene Pesek, Chicago Sun-Times
340-341 National Oceanic and Atmospheric
Administration
342 ALCOA
344 Bell Telephone Laboratories
345 J. Caulkins and C. Leonard
348 National Accelerator Laboratory
350 Los Alamos Scientific Laboratory
352 Reprinted by permission of the Bulletin of
the Atomic Scientists, 1971
354 University of Chicago
356 R. K. Davenport, Yerkes Primate Research
Center, Atlanta
358 The White House
359 Drawing by Weber © 1971 The New Yorker
Magazine, Inc.
360 Wide World
363-364 TRW, Inc.
367-368 Sovfoto
369 NASA
371 Monocab, Inc; The Bendix Corporation;
Transportation Technology, Inc.
374 B. F. Goodrich Company
375 The New York Times; Western Foundation of
Vertebrate Zoology
376 Drawing by Richter © 1972 The New Yorker
Magazine, Inc.
377 Wide World
378 Michael Festing, MRC Medical Research
Council, Surrey, England
379 Chicago Zoological Society
380 Gilbert L. Meyers*; Wide World

Men of Science

382-387 Declan Haun*
388 Svenska Dagbladt; Declan Haun*
389-390 Declan Haun*
392-396 Lee Balterman*
398 Marc St. Gil, Black Star*
400-401 NASA
403-411 Marc St. Gil, Black Star*
415 Wide World
416 Harvard University; UCLA; Cancer
Research Institute
417 Eugene P. Wigner; University of Wisconsin
418 Stanford University; Columbia
University; The Johns Hopkins University
419 The Johns Hopkins University; Oregon State
University
420 United States Air Force
421 Jay Scarpetti, Polaroid Corporation;
Ken Heyman*
423 Wide World; Pictorial Parade; Pictorial Parade
424 Wide World; Wide World; United Press Int.

Typography

Display—Univers
Monsen Typographers, Inc., Chicago
Text—Baskerville Fototronic
The Poole Clarinda Co., Chicago
Text—Baskerville monotype (modified)
LST Typography, Chicago
Monsen Typographers, Inc., Chicago
R. R. Donnelley & Sons Co., Chicago

Offset Positives

Process Color Plate Company, Chicago
Jahn & Ollier Engraving Co., Chicago

Printing

Kingsport Press, Inc., Kingsport, Tenn.

Binding

Kingsport Press, Inc., Kingsport, Tenn.

Paper

Text
White Field Web Offset (basis 60 pound)
Westvaco, Luke, Md.

Cover Material

White Offset Blubak
Holliston Mills, Inc., Kingsport, Tenn.